COMPUTATIONAL GEOMETRY

AF606443

WITHDRAWN

Machine Intelligence and Pattern Recognition

Volume 2

Series Editors

L. N. KANAL

and

A. ROSENFELD

University of Maryland
College Park
Maryland
U.S.A.

NORTH-HOLLAND
AMSTERDAM · NEW YORK · OXFORD

3 8006 10096 1161

514.19 Copy B

TOUSSAINT, G.T.

Computational geometry
Machine Intelligence and Pattern Recognition 2.

10. JAN 90

25/6/93

11. OCT

CRANFIELD INSTITUTE OF TECHNOLOGY LIBRARY
BEDFORD MK43 0AL

Computational Geometry

Edited by

Godfried T. TOUSSAINT

School of Computer Science
McGill University
Montreal
Canada

1985

NORTH-HOLLAND
AMSTERDAM · NEW YORK · OXFORD

© ELSEVIER SCIENCE PUBLISHERS B.V., 1985

All rights reserved. No part of this publication may be reproduced, stored in a retrieval system, or transmitted, in any form or by any means, electronic, mechanical, photocopying, recording or otherwise, without the prior permission of the copyright owner.

ISBN: 0 444 87806 8

Publishers:

ELSEVIER SCIENCE PUBLISHERS B.V.
P.O. Box 1991
1000 BZ Amsterdam
The Netherlands

Sole distributors for the U.S.A. and Canada:

ELSEVIER SCIENCE PUBLISHING COMPANY, INC.
52 Vanderbilt Avenue
New York, N.Y. 10017
U.S.A.

Library of Congress Cataloging-in-Publication Data

Main entry under title:

Computational geometry.

(Machine intelligence and pattern recognition ; v. 2)
1. Geometry--Data processing. I. Toussaint, Godfried T., 1944- . II. Series.
QA447.C57 1985 516'.028'54 85-13639
ISBN 0-444-87806-8 (U.S.)

PRINTED IN THE NETHERLANDS

TO ADELINE AND FELIX
my progenitors

PREFACE

Computational geometry is having an ever-broadening influence on the science of pattern recognition. Procedures in almost every area of pattern recognition can be made more rigorous and the efficiency of algorithms can be improved by the application of principles discovered in computational geometry. Two recent surveys attesting to this fact are: "Pattern Recognition and Geometrical Complexity" by G.T. Toussaint in the Proceedings of the Fifth International Conference on Pattern Recognition, Miami Beach, December 1980; and "Computational Geometric Problems in Pattern Recognition" by G.T. Toussaint in the Proceedings of the NATO Advanced Study Institute, Oxford University, April 1981.

This book contains a collection of fifteen papers covering a wide variety of topics relevant to pattern recognition to a varying extent. Some of the papers are of a tutorial-survey nature and the collection should serve well as a graduate text in computational geometry, as well as provide researchers in the area with some of the latest developments.

Convex hulls have been used extensively in many pattern recognition problems and algorithms for this problem are closely related to sorting sequences of numbers. Selim Akl describes new optimal parallel algorithms for selection, sorting, and computing convex hulls. It is often required to compute the convex hull in an on-line manner where a total of n points arrive one-at-a-time and it is desired to insert points into a "growing" convex hull. A similar problem of computing the intersection of half-planes occurs when we wish to insert a half-plane into a convex polygon. David Avis, Hossam ElGindy, and Raimund Seidel present a new simple data structure that solves both of these problems in $0(n \log n)$ time.

Although the *nearest-point* Voronoi diagram is well-known in pattern recognition circles, little is known about the *furthest-point* Voronoi diagram. Binay Bhattacharya and Godfried Toussaint show how the furthest-point Voronoi diagram can be used to solve several geometric problems efficiently.

Another inherently geometric problem that occurs in syntactic pattern recognition concerns decomposing simple polygons into simpler components. Bernard Chazelle and David Dobkin solve the problem of decomposing a non-convex simple polygon into a minimum number of convex polygons. Mark Keil and Jorg Sack survey the area of minimum decompositions.

Although the average or expected complexity of an algorithm is often the most useful measure of performance in practical situations, most of the complexity results in computational geometry are concerned with worst-case situations. This is in part due to the difficulty involved in geometric probability. Luc Devroye gives a survey of some recent results that can be helpful in the expected time analysis of algorithms in computational geometry and determines the expected complexity of a variety of algorithms for solving different geometric problems.

One of the most effective ways of obtaining fast running times in practice is to use *bucketing* techniques. An in-depth survey of this area is provided by Takao Asano, Masato Edahiro, Hiroshi Imai, Masao Iri, and Kasuo Murota. Specifically, these techniques are applied to the problems of minimum-weight perfect matchings in the plane, two-dimensional Voronoi diagrams, point location and range search in the plane, and shortest paths in networks.

David Kirkpatrick and John Radke outline a new methodology for describing the "internal structure" (or "skeleton") of planar point sets. The methodology, which is based on parameterized measures of neighbourliness, gives rise to a spectrum of possible internal shapes. Applications to the analysis of both point set and network patterns are described.

D.T. Lee and I.M. Chen present an efficient algorithm for computing the visible boundary edges of a set of nonoverlapping convex polygons. The method can be applied to solve three dimensional hidden-surface problems based on a scan-line approach, namely, each scan-line plane intersects a set of objects in three-dimensional space and yields a set of nonoverlapping convex polygons.

A problem that is well-known in the Operations Research literature is the so-called "Euclidean Distance Facility Location Problem" or the minimum spanning circle problem. Robert Melville presents the results of an implementation study of two algorithms for this problem.

The signature is a function that can be associated with any continuous plane curve. The signature value at arc length s is the length of the curve to the left of the directed tangent to the curve at the point p determined by s. The signature is invariant under translation, rotation, and scale changes to its curve, and nearly invariant under slanting transformations. These properties make it useful as the basis of a curve similarity measure. Joseph O'Rourke and Richard Washington demonstrate with several examples of experimental results that the *signature distance* between two curves has appealing behavior, especially when applied to slanted handwriting samples.

Lower bounds on the complexity of a geometric problem are theoretical results with very practical consequences: they allow us to stop looking for algorithms that do not exist. Raimund Seidel presents a new powerful methodology for proving lower bounds for certain geometric problems.

Spurred by developments in spatial planning in robotics, computer graphics, and VLSI layout, considerable attention has been devoted recently to the problem of moving sets of objects, such as line segments and polygons in the plane or polyhedra in three dimensions, without allowing collisions between the objects. One class of such problems considers the *separability* of sets of objects under different kinds of motions and various definitions of separation. Godfried Toussaint surveys this new area of research in a tutorial fashion, presents new results, and provides a list of open problems and suggestions for further research. Along similar lines, research and development work in robotics and industrial automation has created a need for good motion planning algorithms for collision avoidance; this in turn has stimulated research on the inherent computational complexity of motion planning. The purpose of Sue Whitesides' paper is to survey some recent algorithms and complexity results for motion planning, placing particular emphasis on computational geometry issues.

Often in image processing the polygons being processed have their edges parallel to two orthogonal directions, i.e., they are *isothetic*. This additional structure can usually lead to more efficient algorithms. Derick Wood surveys this area of isothetic computational geometry.

In September 1983, I was asked to arrange a special session on computational geometry for the 1983 International Symposium on Information Theory, held in St. Jovite, Quebec, Canada. Laveen N. Kanal, who was attending the session, suggested collecting those papers and making them into a book. In order to give a more complete treatment of the topic, additional authors were subsequently invited to contribute papers. I would like to express my sincere thanks to all the authors for their contribution and a special thanks to Laveen N. Kanal for giving me the opportunity of putting this volume together.

Montreal
April 16, 1985

Godfried T. Toussaint

TABLE OF CONTENTS

COMPUTATIONAL GEOMETRY, G.T. Toussaint (Editor)
© Elsevier Science Publishers B.V. (North Holland), 1985

OPTIMAL PARALLEL ALGORITHMS FOR SELECTION, SORTING AND COMPUTING CONVEX HULLS

Selim G. Akl

Department of Computing and Information Science
Queen's University
Kingston, Ontario, Canada

Parallel algorithms are described for three problems that are often to be solved in computational geometry, namely selection, sorting and computing convex hulls. For a problem of size n, all three algorithms use n^{1-e} processors where $0<e<1$ and depends on the number of available processors. The cost of each of the algorithms, i.e. the number of processors multiplied by the worst-case running time, matches an existing lower bound on the number of operations required in the worst case to solve the given problem and is therefore optimal to within a constant factor.

1. INTRODUCTION

With the growing number of areas in which computers are being used, there is an ever increasing demand for more computing power than today's machines can deliver. Extremely fast computers are being sought by many applications to process enormous quantities of data in reasonable amounts of time. However, it is becoming apparent that significant increases in speed will very soon be impossible to achieve by simply using faster electronic devices as was done in the past three decades. This is due, on the one hand, to the fact that with today's super fast circuit elements, more time is needed for a datum to travel between two devices than it takes for it to be processed by either of them. On the other hand, the reduction of distance between devices through very high scale integration is quickly reaching a limit beyond which not only optical resolution is impossible but also the reliability and speed of circuit elements decrease.

An alternative route to the attainment of very high computational speeds is to use a parallel computer, i.e. one which possesses several processing units (or processors). Here the problem is broken into smaller parts which are solved simultaneously, each on a different processor. This approach becomes truly attractive when one considers the rapidly decreasing cost of computer components. Hundreds or even thousands of processors can thus be assembled to dramatically reduce the solution time for a problem. Today, an important and very active field of study is developing which concerns itself with the design and analysis of algorithms for parallel computation.

There are a number of properties that it is desirable to have in a parallel algorithm. Assume that the algorithm is intended for solving a problem of size n. The first two properties concern the number of processors used by the algorithm, which is a function P(n) of n. Firstly, it is important that P(n) be sublinear in n: no matter how inexpensive computers become, in most practical situations it is unrealistic on the part of the algorithm designer to assume that the number of processors he has at his disposal is bigger than or equal to n, especially when n is extremely large. Secondly, P(n) should adapt to the actual number of processors on the available parallel computer: thus functions such as log n or $n^{1/2}$ (although sublinear) will not do because of their inflexibility.[1] The next two properties concern T(n) the worst-case parallel running time of the algorithm, which is also a function of n. Obviously, it is required that T(n) be significantly smaller than the time required by the best sequential algorithm for the problem. Furthermore, T(n) should vary inversely with P(n): the larger P(n) the smaller T(n) and vice versa. Finally, the last and ultimate goal of the algorithm designer is that the cost C(n) of the parallel algorithm, i.e. the product of P(n) and T(n), be optimal in the sense that it matches a known lower bound on the number of sequential operations required in the worst case to solve the problem at hand. It is interesting to note in this context that if the cost is smaller than the running time of the best (but not necessarily optimal) sequential solution known, then a better sequential algorithm can be obtained by simulating the parallel algorithm on a single processor!

In this paper we describe three parallel algorithms. The first two algorithms are for selection and sorting, respectively, two operations which are often used in the solution of computational geometric problems, while the third is for one of the most celebrated and useful constructions in computational geometry, namely determining the convex hull of a set of planar points. Each of these parallel algorithms satisfies the properties listed above: (i) it uses n^{1-e} processors, where $0<e<1$ is a parameter which depends on the number of processors on the parallel computer, and (ii) runs in $O(f(n)/n^{1-e})$ time where f(n) is the minimum number of sequential operations required to solve the problem in the worst case.

Unlike the case with uniprocessor computers which generally follow the conventional model of computation originally described by von Neumann [1], several different architectures exist for parallel computers. Our chosen computational model, the shared-memory single-instruction stream multiple-data stream computer, is described in section 2. In section 3, a parallel algorithm is presented for selecting the kth smallest element of a totally ordered (but not sorted) sequence of n elements, $1 \leq k \leq n$. The algorithm uses n^{1-e} processors, $0<e<1$, and runs in

$O(n^e)$ time, for an optimal cost of O(n). It is used in section 4 to derive a parallel algorithm for sorting a totally ordered sequence of n elements into increasing order using n^{1-e} processors, $0<e<1$, in $O(n^e \log n)$ time, for an optimal cost of O(n log n). Section 5 is concerned with computing the convex hull of a set of n points in the plane and a parallel algorithm is described for the problem which uses n^{1-e} processors, $0<e<1$, and runs in $O(n^e \log h)$ time, where h is the number of edges on the convex hull. It should be noted that this performance matches that of the best currently known sequential convex hull algorithm which runs in O(n log h) time. In addition, since h = n in the worst case, the algorithm is cost-optimal. Finally, section 6 offers some conclusions and a number of open problems for further research.

2. COMPUTATIONAL MODEL

Our model of a parallel computer is the shared-memory single-instruction stream multiple-data stream (SIMD) machine [2]. In this model, N processors numbered 1 to N share a common memory and operate under the control of a single instruction stream issued by a central control unit. In addition to the shared memory, each processor possesses a local memory in which programs and data are stored. The processors operate synchronously: during a given time unit a selected number of processors are active and execute the same instruction each on a different data set; the remaining processors are inactive. When two processors wish to communicate they do so through the shared memory: one processor writes a datum in the shared memory which is subsequently read by the other processor. Different processors can access the shared memory at the same time. However, no two processors are allowed to simultaneously read from or write into the same memory address. When more than one processor need the same datum at the same time, a broadcast operation is performed which is best-described by the following ALGOL-like procedure:

procedure BROADCAST(m,N,B)

Input: A datum m in shared memory to be broadcast to N processors numbered 1 to N using a shared memory array B of length N which is initially empty.

Output: Each of the N processors has received m.

Step 1: processor 1 copies m in its own memory and then writes it into B[1].

Step 2: **for** i=0 **to** (logN - 1) **do**

for $j=2^i+1$ **to** 2^{i+1} **do in parallel**

processor j copies $B[j-2^i]$ in its own memory and then writes it into B[j]

end for

end for. []

It is clear that when this procedure which requires O(log N) time terminates each of the N processors is in possession of the datum m.

A particularly useful application of broadcasting in the context of this paper is in letting the processors know the value of the important parameter e when a problem of size n is to be solved. Initially, each of the N processors knows its own number i, $1 \le i \le N$, but not the value of N. Thus, before executing any algorithm, the values of n and N are broadcast to all processors, each of which computes e from $N = n^{1-e}$. We will therefore assume henceforth that e is initially known to all processors at the beginning of computation.

We conclude this section by presenting another parallel procedure which will be often invoked by the algorithms of this paper. Assume that processor i, $1 \le i \le N$, contains in its local memory a number s_i. The following procedure ALLSUMS replaces the s_i in each processor by the sum $s_1 + s_2 + ... + s_i$.

procedure ALLSUMS ($s_1, s_2, ..., s_N$)

Input: N processors, numbered 1 to N, with processor i containing a number s_i.

Output: The number contained in processor i is now $s_1 + s_2 + ... + s_i$, $1 \le i \le N$.

for j=0 **to** (logN - 1) **do**

for i = 2^j+1 **to** N **do in parallel**

processor i obtains s_{i-2^j} through shared memory then computes $s_i = s_i + s_{i-2^j}$

end for

end for. []

This procedure also requires O(log N) parallel time.

3. A PARALLEL ALGORITHM FOR SELECTION

Given a sequence S of n elements drawn from a totally ordered set, and an integer k, $1 \leq k \leq n$, the selection problem calls for finding the kth smallest element in S. This problem has been well studied in the literature on sequential computation and an optimal divide-and-conquer algorithm exists for its solution whose worst-case running time is O(n) [3].

In parallel computation the obvious method for solving this problem is to sort the sequence in nondecreasing order (using any of the existing efficient parallel sorting algorithms described, for example, in [4] or [5]) and then to pick the kth element of the sorted sequence. Evidence from sequential computation, however, suggests that this may be an unduly expensive approach to parallel selection (regardless of how efficiently we can sort in parallel). Indeed when one processor is used, sorting requires $\Omega(n \log n)$ comparisons [6], whereas selection can be done in linear time as pointed out above; by analogy a better parallel selection algorithm must exist. Furthermore, sorting in order to select is absurd if we are to use selection as a step in a parallel sorting algorithm, as will be done in section 4.

We now propose a parallel selection algorithm which is an adaptation of the sequential algorithm of [3] to run on a share-memory SIMD model of computation. The following procedure PARALLEL SELECT assumes the existence of n^{1-e} processors numbered 1 to n^{1-e} and operating in parallel where $0<e<1$. The presence of a sequential selection procedure SELECT is also assumed.

procedure PARALLEL SELECT (S,k)

Input: An unsorted sequence S of n elements and an integer k, $1 \leq k \leq n$.

Output: The kth smallest element of S.

Step 1: If $|S| < 3$ then using one processor (and at most one comparison) return the kth element else subdivide S into n^{1-e} subsequences of n^e elements each and assign a subsequence to each processor.

Step 2: With all processors operating in parallel, processor i finds the median m_i of its associated subsequence using procedure SELECT. Let M be the sequence of medians thus found: the ith position of M holds m_i.

Step 3: Find the median m of M by calling PARALLEL SELECT (M, $|M|/2$).

Step 4: Subdivide S into three subsequences, S_1, S_2 and S_3 of elements smaller than, equal to and larger than m, respectively.

Step 5: If $|S_1| \geq k$ then PARALLEL SELECT (S_1,k)
else if $|S_1| + |S_2| \geq k$ then return m
else PARALLEL SELECT $(S_3, k-|S_1|-|S_2|)$. []

A step-by-step analysis of PARALLEL SELECT follows. In it we assume that T(n) is the parallel running time required by the algorithm.

(1) The beginning address A of sequence S in the shared memory and its size n as well as the value of k are broadcast to all processors; this can be done in $O(\log n^{1-e})$ time using procedure BROADCAST. Processor i computes, in constant time, the address of the first and last elements of its associated subsequence from $(A + (i-1)n^e)$ and $(A + i\,n^e-1)$, respectively. Hence step 1 requires $c_1 \log n$ time units where c_1 is a constant.

(2) If procedure SELECT is the sequential selection algorithm of [3], then the median of each subsequence can be found in $O(n^e)$ operations. Hence step 2 requires c_2n^e time units where c_2 is a constant.

(3) Since $|M| = n^{1-e}$, step 3 requires $T(n^{1-e})$ time.

(4) Subdividing S into S_1, S_2 and S_3 can be done by letting each processor i split its associated subsequence into three lists S_1^i, S_2^i and S_3^i, and then merging the S_1^i's, the S_2^i's and the S_3^i's to form S_1, S_2 and S_3, respectively. Broadcasting m to the n^{1-e} processors can be done in $O(\log n^{1-e})$ time using procedure BROADCAST. The time required by each processor to split its subsequence is linear, i.e. $O(n^e)$. Merging the S_1^i's can be done in $O(n^e)$ time by the following procedure (similar procedures, with the same running time, can be used to merge the S_2^i's and S_3^i's, respectively). Let s_i denote the size of S_1^i, i.e. $s_i = |S_1^i|$. For each i, $1 \leq i \leq n^{1-e}$, the sum

$$z_i = \sum_{j=1}^{i} s_j$$

is computed. All such sums can be obtained by n^{1-e} processors using procedure ALLSUMS in $O(\log n^{1-e})$ time. Taking $z_0 = 0$, all processors simultaneously write their lists in S_1, with processor i starting to copy S_1^i in position $z_{i-1}+1$ of array S_1. The time elapsed during this step is proportional to the length of the longest S_1^i, which cannot exceed n^e. Notice also that $|S_1|$, the size of S_1 needed in step 5, has already been obtained through the computation of $z_{n^{1-e}}$. Hence the time required by step 4 is dominated by c_3n^e where c_3 is a constant.

(5) Since m is the median of M, $n^{1-e}/2$ elements of S are guaranteed to be larger than it. Furthermore, every element of M is smaller than at least $n^{e}/2$ elements of S. Thus $|S_1| \leq 3n/4$. Similarly, $|S_3| \leq 3n/4$. Hence step 5 requires at most T(3n/4) time.

From the above we have

$$T(n) = c_1 \log n + c_2 n^e + c_3 n^e + T(n^{1-e}) + T(3n/4)$$

whose solution is

$$T(n) = O(n^e), \text{ for } n > 4. \text{ Hence,}$$

$$C(n) = n^{1-e} . O(n^e)$$

$$= O(n).$$

PARALLEL SELECT is therefore cost-optimal, to within a constant multiplicative factor, in view of the $\Omega(n)$ lower bound on sequential selection [7].

It is useful to note here that the best available parallel sorting algorithms achieve a total cost of $O(n \log^2 n)$ on the shared-memory SIMD model, which is not optimal. For example, the algorithm of [4] uses n/2 processors and runs in $O(\log^2 n)$ time.[2] On the other hand, a parallel running time for sorting of O(log n) is obtained in [5] by using O(n log n) processors.[3] Both of these algorithms therefore use more processors in general than PARALLEL SELECT. However, for any value of e, PARALLEL SELECT is asymptotically slower than either of these two algorithms. Hence if enough processors are available and speed of execution is the only matter of concern, then sorting, by [4] or [5], say, as a solution to the parallel selection problem may be preferable to PARALLEL SELECT.

4. A PARALLEL ALGORITHM FOR SORTING

Let $S = \{d_1, d_2, ..., d_n\}$ be a sequence of n real numbers to be sorted in nondecreasing order. Several algorithms exist for solving this problem on a sequential model of computation in O(n log n) time [6]. These algorithms are optimal in view of the $\Omega(n \log n)$ lower bound on the number of operations required to sort in the worst case [6]. In this section we describe a parallel sorting algorithm which is an adaptation of the sequential algorithm QUICKSORT [6] to run on the shared-memory SIMD model of computation. As mentioned in the introduction, the algorithm uses n^{1-e} processors, $0<e<1$, runs in $O(n^e \log n)$ time and is therefore cost-optimal. To simplify the presentation we assume that all elements of S are distinct.

It may be helpful to first recall how QUICKSORT operates. This is best done through the following recursive procedure.

procedure QUICKSORT (S)

Input: A sequence S of numbers in random order.
Output: The sequence S sorted into increasing order.

Step 1: If $|S|<3$ then sort S directly using at most one comparison and return else find the median m of S and place it in its final position in the sorted sequence.
Step 2: Create the two subsequences S_1 and S_2 of S, of numbers smaller than and larger than m, respectively.
Step 3: QUICKSORT (S_1)
Step 4: QUICKSORT (S_2). []

Ideally in PARALLEL SORT, the parallel version of QUICKSORT, steps 3 and 4 of QUICKSORT would be executed simultaneously, since each of the two subproblems S_1 and S_2 has the same structure as the original problem S, and contains at most half as many points as S. Unfortunately, this is not possible here: if n^{1-e} processors solve a problem of size n in time T(n), then $(n/2)^{1-e}$ processors would be needed to solve a problem of size n/2 in time T(n/2). In our case, only $n^{1-e}/2$ processors would be assigned to each of S_1 and S_2, which is clearly less than the number required. On the other hand, implementing steps 3 and 4 sequentially would mean that we are solving a problem of size n/2 using n^{1-e} processors (more than the $(n/2)^{1-e}$ required) and many processors would be idle. But more importantly, such an implementation would not achieve the desired performance, as can be easily verified.

We now show how this difficulty can be surmounted. First, the $(2^{1/e}-1)$ percentiles of S are found which divide S into $2^{1/e}$ subsequences of size $(n/2^{1/e})$ each, and these subsequences created. Observe that the union of these subsequences denoted by S_1^j and S_2^j, for j = 1,2,..., $2^{1/e-1}$, forms S_1 and S_2, respectively. Procedure PARALLEL SORT is now applied in parallel to half of these subsequences (namely S_1) using $(n^{1-e}/2^{1/e-1})$ processors per subsequence. The same is then done for the other half (i.e. S_2). It should be noted that the number of processors used to solve each subsequence of size $(n/2^{1/e})$ is exactly the $(n/2^{1/e})^{1-e}$ required for a proper recursive application of the algorithm.[4]

In order to convey the main idea behind the algorithm in the simplest and most intuitive way possible, a formal description of PARALLEL SORT is given below for the case e = 1/2.

procedure PARALLEL SORT (S)

Input: A sequence S of n numbers in random order.

Output: The sequence S sorted into increasing order.

Step 1: If $|S| < 3$ then sort S directly using one processor (and at most one comparison) and return
else find the first, second and third quartiles of S; these elements, denoted by m_1, m_2 and m_3, respectively, are placed in their final positions in the sorted sequence.

Step 2: Split the remaining elements of S into four subsequences of approximately equal size, namely:

$S_1^1 = \{d_i : d_i < m_1\}$,

$S_1^2 = \{d_i : m_1 < d_i < m_2\}$,

$S_2^1 = \{d_i : m_2 < d_i < m_3\}$ and

$S_2^2 = \{d_i : m_3 < d_i\}$.

Step 3: Do steps 3.1 and 3.2 in parallel

3.1 PARALLEL SORT (S_1^1)

3.2 PARALLEL SORT (S_1^2)

Step 4: Do steps 4.1 and 4.2 in parallel

4.1 PARALLEL SORT (S_2^1)

4.2 PARALLEL SORT (S_2^2). []

Let T(n) be the parallel running time of PARALLEL SORT. From section 3 we know that steps 1 and 2 can be implemented in $O(n^e)$ time. Thus, for some constant c,

$$T(n) = cn^e + 2\,T(n/2^{1/e})$$

whose solution is $T(n) = O(n^e \log n)$. Hence,

$$C(n) = n^{1-e} \,.\, O(n^e \log n)$$

$$= O(n \log n).$$

PARALLEL SORT is therefore cost optimal, to within a constant multiplicative factor, in view of the $\Omega(n \log n)$ lower bound on the number of sequential operations required to sort [6].

Table 1 provides a comparison of PARALLEL SORT with other parallel sorting algorithms for the shared-memory SIMD model. The table is self-explanatory. It is perhaps worthwhile to point out that the algorithms of [4], [9], [10], [11], [13] and [15] were originally proposed for parallel computers where the processors communicate through a special interconnection network. As pointed out in section 3, however, these algorithms can be simulated on the shared-memory SIMD machine while retaining their asymptotic performance. As the table shows, all the algorithms that are asymptotically faster than PARALLEL SORT use more processors and are not cost-optimal. Hence, if enough processors are available and speed of execution is the only matter of concern then sorting by either one of [4] or [13], say, may be preferable to PARALLEL SORT. The algorithm of [12] has a performance comparable to that of PARALLEL SORT except for the fact that it has memory fetch conflicts and is optimal only for a restricted range of values of N, the number of processors.

Property / Algorithm	P(n)	T(n)	Cost-Optimal?	Adaptive?	Free of memory fetch conflicts?
Todd[9]	log n	n	yes	no	yes
Thompson [10] Nassimi[11]	n	$n^{1/2}$	no	no	yes
Stone [4]	n	$\log^2 n$	no	no	yes
PARALLEL SORT	n^{1-e} $0<e<1$	$n^e \log n$	yes	yes	yes
Shiloach [12]	$1 \leq N \leq n$	$(n/N) \log n + \log n \log N$	yes,if $N \leq n/\log n$	yes	no
Preparata [5] Nassimi [13]	$n^{1+1/k}$ $1 \leq k \leq \log n$	k log n	no	yes	yes
Preparata [5] Borodin [14] Ajtai [15]	n log n	log n	no	no	no

Table 1

5. A PARALLEL ALGORITHM FOR COMPUTING CONVEX HULLS

Given a set $S = \{p_1, p_2, ..., p_n\}$ of n points in the plane, the convex hull of S, denoted by CH(S), is a convex polygon whose corners are points of S such that every point of S is either a corner of CH(S) or lies inside CH(S). The problem of computing the convex hull on a sequential model of computation has been thoroughly studied over the past decade and a number of algorithms exist for its solution whose running time is O(n log n) [16-19]. These algorithms are worst-case optimal as their complexity matches the Ω(n log n) lower bound on the convex hull problem [20]. Recently, a new algorithm was proposed in [21] whose complexity is O(n log h) where h is the number of edges of CH(S). This algorithm is superior to the previous ones in the sense that its running time is sensitive to the size of the output. In the worst-case, of couse, h=n and the algorithm's performance matches that of its predecessors.

In this section we describe a parallel convex hull algorithm which is an adaptation of the algorithm in [21] to run on a shared-memory SIMD model of computation. The algorithm uses n^{1-e} processors and runs in $O(n^e \log h)$ time, where $0<e<1$. This performance matches the best current bound on sequential computation of the convex hull achieved by the algorithm of [21]. Since in the worst case the algorithm's running time is $O(n^e \log n)$, its cost is O(n log n) which is optimal. In that sense the new algorithm is superior to the best available parallel algorithms for this problem as will be shown later.

In what follows we begin by briefly reviewing the sequential convex hull algorithm of [21]. A parallel implementation of the algorithm and its analysis are then presented. We conclude this section by comparing the new parallel algorithm with existing ones.

5.1 The Sequential Convex Hull Algorithm

We start by presenting a simplified version of the algorithm of [21]. Let x_i and y_i denote the x and y cartesian coordinates, respectively, of point p_i. We will assume for ease of presentation that no two points have the same x or y coordinates and that no three points fall on a straight line.

Let p_{min} and p_{max} denote the points with minimum and maximum x coordinates, respectively. Then CH(S) consists of two parts: an upper convex polygonal path from p_{min} to p_{max}, and a lower one, as shown in Fig. 1. Given these two polygonal paths, they can be concatenated to yield CH(S). Thus the algorithm for finding the convex hull of S can be described as follows:

procedure CONVEX HULL(S)

Input: A set S of n points in the plane

Output: The convex hull of S.

Step 1: Determine min and max, the indices of the points of S with minimum and maximum coordinates, respectively.

Step 2: Find the upper convex polygonal path.

Step 3: Find the lower convex polygonal path.

Step 4: Concatenate the two paths found in steps 2 and 3. []

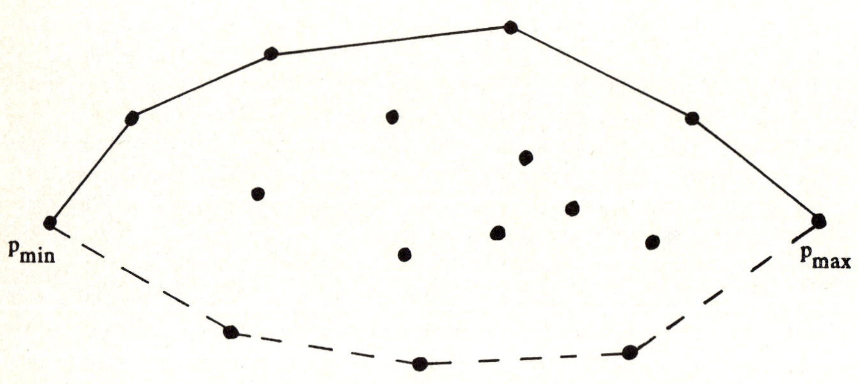

Fig. 1

Since steps 1 and 4 of the above procedure are obvious, we now concentrate on the description of an algorithm for step 2. This algorithm can be easily modified to yield an algorithm for step 3. The following procedure UPPER HULL (S,k,m,) determines the upper convex polygonal path from p_k to p_m. It first places a vertical line L dividing S into two sets S_{left} ad S_{right} of approximately the same size and then determines the unique edge of the upper convex polygonal path intersecting this line as shown in Fig. 2. The procedure is then applied recursively to S_{left} and S_{right}. Initially, the procedure is invoked from CONVEX HULL(S) with k = min and m = max.

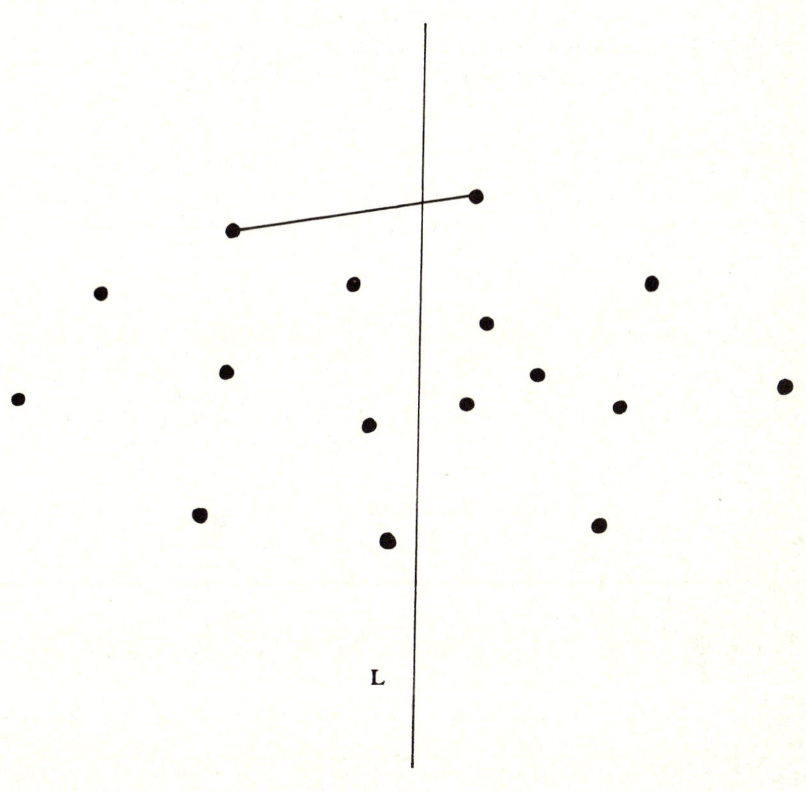

Fig. 2

procedure UPPER HULL (S,k,m)

Input: A set S of n points in the plane and the indices k and m of two selected points of S.

Output: The upper convex polygonal path of S from p_k to p_m.

Output: The upper convex polygonal path of S from p_k to p_m.

Step 1: Find the median of the set $\{x_1, x_2, ..., x_n\}$, i.e. find a real number A which splits S into two subsets of approximately equal size S_{left} and S_{right} containing points of S with an x coordinate smaller than A and larger than A, respectively.

Step 2: Find the unique edge (p_i, p_j) of the upper convex polygonal path intersecting the vertical line x = A.

Step 3: If i = k then output i

else UPPER HULL (S_{left},k,i).

Step 4: If j = m then output j

else UPPER HULL (S_{right},j,m). []

As steps 1, 3 nd 4 are straightforward, we conclude our description by showing how step 2 can be implemented. The following procedure BRIDGE (S,A) returns the indices of the two points p_i and p_j where (p_i, p_j) is the unique edge of the polygonal path intersecting the vertical line x = A.

procedure BRIDGE (S,A)

Input: A set S of n points in the plane and a vertical line x = A dividing the set.

Output: The edge of the upper convex polygonal path of S crossing the vertical line.

Step 1: The points of S are paired up into couples (p_a, p_b) such that $x_a < x_b$. These ordered pairs define $\lfloor n/2 \rfloor$ straight lines whose slopes are $\{s_1, s_2, ..., s_{\lfloor n/2 \rfloor}\}$.

Step 2: Find the median K of the set $\{s_1, s_2, ..., s_{\lfloor n/2 \rfloor}\}$.

Step 3: Find a straight line Q of slope K which contains at least one point of S but has no point of S above it.

Step 4: If Q contains two points of S, one on each side of L, then return their indices. Else if Q contains no points of S_{right} then for every straight line defined by (p_a, p_b) whose slope is larger than or equal to K, $S \leftarrow S - \{p_a\}$. Else if Q contains no points of S_{left} then for every straight line defined by (p_a, p_b) whose slope is smaller than or equal to K, $S \leftarrow S - \{p_b\}$.

Step 5: BRIDGE (S,A). []

A detailed description of CONVEX HULL(S), together with a correctness proof and analysis are provided in [21], where it is shown that the algorithm runs in O(n log h) time when CH(S) has h edges.

5.2 A Parallel Implementation of CONVEX HULL

We now turn to the main theme of this section, namely how CONVEX HULL(S) can be implemented in parallel. Assume that n^{1-e} processors are available where $0 < e < 1$. In what follows we present a step-by-step description and analysis of the parallel implementation.

PARALLEL CONVEX HULL (S)

Step 1 of this procedure can be implemented using PARALLEL SELECT which, as we know from section 3, runs on n^{1-e} processors in $O(n^e)$ time. Steps 2 and 3 are discussed below in the analysis of UPPER HULL. Let h_U and h_L be the number of edges of the upper and lower convex polygonal paths, respectively. We denote the complexities of steps 2 and 3 by $F_U(n, h_U)$ and $F_L(n, h_L)$, respectively. Step 4 can be executed in constant time by a single processor. Thus, if t(n,h) is a function describing the running time of PARALLEL CONVEX HULL(S), where $h=h_U + h_L$, we have

$$t(n,h) = c_1 n^e + F_U(n, h_U) + F_L(n, h_L) + c_2,$$

where c_1 and c_2 are constants.

PARALLEL UPPER HULL (S,k,m)

Ideally, in a parallel implementation of UPPER HULL, steps 3 and 4 should be executed simultaneously since each of the two subproblems S_{left} and S_{right} has the same structure as the original problem S. Unfortunately, this is not possible for the same reasons presented in section 4 and we therefore resort to a solution similar to the one used in the case of PARALLEL SORT. First, $(2^{1/e} - 1)$ points are found in the plane whose x coordinates divide S into $2^{1/e}$ subsets of size $(n/2^{1/e})$ each, and vertical lines through them drawn. Note that the union of these subsets thus created and denoted by S^j_{left} and S^j_{right}, for $j = 1,2, ..., 2^j$, forms S_{left} and S_{right}, respectively. The upper hull edges crossing these vertical lines are now computed through repeated application of procedure BRIDGE. Procedure PARALLEL UPPER HULL is now applied recursively and in parallel to half of these subsets (namely S_{left}) using $(n^{1-e}/2^{1/e}-1)$ processors per subset.

The same is then done for the other half (i.e. S_{right}). A formal description of PARALLEL UPPER HULL is given below for the case e = 1/2.

procedure PARALLEL UPPER HULL (S,k,m)

Input: A set S of n points in the plane and the indices k and m of two selected points of S.

Output: The upper convex polygonal path of S from p_k to p_m.

Step 1: Find the first, second and third quartiles of the set $\{x_1, x_2, ..., x_n\}$ i.e. find real numbers A_1, A_2 and A_3 which split S into four subsets of approximately equal size, namely

$S^1_{left} = \{x_i : x_i < A_1\}$,

$S^2_{left} = \{x_i : A_1 < x_i < A_2\}$,

$S^1_{right} = \{x_i : A_2 < x_i < A_3\}$ and

$S^2_{right} = \{x_i : A_3 < x_i\}$.

Step 2: For each of the vertical lines $x = A_1$, $x = A_2$ and $x = A_3$, find the unique edge of the upper convex polygonal path intersecting it. Let these edges be (p_{i_1}, p_{j_1}), (p_{i_2}, p_{j_2}) and (p_{i_3}, p_{j_3}), respectively.

Step 3: Do 3.1 and 3.2 in parallel

3.1: If $i_1 = k$ then output i_1
else PARALLEL UPPER HULL (S^1_{left}, k, i_1).

3.2: If $j_1 = i_2$ then output j_1

else if $i_1 \neq i_2$ then PARALLEL UPPER HULL (S^2_{left}, j_1, i_2).

Step 4: Do 4.1 and 4.2 in parallel

4.1: If $j_2 = i_3$ then output j_2

else if $i_2 \neq i_3$ then PARALLEL UPPER HULL (S^1_{right}, j_2, i_3).

4.2: If $j_3 = m$ then output j_3

else PARALLEL UPPER HULL (S^2_{right}, j_3, m). □

Step 1 of PARALLEL UPPER HULL can be implemented using PARALLEL SELECT and hence runs in $O(n^e)$ time. Step 2 is discussed below in the analysis of PARALLEL BRIDGE; let its complexity function be denoted by B(n). Thus, for some constants c_3 and c_4,

$$F_U(n,h_U) = c_3 n^e + c_4 B(n) +$$

$$\max_{h_1+h_2=h_U-1} \{\max_{h_1^j} [F_U(|S^j_{left}|, h_1^j)] + \max_{h_2^j} [F_U(|S^j_{right}|, h_2^j)]\},$$

where h_1, h_1^j, h_2, h_2^j are the number of edges on the upper convex polygonal paths associated with S_{left}, S^j_{left}, S_{right} and S^j_{right}, respectively.

<u>PARALLEL BRIDGE (S,A)</u>

Step 1 of this procedure is performed in parallel by assigning different subsets of S of size n^e to the n^{1-e} processors each of which creates $\lfloor n^e/2 \rfloor$ pairs of points (p_a, p_b) and computes the slopes of the straight lines they form. Step 1 thus requires $O(n^e)$ time. Step 2 can be implemented using PARALLEL SELECT in $O(n^e)$ time. Step 3 is executed by finding the (at most two) points maximizing the quantity $(y_i - Kx_i)$. This quantity can be obtained for all values of i by having each processor compute it for the points in its assigned subset of S. The maximum of these quantities is found using PARALLEL SELECT. Hence step 3 is also $O(n^e)$. Finally, in step 4, determining whether Q contains the required edge can be done by one processor in constant time. Otherwise, the value of K is broadcast to all n^{1-e} processors in $O(\log n^{1-e})$ time using procedure BROADCAST. Each processor compares K to the $\lfloor n^e/2 \rfloor$ slopes it has computed in step 1, and updates S accordingly; this requires $O(n^e)$ time. The overall time requirement of step 4 is therefore $O(n^e)$. Since one-quarter of the points are discarded in step 4, the complexity of step 5 is B(3n/4). Thus, for some constant c_5,

$$B(n) = c_5 n^e + B(3n/4)$$

whose solution is $B(n) = O(n^e)$.

Therefore, $F_U(n, h_U) = O(n^e \log h_U)$ and,

similarly $F_L(n, h_L) = O(n^e \log h_L)$.

It follows that $t(n,h) = O(n^e \log h)$.

The total number of operations required by PARALLEL CONVEX HULL is thus $O(n \log h)$ which matches the best currently known upper bound for computing convex hulls sequentially [21]. Furthermore, when h=n, the worst-case running time of the algorithm is $T(n) = O(n^e \log n)$. Thus,

$$\begin{aligned} C(n) &= n^{1-e} \,.\, O(n^e \log n) \\ &= O(n \log n) \end{aligned}$$

and hence PARALLEL CONVEX HULL is cost-optimal, to within a constant multiplicative factor, in view of the $\Omega(n \log n)$ worst-case lower bound on sequential convex hull computation [20].

5.3 Comparison with other parallel convex hull algorithms

Table 2 compares PARALLEL CONVEX HULL to existing parallel convex hull algorithms for the shared-memory SIMD model of computation. Two points are worth making here:

(1) The algorithms of [22] and [24] were proposed for parallel processor networks that can be easily simulated by the shared-memory SIMD model.

(2) The algorithm of [22], as described therein, computes the convex hull in constant time using n^3 processors on a model of computation which assumes unbounded fan-in for logical gates [25]. If this assumption is removed then the algorithm runs in O(log n) time on the shared-memory SIMD model as indicated in the table.

As the table shows, the first three algorithms are asymptotically faster than PARALLEL CONVEX HULL but use more processors and are not cost-optimal. Hence if enough processors are available and speed of execution is the only matter of concern then computing convex hulls by either one of [22] or [23], say, may be preferable to using PARALLEL CONVEX HULL. The performance of the fourth algorithm is comparable to that of PARALLEL CONVEX HULL except for the fact that the algorithm is cost-optimal only for a restricted range of values of the number of processors.

Property / Algorithm	P(n)	T(n)	Cost-optimal?	Adaptive?
Akl [22]	n^3	log n	no	no
Nath [23]	$n^{1+1/k}$ $1 \le k \le \log n$	k log n	no	yes
Nath [23] Chow [24]	n	$\log^2 n$	no	no
Nath [23]	$1 \le N \le n$	(n/N) log n + logn logN	yes, if $N \le n/\log N$	yes
PARALLEL CONVEX HULL	n^{1-e} $0<e<1$	$n^e \log n$	yes	yes

Table 2

6. CONCLUSION

Parallel algorithms were proposed for three problems which permeate computational geometry, namely, selection, sorting and computing convex hulls. The algorithms possess the following three properties:

(1) the number of processors they use is sublinear in the size of the problem to be solved and adapts to the degree of parallelism on the available parallel computer;

(2) their running time is significantly smaller than that of their best sequential counterparts and varies inversely with the number of processors;

(3) they are cost-optimal in the sense that the number of processors they use multiplied by their worst-case running time is equal to a known lower bound on the number of sequential operations required to solve the problem each of them addresses.

The three algorithms thus provide good examples of the number of processors versus running time tradeoff.

For each of the new algorithms, it was pointed out that at least one asymptotically faster algorithm already exists in the literature which uses more processors and is not cost-optimal. This leads us in conclusion to pose the following open problems for any model of parallel computation: For some u, $0 \leq u \leq 1$,

(1) Is there an algorithm for parallel selection which uses $O(n/\log^u n)$ processors and runs in $O(\log^u n)$ time?

(2) Are there parallel algorithms for computing convex hulls and for sorting which use $O(n/\log^{u-1} n)$ processors and run in $O(\log^u n)$ time?

A positive answer to any of these questions would provide an algorithm superior in all respects to the corresponding algorithms discussed in this paper.

7. FOOTNOTES

1. All logarithms in this paper are base 2 and all real function values are rounded to the nearest integer.

2. Note that the algorithm of [4] (in reality an adaptation of a parallel sorting algorithm first described in [8]) is designed for a parallel computer where processors communicate through an interconnection network known as the perfect shuffle. Modifying the algorithm of [4] to run on the shared-memory SIMD model (while keeping its number of processors and time requirements unchanged) is however straightforward if the direct links connecting the processors are simulated by alternated write and read operations by the processors, into and from the common memory, respectively.

3. Note, however, that the algorithm of [5] runs on a shared-memory SIMD model of computation which, unlike ours, allows two or more processors to simultaneously read from the same memory location. Such an algorithm is said to have memory fetch conflicts.

4. In practice, of course, if e is less than 0.5 and is not such that

(i) $1/e$ is an integer less than or equal to 10, say, and
(ii) $n \geq 2^{1/e}$

then the smallest real number larger than e and satisfying (i) and (ii) is taken as e.

8. REFERENCES

[1] Taub, A.H., Ed., The Collected Works of John von Neumann 5 (Macmillan, New York, 1963) 34-79.

[2] Flynn, M.J., Very high speed computing systems, Proceedings of the IEEE 54 (1966) 1901-1909.

[3] Blum, M., Floyd, R., Pratt, V., Rivest, R. and Tarjan, J., Time bounds for selection, Journal of Computer and System Science 7 (1972) 448-461.

[4] Stone, H.S., Parallel processing with the perfect shuffle, IEEE Transactions on Computers C-20 (1971) 153-161.

[5] Preparata, F.P., New parallel sorting schemes, IEEE Transactions on Computers C-27 (1978) 669-673.

[6] Knuth, D.E., The Art of Computer Programming 3 (Addison-Wesley, Don Mills, Ontario, 1973).

[7] Hyafil, L., Bounds for selection, SIAM Journal of Computing 5 (1976) 109-114.

[8] Batcher, K.E., Sorting networks and their applications, Proceedings of the AFIPS 1968 SJCC 32 (1968) 307-317.

[9] Todd, S., Algorithms and hardware for a merge sort using multiple processors, IBM Journal of Research and Development 22 (1978) 509-517.

[10] Thompson, C.D. and Kung, H.T., Sorting on a mesh connected parallel computer, Communications of the ACM 20 (1977) 263-271.

[11] Nassimi, D. and Sahni, S., Bitonic sort on a mesh connected parallel computer, IEEE Transactions on Computers C-27 (1978) 2-7.

[12] Shiloach, Y. and Vishkin, U., Finding the maximum, merging and sorting in a parallel computation model, Journal of Algorithms 2 (1981) 88-102.

[13] Nassimi, D. and Sahni, S., Parallel permutation and sorting algorithms and a new generalized connection network, Journal of the ACM 29 (1982) 642-667.

[14] Borodin, A. and Hopcroft, J.E., Routing, merging and sorting on parallel models of computation, Proceedings of the 14th Annual ACM Symposium on Theory of Computing, San Francisco, California (1982) 338-344.

[15] Ajtai, M., Komlos, J. and Szemeredi, E., An O(n log n) sorting network, Proceedings of the 15th Annual ACM Symposium on Theory of Computing, Boston, Massachussetts (1983) 1-9.

[16] Graham, R.L., An efficient algorithm for determining the convex hull of a finite planar set, Information Processing Letters 1 (1972) 132-133.

[17] Shamos, M.I. and Hoey, D., Closest-point problems, Proceedings of the 16th Annual IEEE Symposium on Foundations of Computer Science, New York (1975) 151-162.

[18] Preparata, F.P. and Hong, S.J., Convex hulls of finite sets of points in two and three dimensions, Communications of the ACM 20 (1977) 87-93.

[19] Akl, S.G. and Toussaint, G.T., A fast convex hull algorithm, Information Processing Letters 7 (1978) 219-222.

[20] Yao, A.C.-C., A lower bound to finding convex hulls, Journal of the ACM 28 (1981) 780-787.

[21] Kirkpatrick, D.G. and Seidel, R., The ultimate planar convex hull algorithm?, Proceedings of the 20th Allerton Conference on Communication, Control and Computing, Monticello, Illinois (1982) 35-42.

[22] Akl, S.G., A constant-time parallel algorithm for computing convex hulls, BIT 22 (1982) 130-134.

[23] Nath, D., Maheshwari, S.N. and Bhatt, P.C.P., Parallel algorithms for the convex hull problem in two dimensions, Technical Report EE 8005, Department of Electrical Engineering, Indian Institute of Technology, New Delhi, India (October 1980).

[24] Chow, A.L., A parallel algorithm for determining convex hulls of sets of points in two dimensions, Proceedings of the 19th Allerton Conference on Communication, Control and Computing, Monticello, Illinois (1981) 214-223.

[25] Chandra, A.K., Stockmeyer, L.J. and Vishkin, U., A complexity theory for unbounded fan-in parallelism, Proceedings of the 23rd Symposium on Foundations of Computer Science, Chicago, Illinois (1982) 1-13.

COMPUTATIONAL GEOMETRY, G.T. Toussaint (Editor)
© Elsevier Science Publishers B.V. (North Holland), 1985

SIMPLE ON-LINE ALGORITHMS FOR CONVEX POLYGONS

David Avis and *Hossam ElGindy*

School of Computer Science
McGill University
805 Sherbrooke Street West
Montreal, Quebec, H3A 2K6 Canada

Raimund Seidel

Department of Computer Science
Cornell University
Ithaca NY. 14853 USA

Abstract:

We give simple O(log$n$) algorithms for the point insertion and halfplane intersection problem for convex polygons. Both algorithms use the same data structure for representing convex polygons. Thus they yield an O(n log$n$) on-line algorithm for building up a convex polygon by a sequence of n point insertions and halfplane intersection.

I. Introduction

Among the problems in computational geometry the planar convex hull problem is one of the oldest and best studied. The task is to determine the smallest convex polygon that contains a given finite set of points in the plane. The first asymptotically worst case optimal O(n log$n$) algorithm was given by Graham in 1972 [Gra], a few years before even the term "computational geometry" was actually coined. Subsequently, further optimal algorithms were presented by Shamos [Sha], who was also the first to show that $\Omega(n \log n)$ is a lower bound to certain versions of the convex hull problem, by Akl and Toussaint [A-T], by Preparata [Pre], and by Preparata and Hong [P-H]. Eddy [Edd] and also Jarvis [Jar] developed O(nH) methods, i.e. algorithms whose worst case running time depends not only on n, the number of input points, but also on H, the number of vertices the convex hull is found to have. Finally, Kirkpatrick and Seidel [K-S] presented an O(n log$H$) algorithm and showed that this time bound is asymptotically worst case optimal even if the complexity of the problem is measured in terms of both n and H.

Numerous other papers have appeared dealing with different aspects ([BFP],[B-S], [D-T],[Yao],[Avi],[vEB]) and generalizations ([EKS],[C-K],[Sei]) of the planar convex hull problem. This note deals with updating the convex hull of a finite point set. Given a suitable representation of the convex hull of a point set S, how can one efficiently construct the convex hull of $S \cup \{p\}$ for some new point p (point insertion), or how can one construct the convex hull of $S - \{p\}$, where p is in S (point deletion). Preparata [Pre] proposed an O(log$n$) solution for the insertion problem and later van Leeuwen and Overmars [L-O] as well as Gowda [Gow] gave $O(\log^2 n)$ solutions for the deletion and insertion problem.

The purpose of this note is to present very simple $O(\log n)$ algorithms for point insertion and halfplane insertion problems. The halfplane insertion problem requires to efficiently construct the intersection of a convex hull of a finite point set (i.e. a convex polygon) and a halfplane bounded by some straight line, when a suitable representation of the polygon is given. Both algorithms use the same representation for convex polygons. Thus an intermixed sequence of insertions of points and halfplanes can be executed on-line, i.e. after processing a point or a halfplane in the insertion sequence the representation of the current hull is fully available. Our algorithms are extremely simple and are based on a single geometric primitive, namely the position of a point with respect to a directed line.

II. Convex Hulls, Convex Polygons, and their Representations

There are several equivalent ways of defining the convex hull of a point set. We say that the *convex hull of S*, denoted by conv S, is the intersection of all convex sets in the plane containing S. For a finite set S, conv S is always a convex polygon whose vertices (corners) are elements of S.

We can represent conv S by a circular doubly linked list L with an element corresponding to each vertex of conv S. For each vertex v of conv S there is an element p in L with $p.x$ and $p.y$ containing the cartesian coordinates of v, and $p.next$ and $p.previous$ pointing to the next and previous elements in L that correspond to the vertices of conv S that are next and previous to v, respectively, if conv S is traversed in counterclockwise order. (see Figure 2.1.)

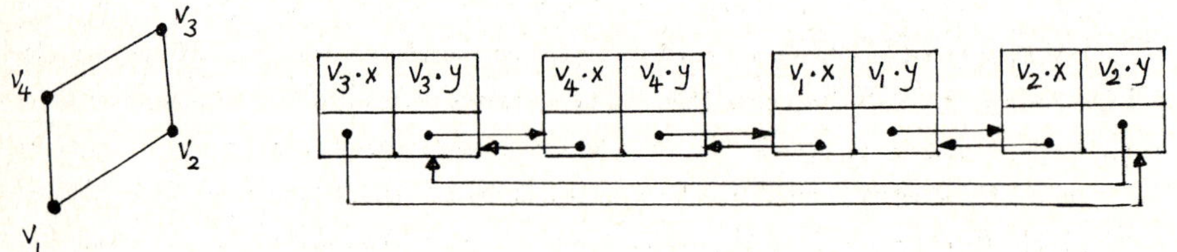

Figure 2.1

This doubly linked list description is not quite adequate since it does not allow rapid access to arbitrary elements in the list. To provide this access, we use a 2-3-tree T whose leaves from left to right are consecutive elements of L, starting at some arbitrary element in L. Every non-leaf node q of T has in addition to the three pointers that can point to sons of q two pointers $q.leftmost$ and $q.rightmost$ pointing to the leftmost and rightmost leaf of the subtree rooted at q, respectively. Furthermore, $q.no\_of\_sons$ denotes the number of sons of q, and $q.height$ denotes the height of q in

T, i.e. the height is 0 for a leafnode and it is one plus the height of any son for a non-leaf node. Note that *q.rightmost* and *q.no_of_sons* could be dispensed with. However, keeping them adds to the clarity of our algorithms. A complete representation for the convex hull of *Figure* 2.1 is given pictorially in *Figure* 2.2. An important property of *2–3–trees* is that they can be "merged" or "concatenated" rapidly. We will take advantage of this property and for completeness we have included a concatenation procedure in Appendix C.

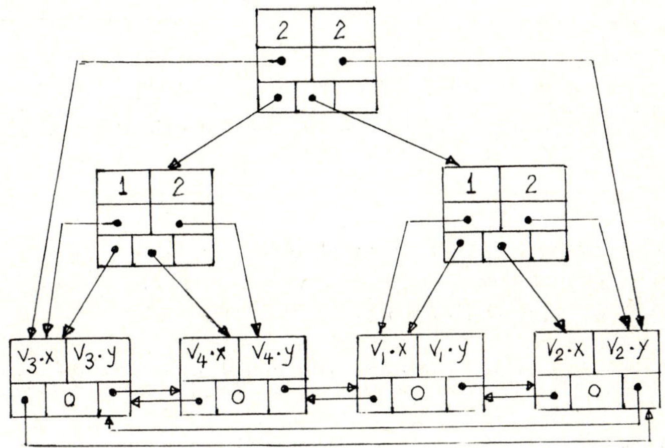

Figure 2.2

III. Point Insertion

In this section we present an algorithm for the point insertion problem. From a 2-3-tree T representing the convex polygon $P = \text{conv} S$, as described in the previous section, one wants to compute the tree T' representing the polygon $P' = \text{conv } S'$, where $S' = S \cup \{p\}$ for some point p.

There are two cases to consider. First, if p lies inside the convex polygon P, then clearly P' equals P and hence T is left unchanged. On the other hand, if p is not contained in P, then we must determine vertices of the convex polygon P which are the left and right "tangent points" for p and replace the chain D of vertices between the left and right tangent points, when P is traversed in counterclockwise order, by the point p itself. A vertex of the convex polygon P is a left (right) tangent point for p if P lies to the left (right) of the directed line joining the vertex to the point p. (see Figure 3.1.)

Let us first concentrate on how it can be determined from our tree representation which vertex of P is a tangent point for p. The following lemma provides the key information for how this operation

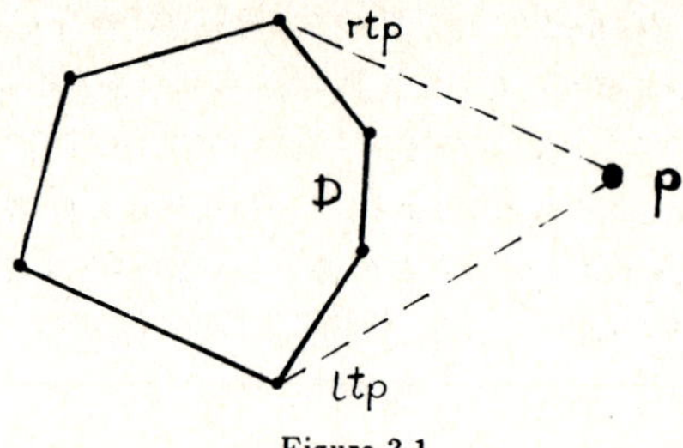

Figure 3.1

can be done efficiently.

Lemma 3.1:

Let T be a tree representing a convex polygon P with n vertices $p_1, p_2, \cdots, p_n$, as described in the previous section, and let p be some point in the plane.

Let q be a node in T that is not the root and let C_q be the chain of vertices of P corresponding to the leaves of the subtree T_q rooted at q.

(1) From q it can be determined in constant time whether, in case a left tangent point ltp for p exists, ltp is a vertex in C_q.

(2) From q it can be determined in constant time whether, in case a right tangent point rtp for p exists, rtp is a vertex in C_q.

Proof:

We outline the proof for the case of the left tangent point. The case of the right tangent point is similar.

Partition the plane into $n+1$ regions $I_0, I_1, \cdots, I_n$, where $I_0 = P$ and for each $k, 1 \leq k \leq n$, region I_k comprises exactly all points in the plane for which p_k is the left tangent point. (see Figure 3.2.)

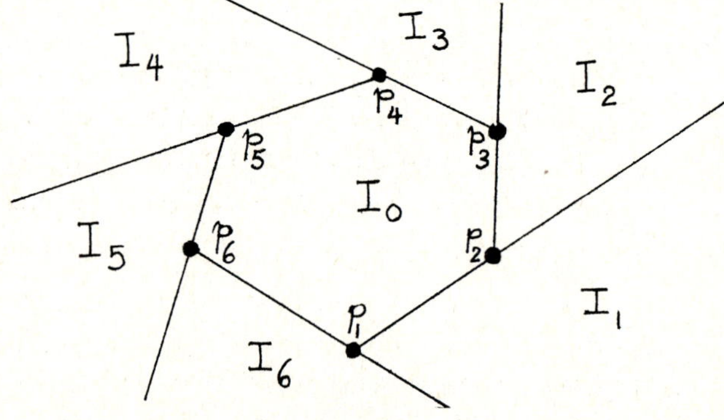

Figure 3.2

Let C_q be the chain $p_1, p_1.next, \cdots, p_m$. If a vertex in C_q is the left tangent point of the point p, then p lies in the union of the regions corresponding to the vertices of the chain C_q, denoted by I_q. However, the boundary of I_q can consist of an arbitrary number of edges and thus membership of p in I_q cannot possibly be tested in constant time. To circumvent this problem enlarge I_q to the region I_q' by including the part of I_0 bounded by the chain of edges of C_q and the edge joining p_1 and $p_m.next$. (see Figure 3.3.) Since the region I_q is completely contained in the enlarged region I_q', p lies in I_q'. As this region is bounded by at most three edges only, membership can clearly be tested in constant time.

Q.E.D.

In Appendix A we give a detailed description of the function CONTAINS_LTP(q,p) for testing whether a vertex of C_q is the left tangent point of the point p. The function CONTAINS_RTP(q,p) is similar.

With Lemma 3.1 it is easy to see that the following function correctly identifies the vertex of a convex polygon P, represented by a tree T, which is the left tangent point of a point p. A function for finding the right tangent point is defined in the same way.

```
FIND_LEFT_TANGENT_POINT(q,p)
begin
   if q is a leafnode then
      q is the left tangent point for p
   else
      for all sons r of q do
         if CONTAINS_LTP(r,p) then
               FIND_LEFT_TANGENT_POINT(r,p)
               return
      print "no left tangent point exists for p"
end
```

The execution time of this procedure is proportional to the height of the tree T, i.e. it is $O(\log n)$, where n is the number of vertices of P. This bound relies heavily on the fact that the test CONTAINS_LTP can be performed in constant time.

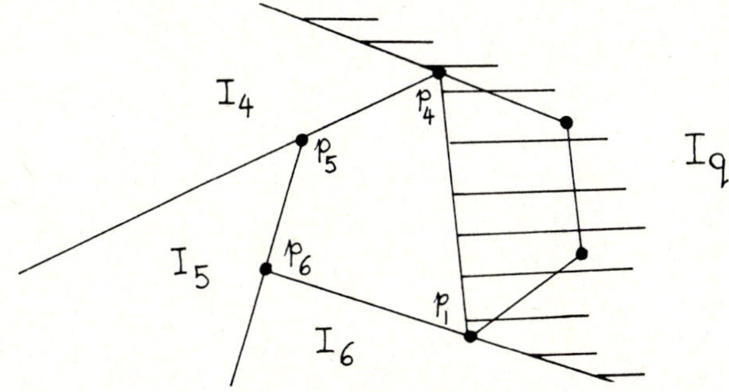

Figure 3.3

So far we have shown how the left and right tangent points for some point p can be determined quickly. Now it remains to show how the tree T representing P has to be modified to a tree T' that represents P' .

Let us assume that the point p lies outside the polygon P and let ltp and rtp denote the left and right tangent point of P for p, respectively. The chain C' of vertices representing P' is obtained from the chain C of vertices representing P by deleting the chain D of vertices between ltp and rtp, when P is traversed in counterclockwise order, and replacing it by the point p. Note that the tree T stores the chain C in counterclockwise order starting at some arbitrary vertex v which could be in D or not. If the leftmost leaf of T happens to be the right tangent point rtp, we say that T is in *normalform*. Updating T becomes particularly simple when it is in normal form. In this case D comprises exactly all the leaves of T that lie to the right of ltp. Thus updating T to T' amounts to locating ltp in T, deleting the chain of leaves to the right of ltp, and inserting p instead. (see Figure 3.4.)

A detailed procedure NORMALIZED_UPDATE(T,p) is given below. It is assumed that it has been established that p lies outside the polygon represented by T and that the right tangent point is the leftmost leaf of T. Also we assume that the tree T is not empty. This case requires a straight forward sequence of steps.

Lemma 3.2:

The running time of the function NORMALIZED_UPDATE(T,p) is proportional to the height of the tree T.

Proof:

The procedure FIND_LTP_AND_COLLECT_LEFT_SIDE first descends on a path of T from the root to some leaf, and then ascends this path again collecting all the subtrees hanging off to the left of this path into a single tree T_1. At each node on the way down, the call to the function

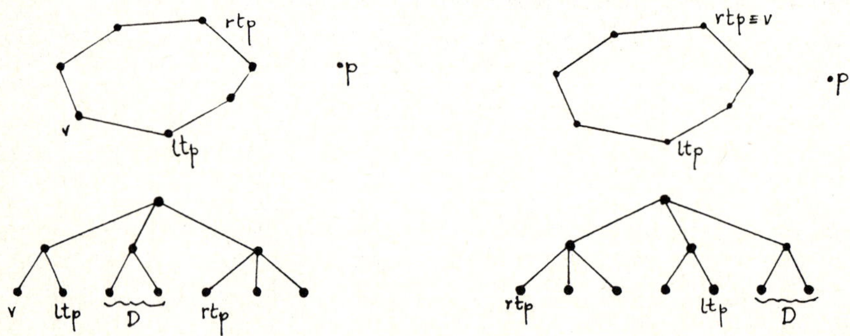

Figure 3.4

```
NORMALIZED_UPDATE(T,p)
    local variable T_1
    FIND_LTP_AND_COLLECT_LEFT_SIDE(q,p)
    begin
       if q.height = 0 then
             T_1 := CONCATENATE(q,p)
             q.next := p
             p.previous := q
             return
       else
             for i:=1 to q.no_of_son do
                if CONTAINS_LTP(q.son[i],p) then
                      FIND_LTP_AND_COLLECT_LEFT_SIDE(q.son[i],p)
                      for j:= i-1 downto 1 do
                            T_1 := CONCATENATE(q.son[j],T_1)
                      return
    end
begin
   FIND_LTP_AND_COLLECT_LEFT_SIDE(T,p)
   T_1.leftmost.previous := T_1.rightmost
   T_1.rightmost.next    := T_1.leftmost
   return (T_1)
end
```

CONTAINS_LTP takes a constant time. Therefore, descending from the root to a leaf of T takes time proportional to the height of T. Using the fact that two 2-3-trees can be concatenated in time proportional to the difference of their heights, it is easy to show with an inductive argument that the overall time taken by the concatenations during the ascend is proportional to the height of T. (cf. [AHU] p.156.)

Q.E.D.

To rearrange the tree T such that the right tangent point *rtp* for a point p becomes the leftmost leaf in T, one needs to locate *rtp* in T, "split" T into two trees T_1 and T_2 such that leaves of T_1 are the leaves of T to the left of *rtp* and leaves of T_2 are the leaves of T to the right of and including *rtp*, and finally concatenate T_2 and T_1, i.e. reverse their relative order. Detailed description of the procedure NORMALIZE(T,p) is given below. Note that the tree is left unchanged if p is discovered to be inside the polygon represented by T. For convenience, we again assume that T is not empty.

```
NORMALIZE(T,p)
        global variable p_outside
        local variable T1,T2

        FIND_RTP_AND_SPLIT(q,p)
        begin
           if q.height = 0 then
                  p_outside := true
                  T1 := NULL
                  T2 := q
                  return
           else
                  for i=1 to q.no_of_sons do
                     if CONTAINS_RTP(q.son[i],p) then
                        FIND_RTP_AND_SPLIT(q.son[i],p)
                        if p_outside then
                           for j=i-1 downto 1 do
                                 T1 := CONCATENATE(q.son[j], T1)
                           for j=i+1 to q.no_of_sons do
                                 T2 := CONCATENATE(T2,q.son[j])
                           return
                  p_outside := false
                  return
        end

begin
   FIND_RTP_AND_SPLIT(T,p)
   if p_outside then
          return ( CONCATENATE(T2,T1) )
   else
          return ( T )
end
```

Lemma 3.3:

The function NORMALIZE(T,p) runs in time proportional to the height of T.

Proof: The proof is identical to the proof of Lemma 3.2 and is therefore omitted.

Now we can completely specify a function INSERT_POINT(p,T) which given a point p and a 2-3-tree T representing the convex hull of a set S returns a 2-3-tree representing the the convex hull of $S \cup \{p\}$.

```
INSERT_POINT(p,T)
      local variables p_outside, T'

begin
      T' := NORMALIZE(T,p)
      if p_outside then
                  return ( NORMALIZED_UPDATE(T' ,p) )
         else
                  return ( T )
end
```

Theorem 3.1:

Let p be a point and let S be a set of n points in the plane.

Using the 2-3-tree representation for convex polygons outlined in Section II and using the function INSERT_POINT given above, the convex hull of $S \cup \{p\}$ can be constructed from the convex hull of S in time $O(\log n)$.

Proof:

The proof follows immediately from lemmas 3.2 and 3.3.

Corollary 3.1:

The convex hull of a sequence of n points $p_1, p_2, \cdots, p_n$ in the plane can be constructed on-line in time $O(n \log n)$ at a cost of $O(\log n)$ per update.

Before closing this section we briefly discuss the space requirements of the algorithm. It is not difficult to see that the concatenation operations in one call to INSERT_POINT can require up to $O(\log n)$ new tree nodes. Of course, a call to NORMALIZED_UPDATE can make a large number of nodes available for re-use. But in order to remain within the $O(\log n)$ update time bound it is in general not possible to put these nodes on the free list individually. However, it is easy to see how these nodes can be put on the free list as $O(\log n)$ groups. Thus it is possible to implement this on-line algorithm such that only $O(n)$ space is used over a sequence of n insertions while the logarithmic update time is maintained.

IV. Intersection of a Convex Polygon and a Halfplane

In this section we present an algorithm for the halfplane insertion problem. We will show that using the representation for convex polygons described in Section II one can construct in $O(\log n)$ time the convex polygon P' that results from intersecting an n-sided convex polygon P and a halfplane H bounded by the straight line L.

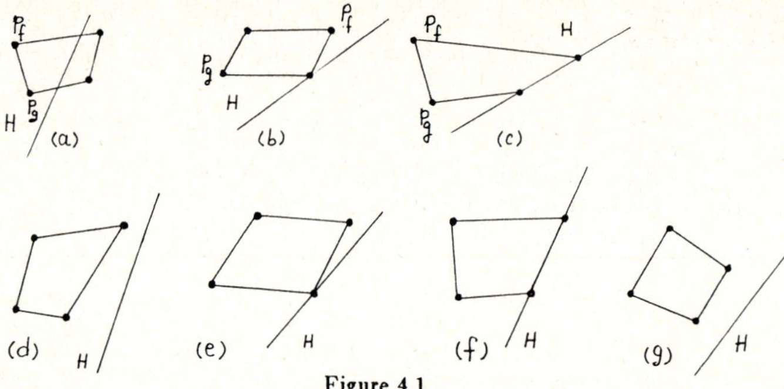

Figure 4.1

Computing the convex polygon P' requires calculating the first vertex p_f and the last vertex p_g of P, when P is traversed in counterclockwise order, that lie in the halfplane H and replacing the chain D of vertices between p_g and p_f by two new vertices g and f, as in cases (a)-(c) in Figure 4.1. A vertex of P is the first vertex with respect to a halfplane H if it lies in the interior of H but its counterclockwise predecessor does not. Similarly, the last vertex of P with respect to H is the vertex that lies in the interior of H but its counterclockwise successor does not. It should be clear that the convexity of P implies that there is at most one first and at most one last vertex and that there exists a first vertex iff there is a last vertex.

The first and last vertices can be undefined, as in cases (d)-(g) in Figure 4.1. In this case P' is equal to either P, as in case (d), or the empty set, as in cases (e)-(g).

Thus the operations necessary to obtain P' are basically identical to the operations for the point insertion algorithm in Section III. The only differences are that the chain D is now replaced by two new vertices instead of one, a minor issue, and that instead of identifying the left and right tangent points one has to determine the first and the last vertices.

Now let us turn to the problem of determining the first and last vertices of a convex polygon with respect a halfplane H. The following lemma provides the key information for performing this operation efficiently.

Lemma 4.1:

Let T be a tree representing a convex polygon P, with n vertices $p_1, p_2, \ldots, p_n$, as described in section II, and let H be a halfplane bounded by the straight line L.

Let q be a node of T that is not the root, and let T_q be the subtree of T rooted at q.

(1) From q it can be determined in constant time whether, in case a first vertex p_f of P with respect to H exists, p_f is a leaf of T_q.

(2) From q it can be determined in constant time whether, in case a last vertex p_g of P with respect to H exists, p_g is a leaf of T_q.

Proof:

We will prove the first part of this lemma. The proof of the second part is similar.

If q is a leaf of T, then the vertex v of P corresponding to q is the first point with respect to H iff v is in the interior of H or on the line L, but its counterclockwise predecessor, w, does not. From q this property can be tested in constant time, since w is available via the pointer q.previous.

Now assume q is a non-leaf and non-root node ot T. Let C_q be the chain of vertices of P corresponding to the leaves of the subtree T_q. Let r be the first vertex and s be the last vertex of C_q, and let a be the counterclockwise predecessor of r and b the counterclockwise predecessor of s (see Figure 4.2). Note that these four points can be determined from q in constant time.

If the straight line L intersects the closed line segment between a and s, then p_f is in C_q if s lies in H and a is not in the interior of H, or in case both s and a lie on L, if b lies in H. This covers all the cases where L intersects the closed line segment between a and s.
Now assume L does not intersect this line segment. We want to decide whether T_q is the subtree to be checked for an intersection between L and the convex polygon P. Using the convexity of this chain we can check for intersection as follows:

Let z be a point on L, let L_a be the straight line parallel to L through point a, and let L_s be the line parallel to L through point s. Then L can intersect the curve formed by the edges between a and s only if a and b are on different sides of L_s but z and b are on the same side of L_s, or if r and s are on different sides of L_a but z and r are on the same side of L_a (see Figure 4.3).

Q.E.D.

Clearly these conditions can be tested in constant time.

In Appendix B we give detailed description of the function CONTAINS_FIRST_VERTEX(q, u_1, u_2) for testing whether a leaf of the subtree rooted at q corresponds to the first vertex with respect to a halfplane H that lies to the left of the directed line L connecting the point u_1 to the point u_2. The function CONTAINS_LAST_VERTEX(q, u_1, u_2) is similar.
We also describe a function TOTALLY_CONTAINED(T, u_1, u_2) for testing in constant time whether, in the case that the line L does not intersect the polygon P represented by T, whether P is totally contained in the halfplane H or not.

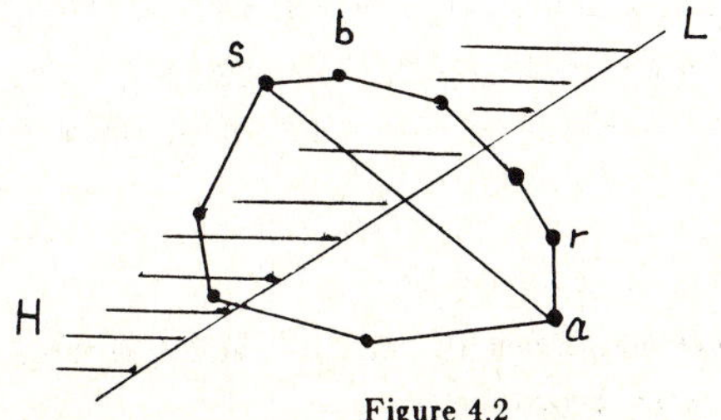

Figure 4.2

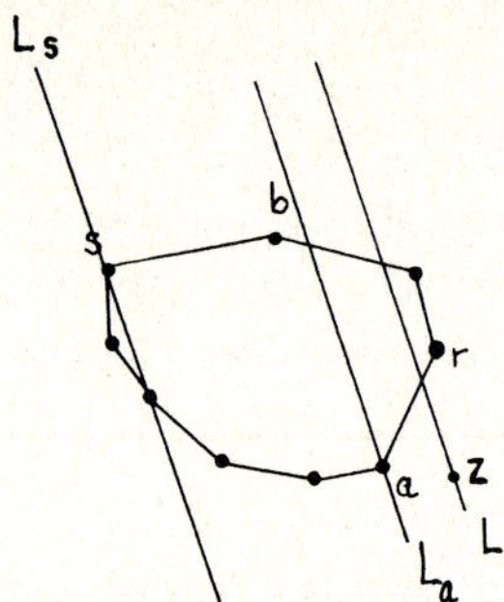

Figure 4.3

Now we describe the function INSERT_HALFPLANE(T,u_1, u_2). Given a 2-3-tree T representing a convex polygon and a halfplane H to the left of the directed straight line joining u_1 to u_2, the function returns a 2-3-tree T' representing the convex polygon that results from the intersection of the polygon represented by T and the halfplane H.

```
INSERT_HALFPLANE (T, u1, u2)
   local variables T' , there_is_an_intersection, new_point1
begin
   T' := NORMALIZE_2(T,p1,p2)
   if there_is_an_intersection then
         return ( NORMALIZED_UPDATE_2(T' , u1, u2) )
   else if TOTALLY_CONTAINED(T,p1,p2) then
         return ( T )
   else
         return ( NULL )
end
```

The functions NORMALIZE_2(T, u_1, u_2) and NORMALIZED_UPDATE_2(T, u_1, u_2), given in detail in Appendix B, differ from their counterparts NORMALIZE and NORMALIZED_UPDATE of section III in a few minor details. Thus we can state the following results without proofs.

Theorem 4.1:

Let T be a 2-3-tree representing an n-sided polygon P as specified in Section II. Let H be a halfplane that lies to the left of the directed line through the two distinct points u_1 and u_2.

Then using the function INSERT_HALFPLANE, the intersection of P and H can be determined in time $O(\log n)$.

Corollary 4.1:

The intersection of n (open) halfplanes $H_1, H_2, \cdots, H_n$ can be constructed on-line in time $O(n \log n)$ at a cost of $O(\log n)$ per update.

Corollary 4.2:

Given a 2-3-tree representation of a n-sided convex polygon P as specified in section II. Any sequence of k arbitrarily interleaved updates of the form "insert a point" or "insert a halfplane" can be processed on-line in time $O(k \log(n+k))$ at the cost of $O(\log(n+k))$ per update.

V. Conclusions

We presented $O(\log n)$ algorithms for inserting a point in a convex polygon and for intersecting a convex polygon with a halfplane. These algorithms naturally yield fast on-line algorithms for the convex hull problem and the problem of constructing the intersection of halfplanes.

Algorithms for solving these problems have been proposed by Preparata [Pre] and by Chazelle and Dobkin [C-D]. So the reader might ask, why other algorithms. There are several reasons. For one thing we believe that our algorithms are particularly simple. They avoid extensive case analysis and use one common representation for convex polygons to efficiently solve both point and halfplane insertion problems.

The used data structure is another reason since it seems to implement naturally both hierarchical representations for convex polygons proposed by Dobkin and Kirkpatrick [D-K]. As a matter of fact, it is an interesting exercise to view the insert algorithm of Section III in terms of the inner hierarchy of [D-K] and the intersection algorithm in terms of the outer hierarchy.

As a last reason, we thought it to be worthwhile to give complete algorithms that treat all degeneracies, are independent of the orientation of the coordinate system, and are based on a single geometric primitive.

References:

[AHU] Aho,V.A.,Hopcroft,J.E.,Ullman,J.D. "The Design and Analysis of Computer Algorithms." Addison-Wesley, Reading, Mass.; 1974.

[A-T] Akl,S.G.,Toussaint G.T. "A Fast Convex Hull Algorithm." Information Processing Letters 7; p.219; 1978.

[Avi] Avis,D. "Comments on a Lower Bound for Convex Hull Determination." Information Processing Letters 11; p.126; 1980.

[BFS] Bentley,J.L.,Faust,G.M.,Preparata,F.P. "Approximation Algorithms for Convex Hulls." Communications of the ACM 25; pp.64-68; 1982.

[B-S] Bentley,J.L., Shamos,M.I. "Divide and Conquer for Linear Expected Time." Information Processing Letters 7; pp.87-91; 1978.

[C-K] Chand,D.R.,Kapur,S.S. "An Algorithm for Convex Polytopes." Journal of the ACM 17; pp.78-86; 1970.

[C-D] Chazelle,B.M.,Dobkin,D.P. "Detection is Easier than Computation." Proc. of the 11th Annual ACM Symposium on Theory of Computing; pp.146-153; 1979.

[D-T] Devroye,L.D., Toussaint,G.T. "A Note on Linear Expected Time Algorithms for Finding Convex Hulls." Computing 26; pp.361-366; 1981.

[D-K] Dobkin,D.P.,Kirkpatrick,D.G. "Fast Detection of Polyhedral Intersections." Proc. of the 9th Colloquium on Automata, Languages, and Programming; Lecture Notes in Computer Science 140, Springer Verlag; pp.154-165; 1982.

[Edd] Eddy,W.F. "A New Convex Hull Algorithm for Planar Sets." ACM Transactions on Mathematical Software 3; pp.398-403 and pp.41-412; 1977.

[EKS] Edelsbrunner,H.,Kirkpatrick,D.G.,Seidel,R. "On the Shape of a Set of Points in the Plane." IEEE Transactions on Information Theory 29; pp.551-559; 1983.

[vEB] Emde Boas,P.v. "On the $\Omega(n \log n)$ Lower-Bound for Convex Hull and Maximal Vector Determination." Information Processing Letters 10; pp.132-136; 1980.

[Gow] Gowda,I.G., "Dynamic Problems in Computational Geometry." Univ. of British Columbia, Dept. of Computer Science, M.Sc.Thesis; 1980.

[Gra] Graham,R.L. "An Efficient Algorithm for Determining the Convex Hull of a Finite Planar Set." Information Processing Letters 1; pp.132-133; 1972.

[Jar] Jarvis,R.A. "On the Identification of the Convex Hull of Finite Set of Points in the Plane." Information Processing Letters 2; pp.18-21; 1973.

[K-S] Kirkpatrick,D.G., Seidel,R. "The Ultimate Planar Convex Hull Algorithm?" Cornell Univ. CS Tech.Report 83-577; 1983.

[O-L] Overmars,M.H.,Leeuwen,J.v. "Dynamically Maintaining Configurations in the Plane." Proc. of the 12th Annual ACM Symposium on Theory of Computing; pp.135-145; 1980.

[Pre] Preparata,F.P. "An Optimal Real-Time Algorithm for Planar Convex Hulls." Communications of the ACM 22; pp.402-405; 1979.

[P-H] Preparata,F.P., Hong,S.J. "Convex Hulls of Finite Sets of Points in Two and Three Dimensions." Communications of the ACM 20; pp.87-93; 1977.

[Sei] Seidel,R. "A Convex Hull Algorithm Optimal for Point Sets in Even Dimensions." Univ. of British Columbia, Dept. of Computer Science, Report 81-14; 1981.

[Sha] Shamos,M.I. "Computational Geometry." Yale University, New Haven, Connecticut; Ph.D.Thesis; 1978.

[Yao] Yao,A.C. "A Lower Bound to Finding Convex Hulls." Journal of the ACM 28; pp.780-787; 1981.

Appendix A:

In this appendix we give detailed description of the function CONTAINS_LTP for testing whether a point p lies in the region I_q' , corresponding to a non-leaf node q which is not the root, defined in section III. Details of the similar function CONTAINS_RTP are left to the reader.

We make use of a single geometric test primitive, namely testing whether a point p is to the left of, on, or to the right of a directed line L which passes through two distinct points w and z. This test can be easily formulated algebraically as:

$$D(p,w,z) < 0 \text{ iff } p \text{ to the left of } L$$
$$D(p,w,z) = 0 \text{ iff } p \text{ on } L$$
$$D(p,w,z) > 0 \text{ iff } p \text{ to the right of } L,$$

where

$$D(p,w,z) \text{ is the determinant } \begin{vmatrix} 1 & p.x & p.y \\ 1 & w.x & w.y \\ 1 & z.x & z.y \end{vmatrix}.$$

The function CONTAINS_LTP(q,p) returns true if and only if the point p lies in the region I_q' , as depicted in Figure A.1.

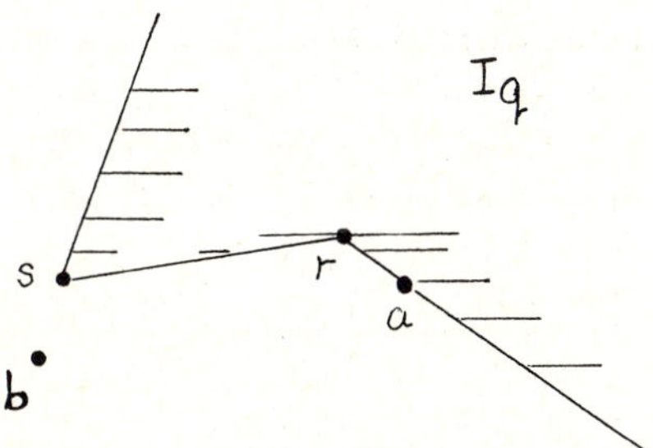

Figure A.1

Obviously, membership of p in I_q' can be tested by extending each of the (at most) three bounding line segments of I_q' to full straight lines and determining the position of p with respect to each of these lines. However, one has to distinguish three cases. Let s and r be the leftmost leaf, respectively rightmost leaf in the subtree rooted at q. Let a be the leaf preceding r, b be the leaf following s, and t be the point $s-(r-a)$. Then the position of point b with respect to the directed line through s and t determines each of the three cases, as shown in Figure A.2.

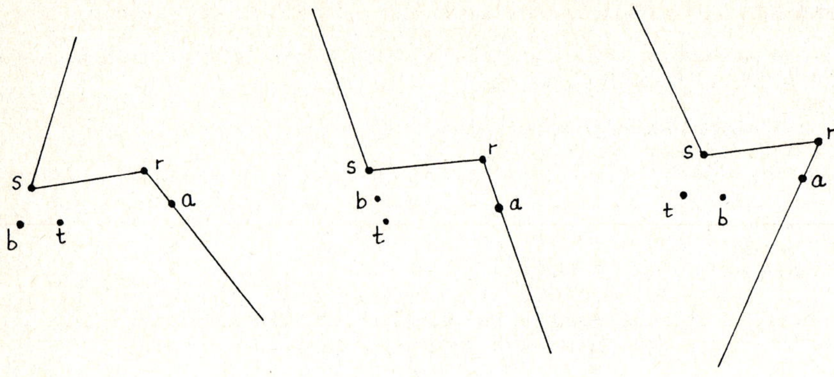

Figure A.2

With these observations the correctness of the following procedure can be easily verified.

```
CONTAINS_LTP(q,p)
begin
    r = if q.height=0 then q else q.leftmost
    s = if q.height=0 then q else q.rightmost
    a = r.previous
    b = s.next
    t = s + a - r
    if D(b,s,t) > 0 then
            return ( ( D(p,a,r) ≤ 0 and ( D(p,s,r) < 0 or D(p,b,s) < 0 ) )
                        or ( D(p,b,s) = 0 and D(p,b,a) < 0 ) )
    else
            return ( ( D(p,b,s) < 0 )
                        or ( D(p,r,s) < 0 and D(p,a,r) ≤ 0 )
                        or ( D(p,b,s) = 0 and D(p,b,a) < 0 ) )

end
```

Appendix B: In this appendix we give detailed description of the functions used in Section IV.

```
CONTAINS_FIRST_VERTEX(q, u1, u2)
begin
   r = if q.height=0 then q else q.leftmost
   s = if q.height=0 then q else q.rightmost
   a = r.previous
   b = s.previous
   S = s + u2 - u1
   A = a + u2 - u1
   return ( ( D(s,u1,u2) < 0 and D(a,u1,u2) ≥ 0 )
                  or ( D(s,u1,u2) = 0 and D(b,u1,u2) < 0 and D(a,u1,u2) > 0 )
                  or ( D(a,s,S)*D(b,s,S) < 0 and D(u1,s,S)*D(b,s,S) > 0 )
                  or ( D(s,a,A)*D(r,a,A) < 0 and D(u1,a,A)*D(r,a,A) > 0 ) )
end
```

```
TOTALLY_CONTAINED(T, u1, u2)
begin
   return ( D(T.leftmost.previous, u1, u2) ≤ 0 and D(T.leftmost.next, u1, u2) ≤ 0 )
end
```

```
NORMALIZE_2(T, u1,u2)
    global variable There_is_an_intersection, new_point1
    local variable T1,T2

    FIND_LAST_VERTEX_AND_SPLIT(q,u1,u2)
    begin
        if q.height = 0 then
                there_is_an_intersection := true
                T1 := NULL
                T2 := q
                Let new_point1.x and new_point2.y be the  coordinates of
                        the intersection of the line defined by q and q.next
                        and the line defined by u1 and u2.
                return
        else
                for i=1 to q.no_of_sons do
                    if CONTAINS_LAST_VERTEX(q.son[i],u1,u2) then
                        FIND_LAST_VERTEX_AND_SPLIT(q.son[i],u1,u2)
                        if there_is_an_intersection then
                            for j=i-1 downto 1 do
                                    T1 := CONCATENATE(q.son[j],T1)
                            for j=i+1 to q.no_of_sons do
                                    T2 := CONCATENATE(T2,q.son[j])
                            return
                there_is_an_intersection := false
                return
    end

begin
    FIND_LAST_VERTEX_AND_SPLIT(T,u1,u2)
    if there_is_an_intersection then
            return ( CONCATENATE(T2,T1) )
    else
            return ( T )
end
```

```
NORMALIZED_UPDATE_2 (T, u1, u2)
    global variable new_point1
    local variable T1, new_point2

    FIND_FIRST_VERTEX_AND_COLLECT_LEFT_SIDE(q,p1,p2)
    begin
      if q.height = 0 then
            let new_point2.x and new_point2.y be the coordinates of the
                  intersection of the line through q and q.previous and the line
                  through p1 and p2.
            T1 := CONCATENATE(new_point1,new_point2)
            T1 := CONCATENATE(T1,q)
            new_point1.next := new_point2
            new_point2.previous := new_point1
            new_point1.previous := q
            q.next := new_point1
            return
      else
            for i:=1 to q.no_of_sons do
               if CONTAINS_FIRST_VERTEX(q.son[i],p1,p2) then
                     FIND_FIRST_VERTEX_AND_COLLECT_LEFT_SIDE(q.son[i],p1,p2)
                     for j:= i-1 downto 1 do
                           T1 := CONCATENATE(q.son[j], T1)
                     return
    end

begin
   FIND_FIRST_VERTEX_AND_COLLECT_LEFT_SIDE(T,p1,p2)
   T1.leftmost.previous := T1.rightmost
   T1.rightmost.next := T1.leftmost
   return (T1)
end
```

Appendix C:

```
CONCATENATE(S,T)
begin
   if S = NULL then return( T )
   if T = NULL then return( S )

   if S.height = T.height then
       let q be a new tree node
       q.height := S.height + 1
       q.no_of_sons := 2
       q.son[1] := S
       q.son[2] := T
       q.leftmost := S.leftmost
       q.rightmost := T.rightmost
       return( q )

   else if S.height < T.height then

      if T.no_of_sons = 2 then
         if S.height+1 = T.height then
             T.no_of_sons := 3
             T.son[3] := T.son[2]
             T.son[2] := T.son[1]
             T.son[1] := S
             T.leftmost := S.leftmost
             return( T )
         else
            return ( CONCATENATE( CONCATENATE(S,T.son[1]) , T.son[2] ) ) )
      else
         return ( CONCATENATE( CONCATENATE(S,T.son[1]) ,
                                        CONCATENATE(T.son[2],T.son[3]) ) )

   else

      if S.no_of_sons = 2 then
         if S.height = T.height+1 then
             S.no_of_sons := 3
             S.son[3] := T
             S.rightmost := T.rightmost
             return( S )
         else
            return ( CONCATENATE( S.son[1] , CONCATENATE(S.son[2],T) ) )
      else
         return ( CONCATENATE( CONCATENATE(S.son[1],S.son[2]) ,
                                          CONCATENATE(S.son[3],T) ) )
end
```

COMPUTATIONAL GEOMETRY, G.T. Toussaint (Editor)
© Elsevier Science Publishers B.V. (North-Holland), 1985

ON GEOMETRIC ALGORITHMS THAT USE THE FURTHEST-POINT VORONOI DIAGRAM*

Binay K. Bhattacharya

Department of Computing Science
Simon Fraser University
Burnaby, B.C. V5A 1S6

Godfried T. Toussaint

School of Computer Science
McGill University
805 Sherbrooke St. West
Montreal, Quebec H3A 2K6
CANADA

In this paper it is shown that the diameter D(P) of a set of n points P on the plane is not necessarily an edge in the dual of the furthest-point Voronoi diagram (FPVD) of P, as previously claimed in [1] and [2]. It is also proved that if P is contained in the disk determined by D(P) then the above property does hold. Furthermore, it is shown that an edge e in the *dual* of the FPVD(P) intersects its corresponding edge in the FPVD(P) if, and only if, P is contained in the disk determined by e . These results invalidate several algorithms for solving the diameter, all-furthest-neighbor, and maximal spanning tree problems proposed in [1] and [2]. A proof of correctness is given for the minimum spanning circle algorithm proposed in [2] and [3]. Finally new O(n log n) algorithms are offered for the minimum spanning circle and all-furthest-neighbor problems.

Index terms: diameter, minimum spanning circle, all-furthest-neighbor problem, maximal spanning tree, Voronoi diagram, convex hull, algorithms, computational geometry.

CR Categories: 5.25, 5.30, 5.5, 3.36, 3.63

*Research supported by N.S.E.R.C. grant No. A-2422 and A-9293.

I. INTRODUCTION

In this paper we are concerned with algorithms that operate on a set of points on the plane $P = \{p_1, p_2, \ldots, p_n\}$ given by their cartesian coordinates. It is further assumed that the points are in *general position* in the sense that no three are collinear and no four are cocircular. A geometric structure which has received attention recently in the design of geometric algorithms is the *furthest-point Voronoi diagram* (FPVD) [1], [2], [3]. The FPVD is a partition of the plane into unbounded convex "polygons" $V(p_k)$ such that $V(p_k)$ is the locus of points further from p_k than from any other point in P. Figure 1 illustrates a set of points and its FPVD in dashed lines. If two polygons $V(p_i)$ and $V(p_j)$

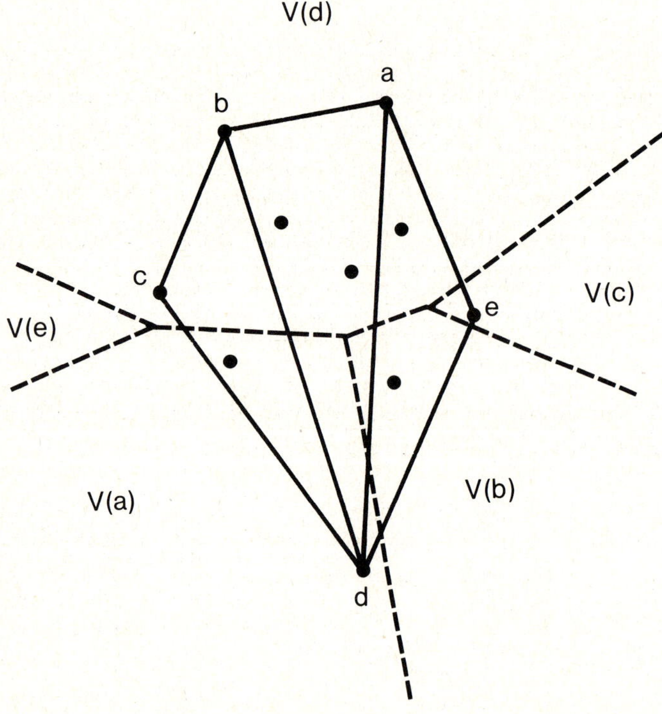

Figure 1: A set of points, the furthest-point Voronoi diagram of the set in dashed lines, and the dual triangulation in solid lines.

share a side then p_i and p_j are *furthest-point Voronoi* (FPV) *neighbors*. If all pairs of points in P that are FPV neighbors are joined by an edge, the resulting *triangulation* is referred to as the *dual* of the FPVD or FPT. In Figure 1 the FPT is shown in solid lines. Analysis of Figure 1 discloses that only points that are on the convex hull of the set (CH(P)) have corresponding FPV "polygons". Points lying in the interior of CH(P) may be ignored without changing the FPVD of P. We now prove this property as we will need this result later on. Although this result is proved in [2] an alternate proof is given here for completeness. The euclidean distance between two points x,y will be denoted by d(x,y), the line segment joining x and y by $\overline{xy}$.

Lemma 1: Given a set P, a point x in the interior of CH(P), and an arbitrary point z on the plane, there always exists a point in P that is further from z than x is.

Proof: Consider the line of support L of P perpendicular to zx such that the extension of zx, in the direction of x, intersects L at w and refer to Figure 2. Let $y \in P$ be the extreme point through which L passes. Since all points in CH(P) other than y lie at some perpendicular distance from L and on one and the same side of it, it follows that $d(z,x) < d(z,w)$. From Pythagoras' theorem we have that $d(z,y) \geq d(z,w)$ and therefore $d(z,x) < d(z,y)$ where $y \in P$. Q.E.D.

Theorem 1: Given a set $P = \{p_1, p_2, \ldots, p_n\}$, if a point p_k, $k \in \{1,2,\ldots, n\}$, lies in the interior of CH(P) then it has no effect on FPVD(P), i.e., $\text{FPVD}(p_1,p_2,\ldots,p_n) = \text{FPVD}(p_1,\ldots,p_{k-1},p_{k+1},\ldots,p_n)$.

Proof: Let P be a convex polygon and let z be any point in the interior of P. Let a be an arbitrary vertex of P. Construct the perpendicular bisector of $\overline{za}$ and refer to Figure 3. If z contributes an edge of V(a), the FPV polygon of a, then the perpendicular bisector of $\overline{za}$ must intersect V(a) to yield a bounded convex polygon (shaded in Figure 3) which is the intersection of V(a) with the closed half-plane which includes a, determined by the bisector of $\overline{za}$. Let x be any point in the interior of this polygon. Since x is on the same half-plane as a, it follows that $d(x,z) > d(x,a)$. Furthermore, by lemma 1 there exists

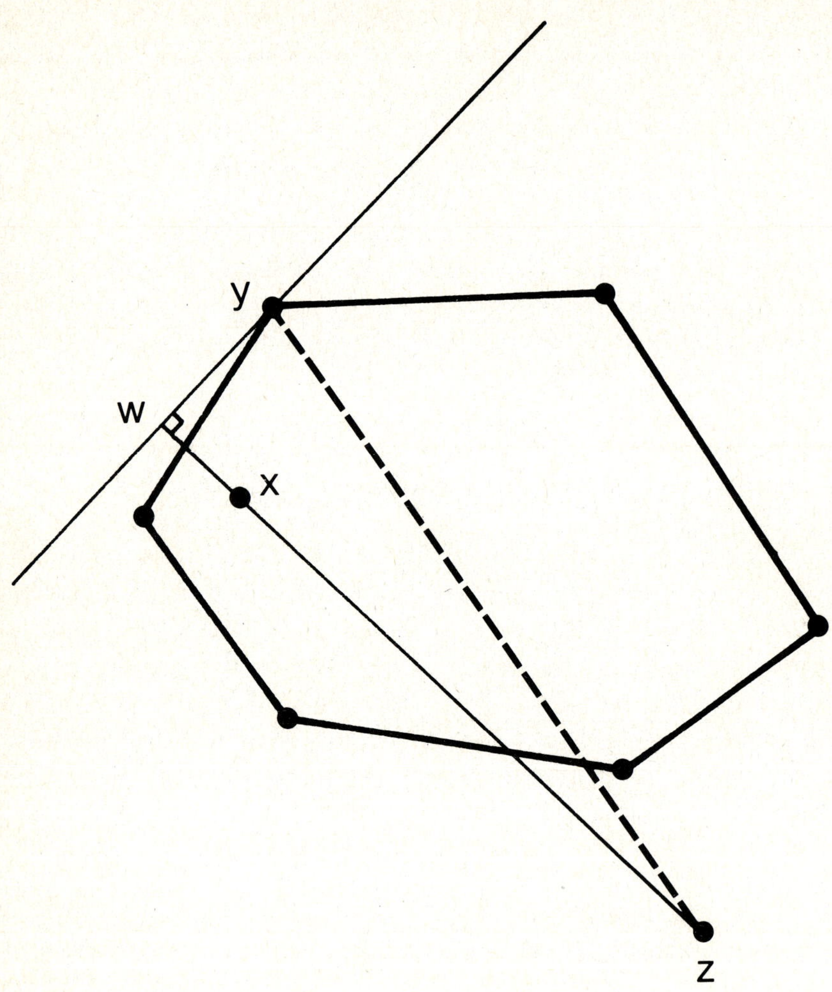

Figure 2: Illustrating the proof of Lemma 1.

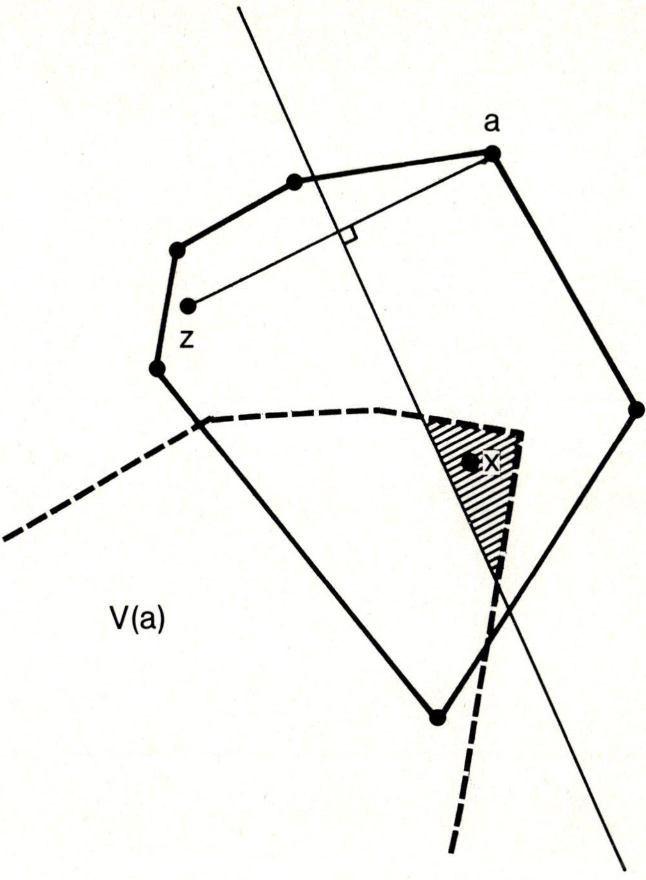

Figure 3: Illustrating the proof of Theorem 1.

some vertex of P, say y, such that $d(y,x) > d(z,x)$. Thus we must have $d(y,x) > d(a,x)$. But by definition, any point in V(a) is further from a than any other vertex of P. In other words $d(a,x) > d(y,x)$, which is a contradiction. Therefore z does not affect V(a). Since a was arbitrary z does not affect any FPV polygon of P. Q.E.D.

II. THE ALGORITHMS THAT USE THE FURTHEST-POINT VORONOI DIAGRAM

2.1. The Diameter of a Set

The diameter of a set of points P is defined as follows:

$$\mathrm{DIAM}(P) = \max_{i,j}\{d(p_i,p_j)\} .$$

The diameter of a set has many applications in the field of pattern recognition [4]. The obvious brute-force approach to computing DIAM(P) leads to an $O(n^2)$ algorithm. In [2] Shamos proposes the following algorithm.

Algorithm D

Step 1: Compute the FPVD(P).

Step 2: Compute the length of every edge in the *dual* of the FPVD(P).

Step 3: Select the longest edge found in step 2.

Since there are $O(n)$ edges in the FPT(P) (the dual of the FPVD(P)), Steps 2 and 3 can be done in $O(n)$ time. Shamos claims that Step 1 can be done in $O(n \log n)$ time via divide-and-conquer which would lead to an $O(n \log n)$ diameter algorithm. The correctness of the algorithm is based on the claim made in [2] that the diameter of a set P is an edge in the FPT(P). We will show in the next section that this claim is false.

2.2. The Minimum Spanning Circle

The minimum spanning circle (MSC) of a set of points P is the smallest circle that encloses P. This problem has received a lot of attention in Operations Research where it is known as the *maximum facilities location problem* [2]. The obvious brute-force approach to this problem leads to an $O(n^4)$ algorithm. It is known that the MSC of P is unique and is determined either by the circumcircle of three points in P, or by the two points in P that determine DIAM(P). In the latter case the MSC is the circle whose diameter equals DIAM(P), [5]. In [3] it is claimed that if the MSC of P is determined by three points, the center of the circle will be a *vertex* of the FPVD(P), i.e., a point where three edges of FPVD(P) meet. The above properties lead Shamos and Hoey [3] to propose the following algorithm for MSC(P).

Algorithm MSC

Step 1: Compute the FPVD(P).

Step 2: Select the longest edge in the *dual* of the FPVD(P) as the *diameter* of P.

Step 3: Test whether the circle determined in Step 2 encloses P; if it does EXIT with the MSC; ELSE continue.

Step 4: For every *vertex* in the FPVD(P) compute the spanning circle and select the smallest circle found as the MSC(P).

Steps 2-4 can be done in $O(n)$ time and Step 1 in $O(n \log n)$ time according to Shamos [2]. Thus the complexity of the algorithm would be dominated by Step 1. However, since Step 2 does not necessarily yield the true diameter of P, the correctness of algorithm MSC is brought into question. In spite of Step 2 it will be shown that the algorithm always finds the correct solution. Preparata [6] offers an alternative $O(n \log n)$ algorithm to solve this problem.

2.3. The All-Furthest-Neighbor Problem

The all-furthest-neighbor problem for a set of points P is to find for each point $p_i \in P$ another point $p_j \in P$, $j \neq i$ such that

$$d(p_i,p_j) = \max_j \{d(p_i,p_j)\} .$$

Alternately we can construct the *furthest neighbor graph* (FNG) by joining two points p_i,p_j with an edge if at least one of p_i,p_j is the *furthest neighbor* of the other. The obvious approach to computing the FNG(P) leads to an $O(n^2)$ algorithm.

In [1] Shamos proposes the following algorithm.

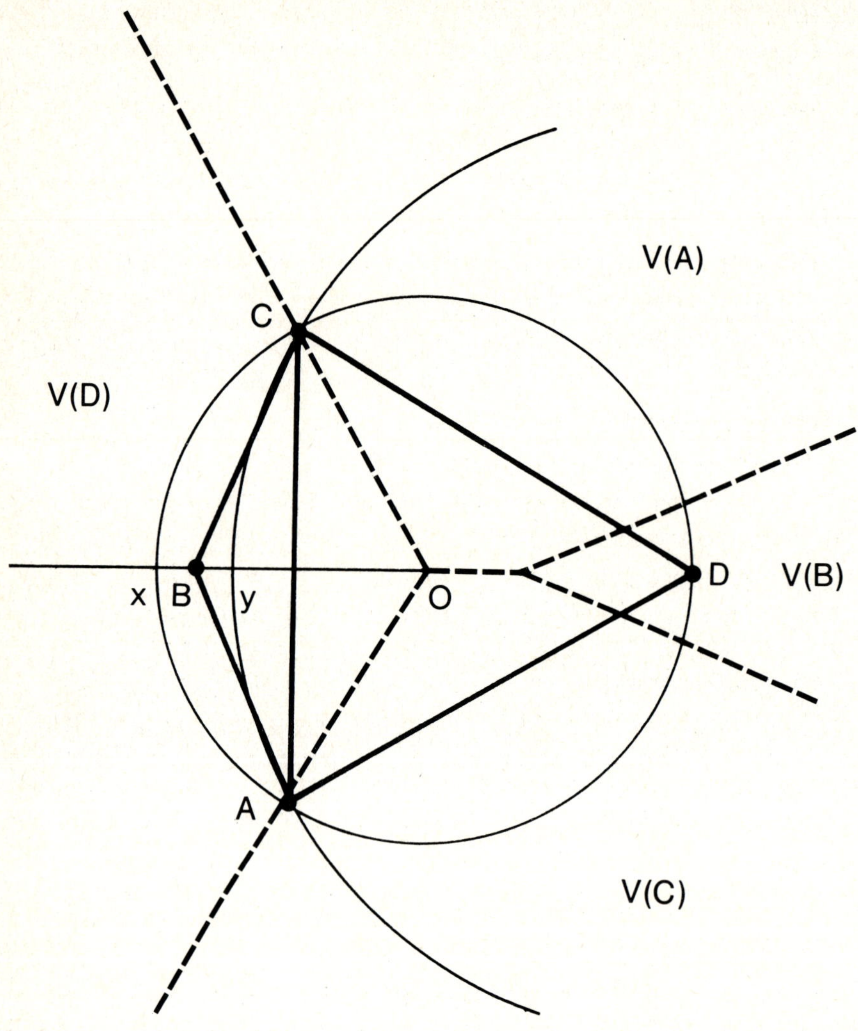

Figure 4: A counterexample to the claim that the diameter of a set is an edge in the dual of the furthest-point Voronoi diagram of the set.

Algorithm FNG

Step 1: Compute the FPVD(P).

Step 2: For each point p_i in P compute the distance between p_i and each FPV *neighbor* of p_i. The maximum distance realized determines the furthest point from p_i.

That this algorithm will not work for arbitrary sets of points is obvious from theorem 1 since points inside the CH(P) are not even accounted for. In the next section we show that the FNG(P) is not a subgraph of the FPT(P) and thus the algorithm fails even for convex polygons.

2.4. The Maximum Spanning Tree

The maximum spanning tree (MXST) problem for a set of point P consists of finding the tree connecting all points of P having the greatest total weight. We are concerned with the case where the weight is the sum of the euclidean edge lengths. In [2] it is pointed out that the MXST(P) is not a subgraph of the FPT(P) and no detailed algorithm is given. However, the following approach is suggested in [1]. First compute the furthest neighbor graph by connecting each point to its furthest neighbor. This graph Shamos [1] claims to be at least half of the MXST(P). While it may be correct that the *true* FNG(P) $\subseteq$ MXST(P), the manner in which the FNG(P) is computed in [1] can lead to a graph which is not a subgraph of MXST(P). Thus this approach is doomed to failure.

III. MAIN RESULTS

3.1. The Counterexample

A counterexample to the algorithms and some of the claims discussed in the previous section is illustrated in Figure 4. Consider a set of four points A,B,C,D. Let the points C,A, and D circumscribe a circle CIRC(C,A,D). Unless otherwise stated CIRC(C,A,D) will include its interior. Also let C,A,D determine an equilateral triangle such that $\overline{CA}$ is a vertical edge. Let O denote the center of CIRC(C,A,D) and let L denote the half-line starting at O and proceeding in the opposite direction of D. Let CIRC(C,A,D) intersect L at x and let the circle with center D and radius = CD = AD intersect L at y. Now place the fourth point B anywhere on the open line segment $\overline{xy}$. The FPVD of this set is shown in dashed lines in Figure 4. Since B lies to the right of x, the intersection of the perpendicular bisectors of $\overline{CB}$ and $\overline{AB}$ intersect at a point to the right of 0. Therefore the "polygons" V(B) and V(D) do not share a side. Hence $\overline{BD}$ is not an edge in the FPT of the set which is indicated with solid thick lines. However, since angle CBA $> 90°$ it follows that $d(C,B) = d(A,B) < d(A,C)$. Furthermore, since B is to the left of y, we have that $d(B,D) > d(C,D) = d(A,D) = d(A,C)$. Therefore $\overline{BD}$ is the diameter of the set. Thus the diameter of a set is not necessarily an edge in the FPT(P). It follows that algorithm D is not guaranteed to work.

Consider next algorithm FNG for computing all the furthest neighbors of the set. Referring to Figure 4 we see that the furthest neighbor of D is B and the furthest neighbor of B is D. However, algorithm FNG will erroneously find either C or A as the furthest neighbor of both B and D since it only considers points as candidates if they are connected by an edge in the FPT.

3.2. New Algorithms for the All-Furthest-Neighbor Problem

We now present two algorithms for the all-furthest-neighbor problem; the first is based on the FPVD(P), the second on CH(P). In the first we consider the problem as a point-location problem. For each point $p_i \in P$ we need to find in what unbounded polygon p_i lies. If p_i lies in $V(p_k)$ then p_k is the furthest neighbor of p_i. Lipton and Tarjan [10] have shown that location of a point in a planar straight-line graph (such as FPVD) with n vertices can be determined in $O(\log n)$ time and $O(n)$ storage using $O(n \log n)$ preprocessing time. Thus once FPVD is obtained all the furthest neighbors can be computed in $O(n \log n)$ time. More efficient algorithms from the point of view of implementation are given in Preparata [14] and Kirkpatrick [15].

Algorithm FNG-1

Step 1: Compute FPVD(P).

Step 2: Preprocess FPVD(P) as in Kirkpatrick [15] or Preparata [14].

Step 3: For each point in P determine its location.

Preparata [6] gives an $O(n \log n)$ algorithm for computing FPVD(*P*). Preprocessing in Step 2 also requires $O(n \log n)$ time [14], [15]. In Step 3 $O(\log n)$ time is needed to locate each point. Thus the complexity of Step 3 is also $O(n \log n)$. Thus algorithm FNG-1 runs in $O(n \log n)$ time.

In practice, if n is not too large the following much simpler algorithm may be preferred. In addition to its simplicity and ease of programming the algorithm has the advantage that for many "reasonable" distributions its expected running time is between $O(n)$ and $O(n \log n)$. Its drawback is an $O(n^2)$ worst-case running time.

Algorithm FNG-2

Step 1: Compute CH(*P*).

Step 2: For each point $p_i \in P$ determine which point $p_k \in CH(P)$ is furthest by computing $d(p_i,p_k)$ for all $p_k \in CH(P)$ and selecting the point that yields the greatest distance.

The correctness of the algorithm follows from theorem 1 since the furthest point among *P* from any z on the plane belongs to CH(*P*).

Let h denote the number of points on CH(*P*). If Jarvis' algorithm [11] is used to find CH(*P*) then both steps run in $O(hn)$ time. For uniform distributions in the unit square the expected value of h is $E\{h\} = O(\log n)$ leading to an expected complexity $E\{C\} = O(n \log n)$. For normal distributions $E\{h\} = O(\sqrt{\log n})$ giving $E\{C\} = O(n\sqrt{\log n})$. Finally, for certain distributions with "heavy tails" [12], [13] $E\{h\} = O(1)$ and thus $E\{C\} = O(n)$.

Looking at the MXST problem for the points in Figure 4 we see that the MXST is not unique. However, any MXST contains $\overline{BD}$ as one of its edges and none of them contains $\overline{BC}$ or $\overline{BA}$ as edges. On the other hand the graph computed by algorithm FNG contains these edges. Therefore algorithm FNG is not guaranteed to yield a subgraph of the MXST.

3.3. New Algorithms for the Minimum-Spanning-Circle Problem

Consider algorithm MSC. We have seen that Step 2 of this algorithm does not necessarily yield the true diameter of the set. It will now be shown that in spite of this erroneous step the algorithm works! We first prove some required theorems. With the counterexample of Figure 4 we have established that the diameter of P is not necessarily an edge in the FPT(P). We now show that if the diametral circle (disk) determined by DIAM(P) encloses all the points of P the above property does hold. Let the disk of two points p_i, p_j, denoted by DISK(p_i,p_j) be defined as follows:

$$\text{DISK}(p_i,p_j) \equiv \{x:[d^2(x,p_i) + d^2(x,p_j)]^{1/2} \leq d(p_i,p_j)\}.$$

The boundary of a set S will be denoted by bd(S).

Theorem 2: Given two points $p_i,p_j \in P$, if $d(p_i,p_j) = \text{DIAM}(P)$ and if $P \subset \text{DISK}(p_i,p_j)$ then $\overline{p_ip_j}$ is an edge in FPT(P).

Proof: From theorem 1 it follows that edges in the FPT(P) exist only between vertices of CH(P). Therefore we need to consider only the case when P is a convex polygon. Let P be a convex polygon with n vertices $p_1,p_2,\ldots,p_n$ in clockwise order and let p_i,p_j determine the diameter of P. The two points p_i and p_j will, in general, partition P into two convex polygonal chains. Without loss of generality, assume p_i and p_j lie on the horizontal line L_h. The two chains are the upper and lower half chains UCHAIN(p_i,p_j) and LCHAIN(p_i,p_j), respectively. To show that $\overline{p_ip_j}$ is an edge in FPT(P) we need only show that the FP "polygons" $V(p_i)$ and $V(p_j)$ share a side. Consider first $V(p_i)$ and any vertex $p_k \in \text{UCHAIN}(p_i,p_j)$ and refer to Figure 5. Extend the line from p_i, through p_k, to intersect bd(DISK(p_i,p_j)) at x_i. Since $\overline{p_ix_i}$ is a chord of bd(DISK(p_i,p_j)) its perpendicular bisector must pass through 0, the center of DISK(p_i,p_j). Since $d(p_i,p_k) < d(p_i,x_i)$, the perpendicular bisector of $\overline{p_ip_k}$ must intersect the vertical line (L_v through 0) at some point c_k below 0. Let $p_k^* \in \text{UCHAIN}(p_i,p_j)$ be the vertex that yields the closest c_k to 0 and let c_k^* denote the corresponding c_k. Similarly, let $p_{k'}^*$ and $c_{k'}^*$ play the corresponding role of LCHAIN(p_i,p_j) with respect to vertex p_i. Note that $c_{k'}^*$ lies on L_v and above 0. Thus $\overline{c_k^*\, c_{k'}^*}$ is an edge of $V(p_i)$. Turning to $V(p_j)$ note that for any p_k, $\overline{p_jp_k}$ is a segment of one and the same circle determined by p_i,p_k, and p_j. Therefore the perpendicular bisector of $\overline{p_jp_k}$ intersects L_v at the center of CIRC(p_i,p_k,p_j) which is precisely c_k. Similarly, for any $p_{k'} \in \text{LCHAIN}(p_i,p_j)$ the perpendicular bisector of $\overline{p_jp_{k'}}$ will intersect L_v at $c_{k'}$. Therefore $\overline{c_k^*\, c_{k'}^*}$ is also an edge of $V(p_j)$ and thus $V(p_i)$ and $V(p_j)$ must share a side. Q.E.D.

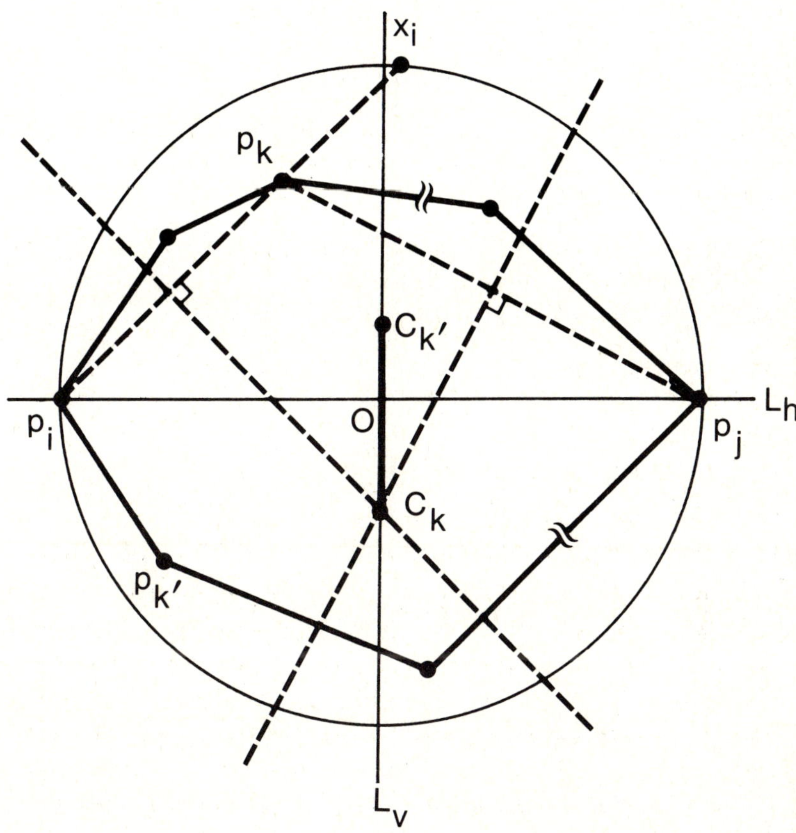

Figure 5: Illustrating the proof of Theorem 2.

Theorem 3: If $d(p_i,p_j) \neq DIAM(P)$ then $P \not\subset DISK(p_i,p_j)$.

Proof: Assume that $P \subset DISK(p_i,p_j)$ and consider any other pair of points $p_r,p_s \in P$, $r \neq s \neq j$. By assumption $p_r,p_s \in DISK(p_i,p_j)$ and since $DISK(p_i,p_j)$ is a convex set $\overline{p_rp_s}$ must also lie in $DISK(p_i,p_j)$. Extend $\overline{p_rp_s}$ in both directions until it intersects $bd(DISK(p_i,p_j))$ at x_r and x_s, respectively. Now $d(p_r,p_s) < d(x_r,x_s)$. Since $\overline{x_rx_s}$ is a segment of $bd(DISK(p_i,p_j))$ it follows that $d(x_r,x_s) < d(p_i,p_j)$ which is after all the diameter of $bd(DISK(p_i,p_j))$. Therefore $d(p_r,p_s) < d(p_i,p_j)$. Since p_r and p_s were arbitrary, $d(p_i,p_j)$ must be $DIAM(P)$. Q.E.D.

Corollary: An edge $\overline{p_ip_j}$ in $FPT(P)$ intersects its corresponding edge in $FPVD(P)$ if, and only if, $P \subset DISK(p_i,p_j)$.

Proof: If $P \subset DISK(p_i,p_j)$ then it follows from theorem 3 that $d(p_i,p_j) = DIAM(P)$ and from theorem 2 it follows (referring to Figure 5) that $V(p_i)$ and $V(p_j)$ share the edge $\overline{c_kc_{k'}}$. Since $c_{k'}$ lies above 0 and c_k below 0, $\overline{p_ip_j}$ must intersect $\overline{c_kc_{k'}}$. If $\overline{p_ip_j}$ is an edge in $FPT(P)$ and it intersects its corresponding edge in $FPVD(P)$ then for all vertices $p_k \in UCHAIN(p_i,p_j)$, c_k must lie below 0. Similarly, for all vertices $p_{k'} \in LCHAIN(p_i,p_j)$, $c_{k'}$ must lie above 0. It follows that $P \subset DISK(p_i,p_j)$. Q.E.D.

Theorem 4: Given three points $p_i,p_j,p_k \in P$, the triangle $\Delta p_ip_jp_k$ is contained in the $FPT(P)$ if, and only if, $P \subset CIRC(p_i,p_j,p_k)$.

Proof: The necessity part of the theorem follows straightforwardly from the definition. Consider a triangle $\Delta p_rp_sp_t$ in $FPT(P)$. This implies that pairs of "polygons" such as $V(p_r)$ and $V(p_s)$ share an edge. The point C_{rst}, at which the three edges in question meet, is a FPV *vertex* and is equidistant from p_r,p_s, and p_t. Furthermore, by definition $d(p_q,c_{rst})$, $q = r,s,t$ is no less than $d(p_u,c_{rst})$ for any $u \neq r,s,t$. Therefore all other points must be included in the circle determined by p_r,p_s, and p_t and center c_{rst}.

We turn now to the sufficiency part of the theorem. Given three points p_i, p_j, p_k with the property that $P \subset CIRC(p_i,p_j,p_k)$ we must show that $V(p_i)$ and $V(p_j)$ share a side, $V(p_i)$ and $V(p_k)$ share a side, and similarly for $V(p_j)$ and $V(p_k)$. Consider an arbitrary edge $\overline{p_ip_k}$. Two cases arise: either $\overline{p_ip_k}$ is a) a *diagonal* of $CH(P)$ or b) it is an *edge* of $CH(P)$.

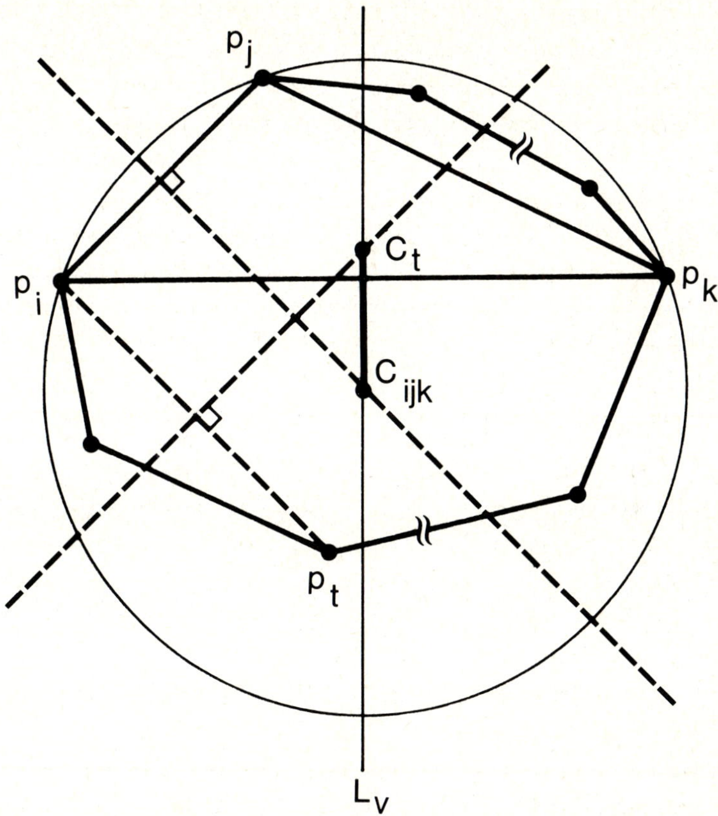

Figure 6: Illustrating the fact that two nonadjacent vertices (of the convex hull) of three vertices that determine a spanning circle must define an edge in the FPT.

Case A: The proof resembles that of theorem 2 and similar terminology is used. Without loss of generality assume $\overline{p_i p_k}$ is horizontal and that p_j lies above $\overline{p_i p_k}$ and refer to Figure 6. Let $p_t \in$ LCHAIN(p_i,p_k) be the vertex such that the perpendicular bisector of $\overline{p_i p_t}$ intersects L_v (the vertical line through c_{ijk}) at c_t closer to c_{ijk} than any other vertex in LCHAIN(p_i,p_k). The corresponding situation for UCHAIN(p_i,p_k) is now slightly different. Since there exists a vertex lying on CIRC(p_i,p_j,p_k), namely p_j it follows that the point corresponding to c_t is c_{ijk} itself. From similar arguments to those in the proof of theorem 2, it follows that V(p_i) and V(p_k) have $\overline{c_{ijk} c_t}$ in common and thus $\overline{p_i p_k} \in$ FPT(P).

Case B: A similar proof holds for the case when $\overline{p_i p_k}$ is an edge of CH(P). In this case the edge shared by V(p_i) and V(p_k) is an infinite half-line starting at c_{ijk}, collinear with the perpendicular bisector of $\overline{p_i p_k}$, and proceeding in the direction away from $p_i p_k$. Q.E.D.

We are now in a position to show that algorithm MSC works in spite of Step 2. Let P be some input to algorithm MSC. Two cases arise: either a) $P \subset$ DISK(DIAM(P)) or b) $P \not\subset$ DISK(DIAM(P)). Consider Case A first. From theore 2 it follows that Step 2 of MSC will find the true diameter of P. Since $P \subset$ DISK(DIAM(P)) the algorithm will stop after executing Step 3 and exit with the correct solution. Now consider Case B. From the counterexample it follows that Step 2 may compute an edge, say $p_u p_v$, which is not the diameter of P. From theorem 3 it follows that if this is the case $P \not\subset$ DISK(p_u,p_v) and thus in Step 3 it is correctly concluded through false reasoning that $P \not\subset$ DISK(DIAM(P)). The algorithm then branches to Step 4 the correctness of which follows from theorem 4.

We offer below two alternative algorithms for computing MSC(P).

Algorithm MSC-1

Step 1: Compute CH(P).

Step 2: Compute DIAM(CH(P))

Step 3: Test whether $P \subset$ DISK(DIAM(P)); if it does EXIT with MSC; ELSE continue.

Step 4: Compute FPVD(P).

Step 5: For every *vertex* in FPVD(P) compute the spanning circle and EXIT with the smallest such circle found.

Step 1 can be done in $O(n \log n)$ time using any one of a variety of algorithms available [4]. Step 2 can be carried out in $O(n)$ time using the "antipodal-pair" algorithm of Shamos [7]. Steps 3-5 are identical to Steps 1, 3, and 4 in algorithm MSC of Shamos and Hoey [3]. Note that the diameter algorithms of Dobkin and Snyder [8] and Snyder and Tang [16] cannot be used in Step 2 since they have recently been shown to fail in some cases [17], [18].

Another approach which is closer in spirit to the algorithm of Shamos and Hoey [3] and which avoids Steps 1 and 2 of algorithm MSC-1, follows from the corollary.

Algorithm MSC-2

Step 1: Compute FPVD(P).

Step 2: For each edge in FPVD(P) test whether it intersects its corresponding edge in FPT(P); if an intersecting edge is found EXIT with the corresponding MSC; ELSE continue.

Step 3: For every *vertex* in FPVD(P) compute the spanning circle determined by the three corresponding points and EXIT with the smallest such circle found.

In Step 2 if an edge $\overline{p_i p_j} \in$ FPT(P) intersects its corresponding edge in FPVD(P) then by the corollary $P \subset$ DISK(p_i,p_j) and from theorem 3 $d(p_i,p_j)$ = DIAM(P). Thus the algorithm correctly exits with DISK(p_i,p_j) as the MSC. If, on the other hand, no intersecting edge is found in Step 2 then no edge in FPT(P) is such that its DISK covers P. Then from theorem 3 we have that no edge in FPT(P) is the diameter, and thus $P \not\subset$ DISK(DIAM(P)). Thus no two-point solution exists and the algorithm correctly branches to Step 3.

Step 1 can be done in $O(n \log n)$ time according to Shamos [2] via divide and conquer. Since Steps 2 and 3 are linear, an $O(n \log n)$ algorithm would result. Which of the two algorithms one would choose probably depends on the application. It seems that computing CH(P) and its diameter is faster (by multiplicative constant) than computing FPVD. Thus, if in a particular application the probability that $P \subset$ DISK(DIAM(P)) is large algorithm MSC-1 may be preferred avoiding the computation of FPVD(P).

IV. CONCLUDING REMARKS

The counterexample described in this paper opened several problems in computational geometry. For the *minimum spanning circle* and *all-furthest-neighbor* problems new O(n log n) algorithms have been presented. For the case of convex polygons an O(n) algorithm for a similar problem has been recently discovered by Toussaint [9]. One interesting problem that remained open for some time is whether there exists an O(n) algorithm for computing the minimum spanning circle of convex polygons. This problem was recently and dramatically solved by Megiddo [19] who showed that even for sets of n points the minimum spanning circle could be obtained in linear time. It should be noted that independently of the work presented here [21], D.T. Lee [20] obtained similar results for the all-furthest-neighbor and minimum spanning circle problems and presents new algorithms for constructing the furthest-point-Voronoi diagram of n points in O(n log n) time.

ACKNOWLEDGEMENT

The authors are grateful to David Avis for several useful discussions on this topic and for reading the manuscript.

REFERENCES

[1] M.I. Shamos, *Problems in Computational Geometry*, Carnegie-Mellon University, 1977.

[2] M.I. Shamos, "Computational geometry", Ph.D. Thesis, Yale University, May 1978.

[3] M.I. Shamos and D. Hoey, "Closest-point problems", *Sixteenth Annual IEEE Symposium on Foundations of Computer Science*, October 1975, pp. 151-162.

[4] G.T. Toussaint, "Pattern recognition and geometrical complexity", *Proc. Fifth International Conference on Pattern Recognition*, Miami Beach, December 1980, pp. 1324-1347.

[5] H. Rademacher and O. Toeplitz, *The Enjoyment of Mathematics*, Princeton University Press, 1957.

[6] F.P. Preparata, "Minimum spanning circle", in *Steps into Computational Geometry*, F.P. Preparata, Ed., University of Illinois, Urbana, Illinois, March 1977, pp. 3-5.

[7] M.I. Shamos, "Geometric complexity", *Proc. 7th ACM Symposium on the Theory of Computing*, May 1975, pp. 224-233.

[8] D.P. Dobkin and L. Snyder, "On a general method for maximizing and minimizing among certain geometric problems", *Proc. 20th Annual Sumposium on Foundations of Computer Science.* San Juan, Puerto Rico, October 1979, pp. 9-17.

[9] G.T. Toussaint, "The symmetric all-furthest-neighbor problem", Computers and Mathematics with Applications, Vol. 9, No. 6, 1983, pp. 747-754.

[10] R.J. Lipton and R.E. Tarjan, "Applications of a planar separator theorem", *18th Annual IEEE Symposium on Foundations of Computer Science,* October 1977, pp. 162-170.

[11] R.A. Jarvis, "On the identification of the convex hull of a finite set of points in the plane", *Information Processing Letters,* Vol. 2, 1973, pp. 18-21.

[12] L. Devroye and G.T. Toussaint, "A note on linear expected time algorithms for finding convex hulls", *Computing*, Vol. 26, 1981, pp. 361-366.

[13] L. Devroye, "Moment inequalities for random variables in computational geometry", *Computing*, Vol. 30, 1983, pp. 111-119.

[14] F.P. Preparata, "A new approach to planar point location", *SIAM Journal on Computing,* Vol. 10, No. 3, August 1981, pp. 473-482.

[15] D. Kirkpatrick, "Optimal search in planar subdivisions", *SIAM Journal on Computing,* Vol. 12, No. 1, February 1983, pp. 28-35.

[16] W.E. Snyder and D.A. Tang, "Finding the extrema of a region", IEEE Trans. Pattern Analysis and Machine Intelligence, Vol. PAMI-2, May 1980, pp. 266-269.

[17] D. Avis, G.T. Toussaint and B.K. Bhattacharya, "On the multimodlity of distances in convex polygons", *Computers and Mathematics with Applications,* Vol. 8, No. 2, 1982, pp. 153-156.

[18] B.K. Bhattacharya and G.T. Toussaint, "A counterexample to a diameter algorithm for convex polygons", IEEE Transactions on Pattern Analysis and Machine Intelligence, Vol. PAMI-4, No. 3, May 1982, pp. 306-309.

[19] N. Megiddo,"Linear time algorithm for linear programming in R^3 and related problems", *SIAM Journal on Computing,* Vol. 12, No. 4, November 1983, pp. 759-776.

[20] D.T. Lee, "Farthest neighbor Voronoi diagrams and applications", Tech. Report. #80-11-FC-04, Northwestern University, November, 1980.

[21] G.T. Toussaint and B.K. Bhattacharya, "On geometric algorithms that use the furthest point Voronoi diagram", Tech. Report No. SOCS-1 81-3, School of Computer Science, McGill University, January 1981.

COMPUTATIONAL GEOMETRY, G.T. Toussaint (Editor)
© Elsevier Science Publishers B.V. (North Holland), 1985

OPTIMAL CONVEX DECOMPOSITIONS

by

Bernard Chazelle and David P. Dobkin

Abstract:

The problem of decomposing a non-convex simple polygon into a minimum number of convex polygons is solved. The decomposition allows for the introduction of Steiner points. Two algorithms are proposed. The first verifies that the problem is doable in polynomial time. The second provides an efficient method. Along the way, numerous results of independent interest in pure geometry as well as geometric complexity are stated.

1. Introduction

The problem of decomposing a simple polygon into its basic components has been a recurrent theme in computational geometry. Interest in this problem comes from its central location in the study of object representation. In the same way that English words benefit "greatly"

This research was supported in part by NSF grants MCS 83-03925 and MCS 83-03926, the Office of Naval Research and the Defense Advanced Research Projects Agency under contract N00014-83-K-0146 and ARPA Order No. 4786.

Authors' addresses: Bernard Chazelle is with the dept. of Computer Science, Brown University, Providence, RI 02912. David Dobkin is with the dept. of Electrical Engineering and Computer Science, Princeton University, Princeton, NJ 08544.

from their expressibility in a 26-letter alphabet, complex geometric structures are more easily handled when decomposed into simpler structures. Think of tool designers pressed for simple, modular designs, for instance. Another good example is the recognition of Chinese characters by matching text data against building blocks, as described in [8]. This operation involves the decomposition of polygonal shapes into convex pieces. For further motivation on decomposition problems, see [8,9,18,20]. The problem we consider is:

Given a simple polygon P, what is the minimum number of convex polygons which form a partition of P ?

This is called the OCD problem (for "optimal convex decomposition"). We will briefly review the known results. The first breakthrough on the OCD problem appeared in the proceedings of the 1979 STOC Symp. [5]. There, these authors proved that the problem was polynomial, thereby frustrating widespread suspicion that it was NP-hard. Interestingly enough, this finding was followed by a stream of NP-hardness results for similar problems [12—17]. For example, it was shown by Lingas that the presence of holes in the polygon P was sufficient to make the OCD problem NP-hard [14].

Variants of the OCD problem were considered in [12,13], where the requirement was made that no new vertices should be introduced in the partition of P. In [15,16] the objective function to minimize was no longer the number of pieces but the total edge-length. For other work, consult [1,3,4,8,9,14,18]. Following the original paper of Chazelle and Dobkin [5], the OCD problem was thoroughly treated in the former author's PhD thesis [3], where an $O(n+c^3)$ time decomposition algorithm was presented (n is the number of vertices of P and c is the number of reflex angles). This result represents a quantitative (but not qualitative) improvement over the $O(n^6)$ time algorithm of [5]. Of course it can be argued that for small values of c, the algorithm in [3] is linear or quasi-linear. Unfortunately this statement must be tempered by the rather intricate nature of the algorithm, which makes it an unlikely candidate for efficient implementation.

The purpose of this paper is to present the main ideas of the algorithms in [5] and [3]. We recognize that the interest of our results is primarily theoretical, so we will devote most of our effort to proving that the OCD problem is polynomial. The rest of the paper will outline the most interesting points of the $O(n + c^3)$ time algorithm, referring to [3] for details and complementary information. The paper will be organized as follows: in this section, we give a brief overview of our method. In the next section, we present the basic geometric facts for the study of the OCD problem. In Section 3, we consider the algorithmic aspect of the problem and describe the polynomial algorithm for the OCD problem. In the following section, we address implementation issues and outline new lines of attack to speed up the algorithm. Finally we give conclusions and outline directions for further research in Section 5.

One methodological note is in order. Given the intricacy of our $O(n+c^3)$ algorithm for the OCD problem, the presentation will follow a top-down approach. We present the main ideas of the method first, and then fill in the blanks left. Our rationale is to separate the essential components of the algorithm and the parts which only contribute to its efficiency.

Two simple facts bound all algorithms for this problem. First, each notch (i.e. vertex displaying a reflex angle) *can* be removed by the addition of a polygon to the decomposition. Second, *at most* two notches can be removed through the addition of a single polygon. Hence, the minimum number of convex parts always lies between $\lceil c/2 \rceil + 1$ and $c + 1$. To extend these simple observations, however, is a difficult mathematical problem. To form minimal decompositions additional (Steiner) points must be introduced as vertices of newly generated polygons. This removes the obvious finiteness of the problem and makes simple enumerative procedures impossible. Furthermore the problem cannot be treated in a local manner. These observations led to the conjecture that the problem was NP-hard.

To circumvent these difficulties, we introduce X-patterns, from which minimal decompositions can be generated. An X_k-pattern is a particular interconnection of k notches which removes all reflex angles at the k notches and creates no new notches. A decomposition obtained by applying p patterns of type $X_{i_1}, \ldots, X_{i_p}$ along with straight-line segments to remove the remaining notches can be shown to yield $c + 1 - p$ convex parts. Clearly, decompositions with the most X-patterns also minimize the number of convex polygons. This can be viewed as a generalized matching problem which might lend itself to a dynamic programming approach. Simple examination shows that there is exactly one type of each X_k-pattern for $k = 2, 3$. Were

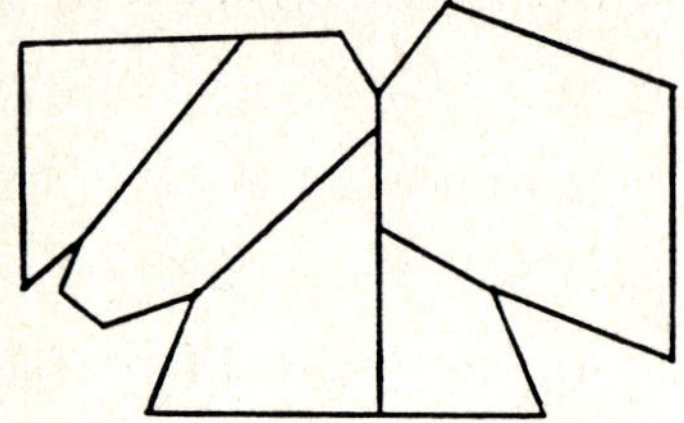

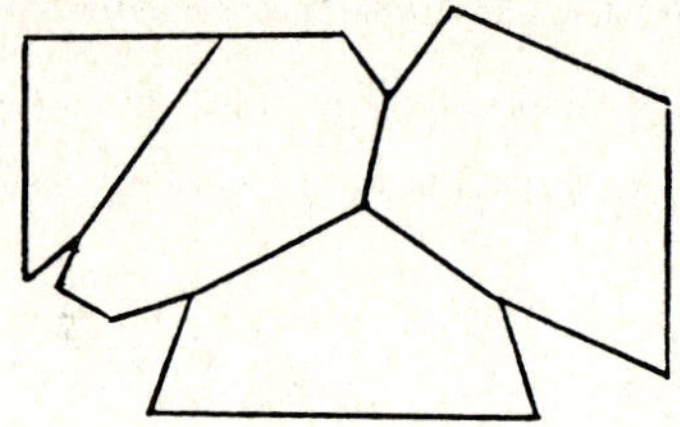

Figure 1 The naive decomposition and an improvement.

this the case for larger k, a polynomial-time algorithm based upon X-patterns would be possible. Unfortunately, determining whether a given set of notches can be interconnected via an X-pattern appears too involved to handle directly.

One solution is to constrain X-patterns in such a way that their detection becomes tractable. This leads to the introduction of Y-patterns, which we can regard as X-patterns endowed with some structural property. We show that with the exception of X_4-patterns any X-pattern can be advantageously replaced by a Y-pattern. Since Y-patterns can be constructed in polynomial time via dynamic programming, we can achieve our first goal, which is to show that the OCD problem is in P. As is shown later on, further geometric analysis leads to substantial gains in the efficiency of the original algorithm.

2. The Geometric Ingredients

In this section, we introduce our notation and show that the OCD problem can be reduced to a form of generalized matching problem. Let P be a simple polygon with n vertices, $p_1, \ldots, p_n$, in clockwise order. As previously mentioned, the vertices of P which display a reflex angle, called *notches*, will play a crucial role in the following. Let $v_1, \ldots, v_c$, be the list of notches in P, given in clockwise order. Throughout this paper, we will use the following convention on the representation of angles. Let ab and ac be two non-collinear line segments. $\angle(ab, ac)$ denotes the angle between 0 and 360 degrees swept by a counterclockwise rotation from ab to ac.

A *decomposition* of P is a set of polygons, $P_1, \ldots, P_k$, whose union gives P, and such that the intersection of any two if non-empty consists totally of edges and vertices. A decomposition is said to be *convex* if all its polygons are convex. We define an *optimal convex decomposition* of P, or OCD for short, as any convex decomposition of minimum cardinality.

We define the *naive decomposition of P* as the set of polygons obtained by removing each notch in turn by means of a simple line segment *naively* drawn from the notch. To be more precise, a naive decomposition of P is obtained by going through each notch $v_1, \ldots, v_c$ in turn, extending a line segment from v_i until we first hit another line already in the decomposition.

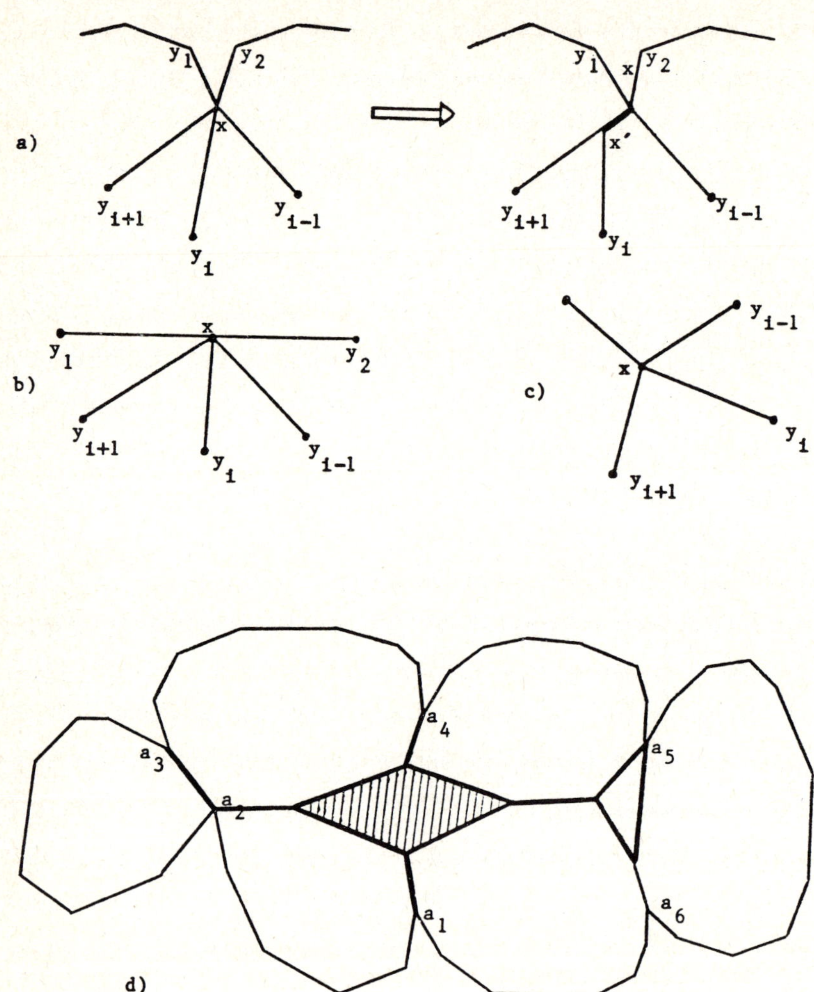

Figure 2 a,b,c) Satisfying degree requirements for X-decompositions.
d) Removing interior polygons.

Of course, the extended line should remove the reflex angle at v_i. Note that in this way we will also guarantee that all lines drawn always lie entirely inside P. To simplify the ensuing analysis, we will ensure that the degree of each vertex in the decomposition does not exceed 3 and that no segment in the naive decomposition connects two notches. These conditions are trivially always satisfiable. Figure 1 illustrates the notion of naive decomposition and shows in particular that it is not always minimal. We have the following (trivial) result.

Lemma 1. Any naive decomposition of P produces exactly $c + 1$ convex parts.

Next we wish to characterize a convenient class of decompositions to which we will restrict our attention in the following. We say that a polygon is *interior* to P if it lies inside P and at most a finite number of its points lie on the boundary of P.

Definition 1. An X*-decomposition* is any convex decomposition containing no interior polygons and such that no vertex is of degree greater than 3, except for the notches, which may be of degree at most 4.

Lemma 2. The class of X-decompositions always contains an OCD.

Proof: Consider an OCD which is not an X-decomposition. We transform its edges to yield an X-decomposition. First, we show how to satisfy the degree requirements. Since this process may introduce interior polygons, we show how to remove interior polygons without increasing the degree of any vertex.

1. We regard the decomposition as a planar graph consisting of *boundary edges* and *added edges* (an edge is said to be added if it does not lie on the boundary of P). Let x be a notch of degree greater than 4 and $y_1, \ldots, y_m$ $(m > 4)$ be its adjacent vertices, with $y_1, y_2 \in$ boundary of P. It is trivial to show that there exists $i > 2$ such that $\angle(xy_{i+1}, xy_{i-1}) < 180$ (Fig.2-a). We can then move xy_i along xy_{i-1} or xy_{i+1} to form a new segment $x'y_i$ with x' chosen close enough to x so as to preserve convexity. We iterate on this process until the notch x becomes of degree 4. The other cases to consider are depicted in Fig.2-b,c. In case b) the vertex x is not a notch but still lies on the boundary of P (it may or may not be a

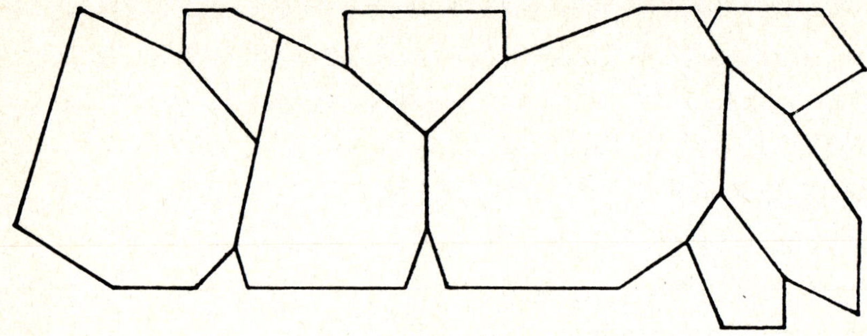

Figure 3 An X-decomposition involving an X3- and an X4-pattern.

vertex of P). In case c) x is a vertex of degree > 3 which does not lie on the boundary of P. The same method used in case a) will reduce the degrees accordingly.

2. Consider the subgraph H formed by the added edges of the OCD under consideration. Pick an interior polygon of the OCD and let G be the connected component in H to which the edges of this polygon belong (Fig.2-d). Let $a_1, \ldots, a_k$ denote the vertices of G (in clockwise order) on the boundary of P and let K be the graph obtained by removing G from the graph of the OCD. Since G lies in P and is a connected component of H, it lies entirely in one face of K, denoted Q. We observe that all the a_i's lie on the boundary of Q. Since G is connected, it determines at least k faces in K, aside from the interior polygon(s). This shows that the OCD has at least $k+1$ faces in Q. The polygon Q may not be convex, but since we had a convex decomposition of P before removing G, all the notches of Q must be notches of P, i.e. must be some of the a_i's. Now, instead of keeping the convex decomposition of Q induced by the OCD, we apply the naive decomposition to it. This will yield at most $k+1$ polygons (exactly $k+1$, actually, since we are dealing with an OCD, and consequently no transformation can *improve* the decomposition). Since the boundary of each of the created polygons contains at least one of the points a_i as a vertex, and each a_i belongs to at most two polygons, none of these polygons can be interior to P. Furthermore it follows from the definition of the naive decomposition that the desired degree constraints will be preserved. Iterating on this process for each of the remaining interior polygons completes the proof. ∎

We are now ready to introduce the important notion of *X-pattern.* Once again we regard the added edges of an X-decomposition as forming a subgraph of the total decomposition. From the definition of X-decompositions, it follows that the subgraph is a forest of trees with each node having degree 1 or 3 except for the notches which may have degree 1 or 2. We will pay special attention to those trees where all vertices of degree 1 or 2 are notches of P.

Definition 2. A planar embedding of a tree lying inside P is called an X-pattern if it is not self-intersecting and:

1. All vertices are of degree 1, 2 or 3.
2. Any vertex of degree 1 or 2 coincides with a notch of P, and its (1 or 2) adjacent edges remove its reflex angle.
3. None of the 3 angles around any vertex of degree 3 is reflex.

An X-pattern with k vertices of degree 1 or 2 is called an X_k-pattern. Vertices of degree 1,2,3 are called N1, N2, N3-nodes, respectively. For simplicity we refer to the vertices of degree 1 or 2 as the notches of the X-pattern. Informally, an X_k-pattern is an interconnection of k notches used to remove them while introducing $k-1$ additional polygons into the decomposition. Figure 3 gives an example of an X-decomposition with one X_3-pattern and one X_4-pattern. We justify the introduction of X-patterns with the following observation.

Lemma 3. An X-decomposition with p X-patterns has at least $c+1-p$ convex parts.

Proof: Let S, t, k be respectively the number of polygons, trees, and tree-vertices lying on the boundary of P. We prove the relation

$$S = k - t + 1 \tag{1}$$

by induction on t. The case $t = 1$ is trivial, so we may assume that the introduction of $t-1$ trees involves k_1 vertices on the boundary of P and creates $S_1 = k_1 - (t-1) + 1$ polygons. Introducing the last tree into the decomposition will account for exactly $k - k_1 - 1$ additional polygons, leading to a total of $S = S_1 + k - k_1 - 1 = k - t + 1$ polygons and proving (1). Each of the $t - p$ trees which are not X-patterns has at least one distinct vertex which lies on the boundary of P and is not a notch. This implies that $t - p \leq k - c$, which alongside (1) establishes the lemma. ∎

Lemma 3 suggests that using p X-patterns saves at most p polygons over the naive decomposition. This leads to the definition of *compatible* X-patterns. A set of X-patterns is said to be *compatible*, if no pair of edges taken from two distinct patterns intersect. For example, the X-patterns in any X-decomposition always form a compatible set. Conversely, we can show that any set of compatible X-patterns can be used to produce a decomposition. The following result is complementary to Lemma 3.

Lemma 4. Any compatible set of p X-patterns can be used to produce an X-decomposition with exactly $c + 1 - p$ convex parts.

Proof: Start the decomposition with the p X-patterns. This will produce a certain number of polygons. If any of them is not convex, apply the naive decomposition to it. Straightforward analysis shows that the final number of polygons will be exactly $c + 1 - p$. ∎

From Lemmas 2, 3 and 4, we are able to express the original OCD problem as a generalized matching problem.

Lemma 5. Let p be the maximum number of compatible X-patterns. An OCD can be obtained by first applying the p X-patterns and then applying the naive decomposition to any remaining non-convex polygon.

Since any X-pattern has at least two notches, the previous result shows that an OCD consists of at least $1 + \lceil c/2 \rceil$ polygons. Lemma 5 suggests a new line of attack for the problem at hand —the sufficiency of computing a maximum set of compatible X-patterns. Unfortunately to do so seems beyond reach, given the excessive number of candidates we might have to consider in the process. X-patterns allow Steiner points (i.e. vertices not on the boundary of P) to be adjacent. Looking at any X-pattern as a mechanical system of extendible arms and joints, this corresponds to a system which is not strongly constrained. We show next that X-patterns can be in general *reduced* to maximally constrained X-patterns, called Y-patterns. The next step will be to prove that Y-patterns (being maximally constrained) can be computed in polynomial time. A rigorous definition of these notions is now in order.

Definition 3. A Y_k-pattern is an X_k-pattern such that

1. no edge joins two nodes of type N3.
2. in any path containing three consecutive nodes of respective type N2,N3,N2, the N2-nodes lie on opposite sides, i.e. the two pairs of edges of P which emanate from the N2-nodes lie on opposite side of the path.

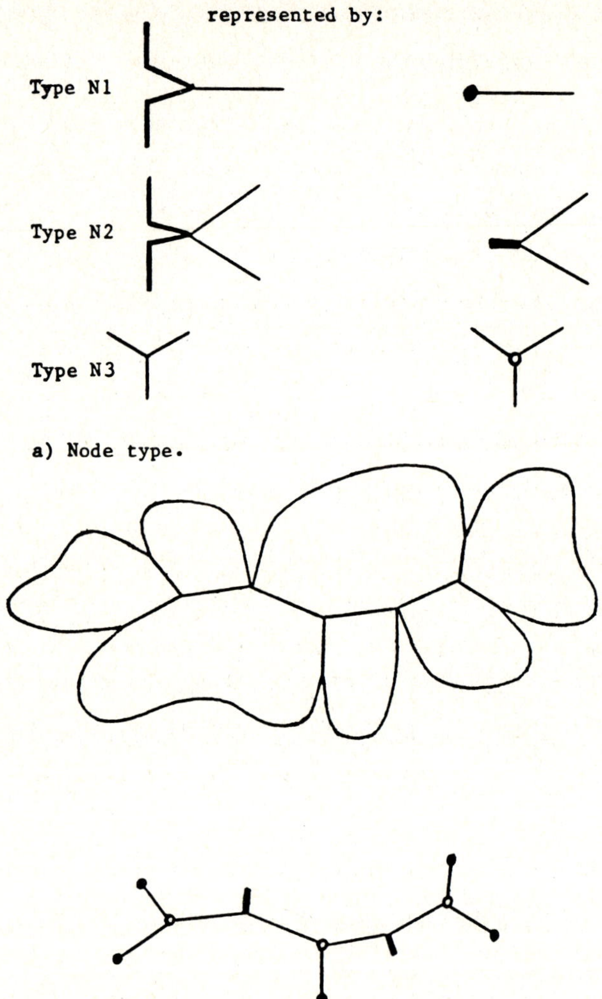
represented by:
Type N1
Type N2
Type N3
a) Node type.
b) a Y7-pattern and its representation.

Figure 4

We will use a special representation for N1, N2 and N3-nodes (Fig.4-a). Note that the N3 nodes of a Y-pattern are its Steiner points. A Y_7-pattern and its representation are given in Fig.4-b. To help visualize Condition 2, observe that if the two N2-nodes were pointing downward, we would not have a Y-pattern.

We next show that all X_k-patterns ($k \neq 4$) can be transformed into Y-patterns. This allows us to limit attention to X-decompositions with only Y and X_4-patterns. The transformations, called *reductions*, involve the stretching, shrinking, and rotating of lines in the original pattern. Reductions involve sequences of steps with each step translating an N3-node from one point to another. All edges in the pattern except for the three edges adjacent to the N3-node remain fixed. Reductions stop before an angle in the pattern becomes reflex or an edge in the pattern strictly intersects an edge on the boundary of P. During a reduction, a tree remains an X-pattern. It may however gain or lose vertices in the process. For example, Figure 5 shows the reduction of an X_3-pattern, which might correspond to one step in the reduction of a more complex X-pattern.

In Figure 5, we see how an X_k-pattern may be reduced to an X_l-pattern ($l < k$). The final tree is still an X_3-pattern; it can also be viewed as an X_2-pattern augmented with a segment provided by the naive decomposition. The X_3-pattern loses a notch thereby being reduced to an X_2-pattern. This is legitimate since from lemma 5 we know that numbers (rather than types) of compatible X-patterns matter. In Fig.5, observe that the notches a and b cannot possibly be interconnected by an X_2-pattern in *any* X-decomposition. However, it is true that an OCD can be obtained by considering the X_2-pattern as the single element of a maximal set of compatible X-patterns. In this regard, the X_2-pattern is of interest to us. If X-patterns can lose notches, they can also gain some, as will be soon shown. This should not surprise us, however, since the previous idea of applying X-patterns and then completing the process with the naive decomposition may also augment the X-patterns with additional edges (while not increasing the number of patterns).

Lemma 6. In an X-decomposition, any X-pattern which is not reducible to an X_4-pattern can be reduced to a Y-pattern.

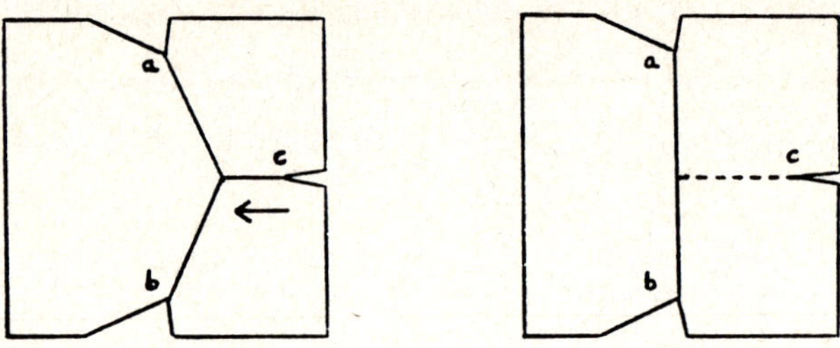

Figure 5 Reductions on X-patterns.

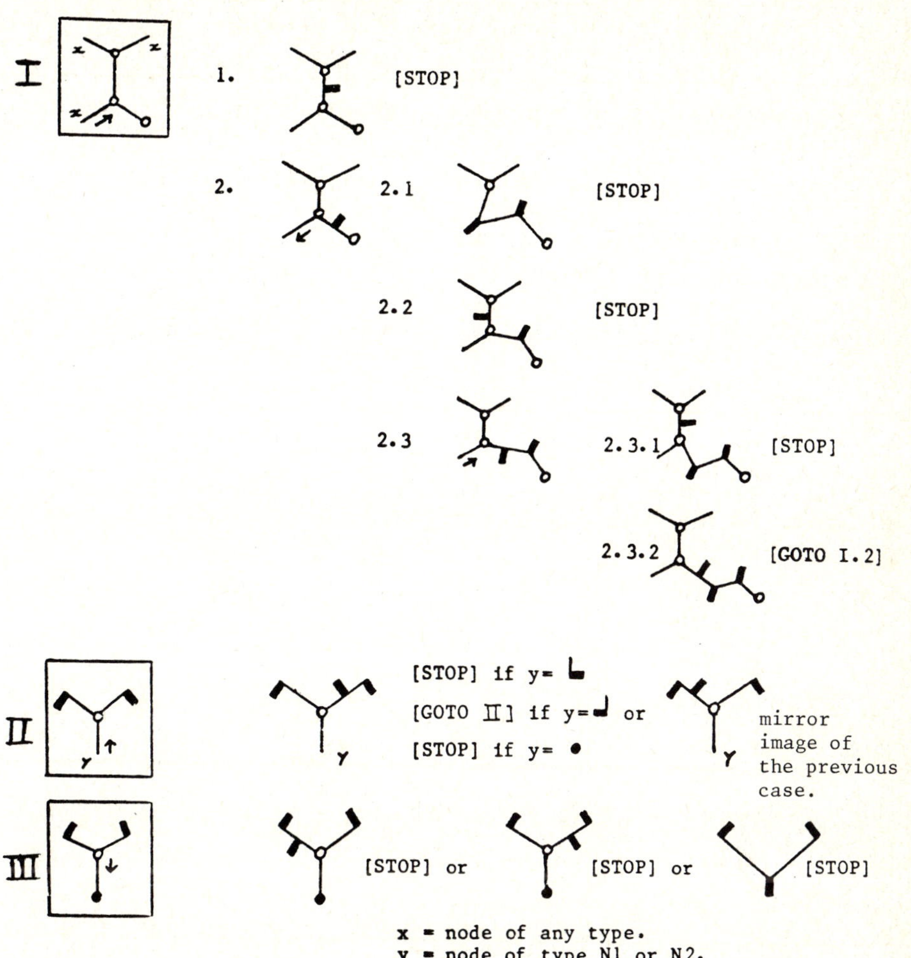

Figure 6 The proof of Lemma 6.

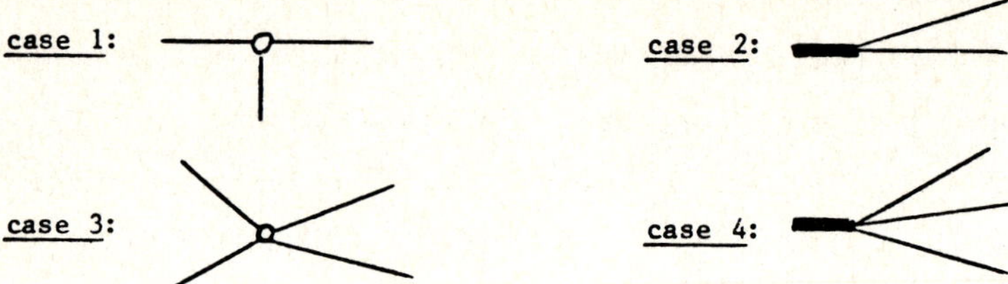

Figure 7 Extreme instances of X-pattern vertices.

Proof: Before describing the appropriate sequence of reductions which will turn X-patterns into Y-patterns, we must ensure that reductions can be carried out freely without merging two patterns in the process, which might increase the number of polygons in the decomposition. If a reduction brings an edge of an X-pattern in contact with any edge of the decomposition not on the boundary of P a reflex angle will have necessarily resulted beforehand. Since the definition of a reduction precludes intersections with the boundary and reflex angles, this case will not occur.

We are now ready to describe the reductions of undesirable X-patterns. We first ensure condition 1 of Definition 3.

Condition 1.

Figure 6-I) illustrates the sequence of actions. As indicated, we assume that the pattern is not of type X_4 but has two N3-nodes adjacent to each other. Recall the notation for node types in Fig.4. We move one of the N3-nodes by translation in the direction indicated by the arrow. The translation continues until either an N2-node results from intersection with the boundary or we fall into one of the "extreme" instances depicted in Fig.7. Assume for the time being that we are in the first case. If the N2-node occurs between the N3-nodes (case 1) we are done. Otherwise (case 2), another reduction leads to 2.1), 2.2) or 2.3). We then iterate on this process until no pair of N3-nodes are adjacent (the label STOP is meant to indicate that the current reduction step is over and that we should check again if more reductions are necessary). Convergence is guaranteed since each step adds a distinct N2-node.

Note that the figure investigates all cases except for those representing extreme instances of X-patterns. These extreme cases are illustrated in Fig.7. Case 2 is to be understood as representing two edges emanating from the same notch, one of which is sufficient to remove the reflex angle. To handle all four cases, we observe that in each of them we can prune one edge from the pattern along with the adjoining subtree and still preserve the non-reflexivity of all angles. Recall that the notches attached to the removed piece, although now unresolved, will be removed later on via the naive decomposition. The crucial observation is that pruning patterns does not affect the overall number of patterns in the decomposition. The pruning

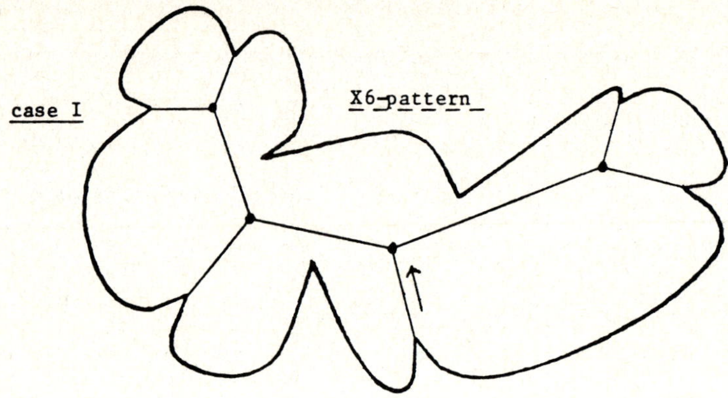

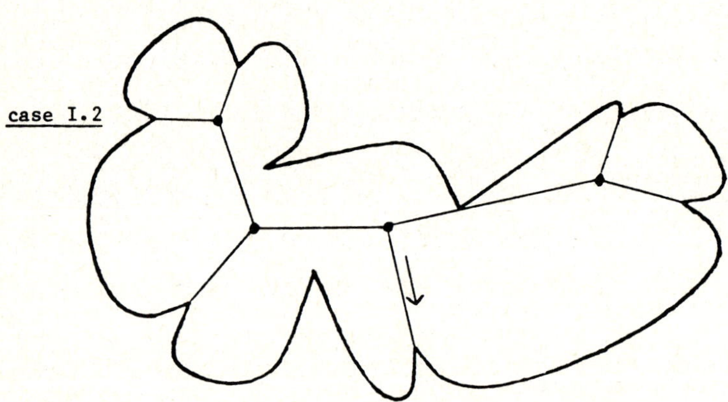

(Figure 8 .../...)

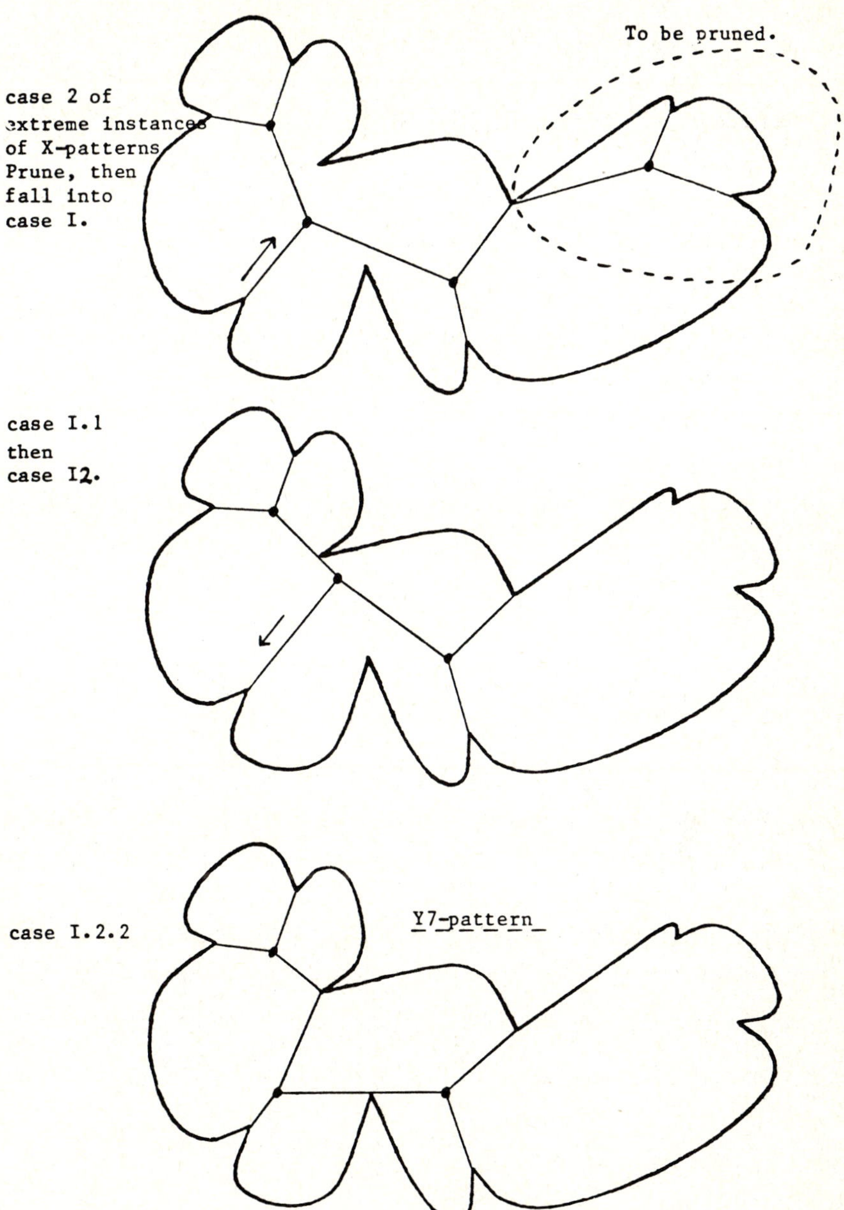

Figure 8 Example of reductions turning an X6-pattern into an Y7-pattern.

process will converge since every removal decreases the current number of N1-nodes by at least one. Unfortunately this statement is a little hasty, since case 2 of Fig.7 surely removes at least one N1-node but, unfortunately, also adds one. To establish convergence, we will have to show that when in case 2 the subtree pruned involves a single notch x, this notch can never be re-introduced by subsequent reductions. Let xy be the unique edge thus pruned. A simple observation shows that no further reduction of the pattern will ever bring any of its edges across the segment xy. For this reason x can never become an N2-node for the pattern. But being an N2-node is a prerequisite for becoming an N1-node again, therefore x is safely *lost* for the pattern once and for all —see [3] for a detailed proof of this fact.

Condition 2.

Once Condition 1 holds, we satisfy Condition 2 by following the instructions outlined in Fig.6-II,III). Proving convergence follows the lines given above and we do not elaborate on it.

The reductions shown in Fig.6 are to be applied iteratively to each X_k-pattern ($k \neq 4$) with adjacent N3-nodes. Convergence is straightforward. We have illustrated the complete reduction of an X_6-pattern in Fig.8. ∎

3. The Polynomial Time Algorithm

The previous section proved the existence of an OCD consisting solely of Y and X_4-patterns. In this section, we present a polynomial time algorithm for constructing such an OCD. For the sake of clarity, we will ignore efficiency issues, merely showing that every routine used in the algorithm runs in polynomial time. Later we will discuss efficient implementation.

We begin by constructing an oracle to answer questions of the form : "*Does there exist a pattern connecting k given notches* ?" in polynomial time. This oracle will be used by the decomposition algorithm.

Let v be an N3-node of a Y-pattern. Remove the three edges vv_i, vv_j, vv_k adjacent to v; the Y-pattern becomes a disconnected set of three subtrees, for which the removed edges play the role of an X_3-pattern. This leads us to introduce the notion of *extended* X-pattern. An *extended* X_l-pattern ($l = 2,3$) is either an X_l-pattern by itself or an X_l-pattern appearing as a subgraph of an X_m-pattern ($m > l$). It is clear that an algorithm for testing the possibility of an X_l-pattern between a given set of notches can also be used to determine the possibility of an extended X_l-pattern, as long as the angles formed at the notches by the subtrees of the X_m-pattern are known in advance.

To see this, we must define the notion of *extended notch* and *extended range*. Let v_i be a notch of the X_l-pattern, and let W be a wedge centered at v_i. We define the *extended range* of v_i as the set of points u such that v_iu lies entirely within both P and W. In essence, the extended range is the intersection of W with the visibility-polygon at v_i [7]. In general the wedge W will be taken as the locus of rays (i.e. half-lines) which emanate from v_i and remove the reflex angle at v_i. When dealing with ordinary X-patterns, the wedge W is simply determined by the edges of P adjacent to v_i. In this case the extended range is simply referred to as *range* of v_i, since it is then defined only with respect to v_i and P. In the case of extended patterns, however, the wedge will take into account the other edges already adjacent to v_i, and will thus be smaller than in the previous case. Later, we consider cases where the wedge W is taken as the entire plane. This is done if we do not wish to remove a reflex angle at a particular notch. In all cases, anyhow, we say that we *extend* the notch according to certain angular specifications.

Lemma 7. Checking for the possibility of an (extended) X_l-pattern between l given notches can be done in polynomial time (for $l \leq 4$).

Proof: For $l = 2,3$ it is clear that an X_l-pattern will be possible if and only if 1) ($l = 2$) the (extended) range associated with each notch contains the other notch; 2) ($l = 3$) the (extended) ranges associated with the l notches have a common intersection point forming three non-reflex angles with respect to the notches. Computing each range can be done in $O(n)$ time by first computing the visibility-polygon [7], and then clipping it along the corresponding wedge. Next we compute the intersections of all ranges, which can be done naively in $O(n^2)$ time. To handle X_4-patterns, we first observe that two different kinds must be considered, as shown in Fig.9. Assume wlog that the two notches v_i and v_j are adjacent to the same N3-node. We successively apply the reductions shown in Fig.9-b-c, with respect to A then B. Assuming wlog that this

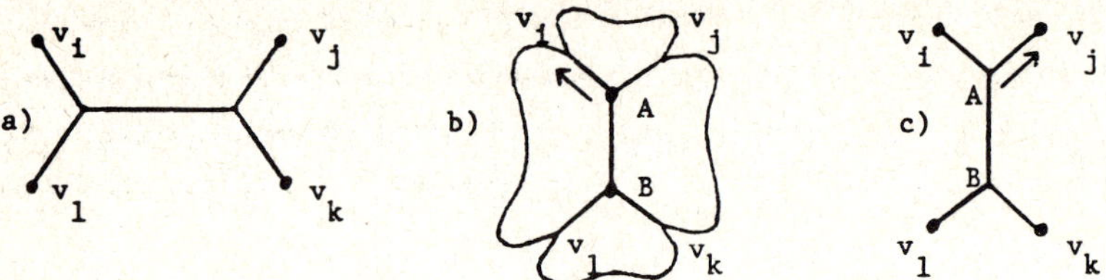

Figure 9 A simple method for computing X4-patterns.

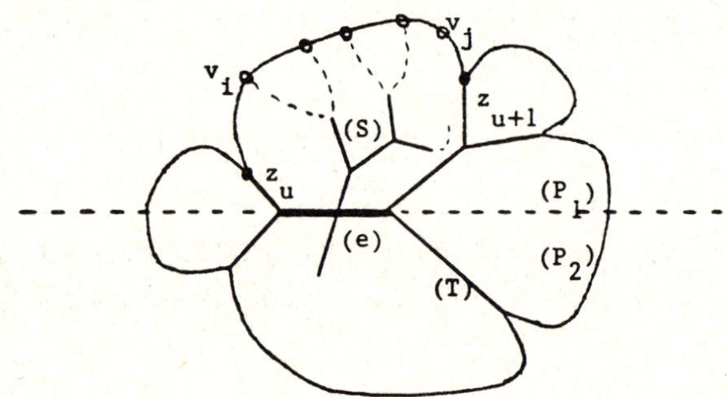

Figure 10 The interaction between X-patterns.

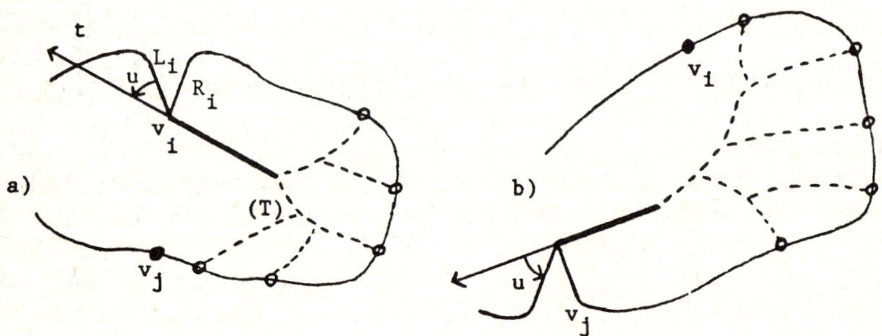

Figure 11 Defining B(i,j) and F(i,j).

does not create an X_3 or a Y_5-pattern with an N2-node between A and B, we simply observe that the four edges emanating from the notches will now be collinear with edges of P. This puts the total number of configurations within $O(c^4)$. We can therefore check whether any of them is feasible, which will lead to an $O(nc^4)$ time algorithm. ∎

We can now turn to the decomposition algorithm. The procedure for determining a maximum compatible set of X-patterns is based on dynamic programming. We rely upon the observation that if a certain X_4 or Y_k-pattern belongs to an OCD of P, it decomposes P into k subpolygons, $P_1, \ldots, P_k$, so that finding an OCD for each P_i yields an OCD for P. We compute maximal compatible sets of patterns for each P_i. Since the notches of P_i are also notches of P, any X-pattern of P_i is also an X-pattern of P. Conversely, we want to show that any X-pattern of P involving only notches in P_i is also an X-pattern of P_i. This is quite important. Dynamic programming proceeds bottom-up, so a maximal set of patterns involving notches of P_i must be found before we know the shape of P_i.

To solve this problem, we define $V(i,j)$ as the set of notches between v_i and v_j in clockwise order, so $V(i,j) = \{v_i, v_{i+1}, \ldots, v_j\}$, with index arithmetic taken modulo c. Let $z_1, \ldots, z_k$ be the notches of an X-pattern, T, given in clockwise order around the boundary of P, and let $V(i,j)$ be the notches of P between z_u and z_{u+1} in clockwise order ($z_u = v_{i-1}, z_{u+1} = v_{j+1}$). We will show that no X-pattern with all its notches in $V(i,j)$ can intersect T. Assume that an X-pattern S intersects an edge e of T. Consider the shortest segment which is collinear with e and has both of its endpoints on the boundary of P. This segment partitions P into two polygons P_1 and P_2 (Fig.10). Since the path of T between z_u and z_{u+1} is a convex polygonal line, it lies entirely in P_1 or P_2 (say, P_1). Since all the notches in $V(i,j)$ are notches of P_1, S must have notches in P_2, a contradiction. This independence result can be understood in combinatorial terms. It states that two X-patterns are intersection-free if and only if their notch sequences are not intermixed, i.e. one sequence falls completely between two consecutive elements of the other sequence.

We next define $S(i,j)$, for every pair of notches v_i, v_j, as a maximum compatible set of X_4 or Y-patterns in $V(i,j)$. To achieve our ultimate goal, i.e. evaluating $S(1,c)$, we compute all values $S(i,j)$ from $\{S(k,l) \mid V(k,l) \subset V(i,j)\}$ using dynamic programming. This can be done directly if v_i and v_j are not connected to the same pattern. We simply test all combinations $\{S(i,k), S(k+1,j)\}$ for all $v_k \in V(i,j-1)$. Otherwise we have to distinguish whether v_i and v_j should be connected together via an X_4 or a Y-pattern.

To handle the latter case, we compute all Y-patterns which might belong to an OCD via dynamic programming. We compute Y-subtrees (i.e. subtrees of Y-patterns) as well as Y-patterns by patching Y-subtrees together. To prevent the number of computations from blowing up, however, we keep only the Y-subtrees that are candidates for belonging to an OCD. A Y-subtree is considered *not* to be a candidate if at the time it is computed we are ensured of the existence of at least one OCD which does not use this Y-subtree (although we may not know this OCD explicitly yet). As a shorthand we say that a pattern or a Y-subtree lies in $V(i,j)$ if all its notches do. It now remains to formalize the intuition given here and describe the polynomial time algorithm.

Consider a Y-pattern which is used in an OCD and has at least one N2-node, v_i. This node splits the Y-pattern into two Y-subtrees, so there exists an index j such that

1. One of the Y-subtrees lies in $V(i,j)$ whereas the other lies in $V(j+1,i)$.
2. All the other patterns in the OCD lie totally either in $V(i,j)$ or in $V(j+1,i)$.

We will consider the candidacy of the Y-subtree in $V(i,j)$ immediately after $S(i,j)$ has been computed. We first observe that if $v_i, v_{i_1}, \ldots, v_{i_m}$ is a list of its notches in clockwise order, we may dismiss the candidacy of the subtree if the following equality is not satisfied:

$$|S(i,j)| = |S(i+1, i_1 - 1)| + \ldots + |S(i_{m-1}+1, i_m - 1)| + |S(i_m + 1, j)|, \tag{2}$$

where $|S(k,l)|$ represents the number of patterns in $S(k,l)$.

Note that the last term in the right-hand side is to be ignored if $i_m = j$. If Relation (2) is not satisfied, the right-hand side is strictly smaller than $|S(i,j)|$. When considering candidate Y-subtrees, we have the idea in mind that only one pattern will have notches in both $V(i,j)$ and $V(j+1,i)$. It would then be unreasonable to use any Y-subtree which does not satisfy (2), since the patterns of $S(i,j)$ would provide a better decomposition altogether.

This simple fact will be consequential in achieving a polynomial time algorithm. We still have to cope, however, with the prospect of keeping an excessive number of candidate subtrees for each pair (i,j). Another geometric observation is in order. Let L_i (resp. R_i) denote the edge of P adjacent to v_i and preceding (resp. following) v_i in clockwise order. Whenever L_i or R_i is used in designating an angle, it is understood as directed outwards with respect to v_i. Let t be the edge adjacent to v_i in the Y-subtree lying in $V(i,j)$. The edge t is called the *arm* of the Y-subtree. When the arm of a Y-subtree enters the expression of an angle, we assume that it is directed towards the notch (here t is directed towards v_i). Among all the Y-subtrees in $V(i,j)$ for which v_i is an N2-node, (2) is true and $u = \angle(L_i, t) < 180$, we may keep the Y-subtree T which minimizes the angle u as the only candidate with respect to $V(i,j)$ (Fig.11-a).

We define $B(i,j)$ as a pointer to the arm of T. If there is no such subtree, $B(i,j)$ is 0. Patterns and subtrees will be represented by linked lists, so $B(i,j)$ will provide access to the entire Y-subtree T whenever necessary. Carrying out the same reasoning counterclockwise in $V(i,j)$ with now v_j as an N2-node, we define $F(i,j)$ in a similar fashion (Fig.11-b).

Having established our notation, we are now in a position to present the decomposition algorithm. We assume a function $\langle ARG \rangle$ for assembling Y-subtrees when computing $S(i,j)$. ARG is in general a pair of Y-subtrees taken from $B(u,v)$ or $F(u,v)$. If the two subtrees can be patched together and form a Y-pattern T, the function $\langle\rangle$ returns (C,T), where C is the maximum number of compatible patterns which can be applied in $V(i,j)$ including T. We return to a discussion of this function after a presentation of the algorithm. Before proceeding with a formal description, a brief overview might be helpful.

After all necessary preprocessing in STEP 1, we use nested loops to implement the dynamic programming scheme. Each step involves computing $S(i,j)$ for a given value of i and j. We start by computing the best Y-pattern which connects v_i and v_j (STEP 2). This involves patching precomputed candidate Y-subtrees. STEP 3 computes a maximum set of compatible patterns in $V(i,j)$, denoted L, assuming that v_i and v_j do not belong to the same pattern. Then we compute M, defined similarly, with the difference that we now allow the presence of an X_4-pattern connecting v_i and v_j. Finally the Y-pattern of STEP 2 (if any) is used to compute N, so the maximal set among L, M, N is finally chosen as $S(i,j)$. STEP 4 computes the Y-subtrees which lie in $V(i,j)$ and are considered as candidates. These subtrees are to be used in further iterations through STEP 2. Once a maximum compatible set of patterns for P has been determined, we finish off the decomposition using the naive decomposition (STEP 5).

Procedure *ConvDec*(P)

STEP 1:

The preprocessing involves checking that P is simple and non-convex. We make a list of the notches $v_1, \ldots, v_c$, and we initialize all $B(i,i)$ and $F(i,i)$ to 0.

for $d = 1, \ldots, c-1$

for $i = 1, \ldots, c$

j:=i+d [mod c]

do STEPS 2,3,4

STEP 2:

Compute the best Y-pattern connecting v_i and v_j as follows:

For each k; $v_k \in V(i+1, j-1)$, compute the set $Q = \bigcup_{1 \le i \le 4} Q_i$, where

$Q_1 = \{\langle F(i,k), B(k,j)\rangle\}$ /* N2-node on path */

$Q_2 = \{\langle B(i,k-1), F(k,j)\rangle\} \bigcup \{\langle B(i,j-1), F(j,j)\rangle\}$ /* no N2 or N3 nodes on path */

$Q_3 = \{\langle B(i,k-1), B(k,j-1)\rangle\}$ /* N3-node on path */

$Q_4 = \{\langle F(i+1,k), F(k+1,j)\rangle\}$ /* N3-node on path */

The elements of Q are pairs of the form (C,T). Let T be the Y-pattern which has the maximum

C value in Q.

STEP 3:

Let $S(i,j)$ be the maximum of L, M, N with respect to cardinality, where (max is taken with respect to cardinality)

$L = \max_{v_k \in V(i,j-1)} \{S(i,k) \bigcup S(k+1,j)\}$

/* corresponds to a patching together of Y-patterns */

$M = \max\{\{x_{i,a,b,j}\} \bigcup S(i+1,a-1) \bigcup S(a+1,b-1) \bigcup S(b+1,j-1)\}$

for all X_4-patterns $x_{i,a,b,j}$ connecting v_i, v_a, v_b, v_j, with $v_a, v_b \in V(i,j)$.

/* corresponds to the use of an X_4-pattern */

$N = \{\text{the } Y\text{-pattern } T \text{ of STEP 2}\} \bigcup S(i+1, i_1 - 1) \bigcup \ldots \bigcup S(i_{p-1}+1, i_p - 1) \bigcup S(i_p + 1, j-1)$

where $v_i, v_{i_1}, \ldots, v_{i_p}, v_j$ are the notches of T in clockwise order.

/* corresponds to the use of the Y-pattern T */

STEP 4:

Compute $B(i,j)$ and $F(i,j)$.

STEP 5:

Finish off the decomposition using the naive decomposition, i.e. adding one polygon for each remaining notch.

The remaining of this section is devoted to explaining the various steps of the algorithm and analyzing its complexity.

1. *Patching Y-subtrees* (STEP 2)

The function $\langle ARG \rangle$ takes two Y-subtrees and constructs a Y-pattern if these two subtrees can be patched together. ARG is any argument of the form: $(F(i,k), B(k,j))$, $(B(i,k-1), F(k,j))$, $(B(i,k-1), B(k,j-1))$ or $(F(i+1,k), F(k+1,j))$, with v_i, v_k, v_j occurring in clockwise order.

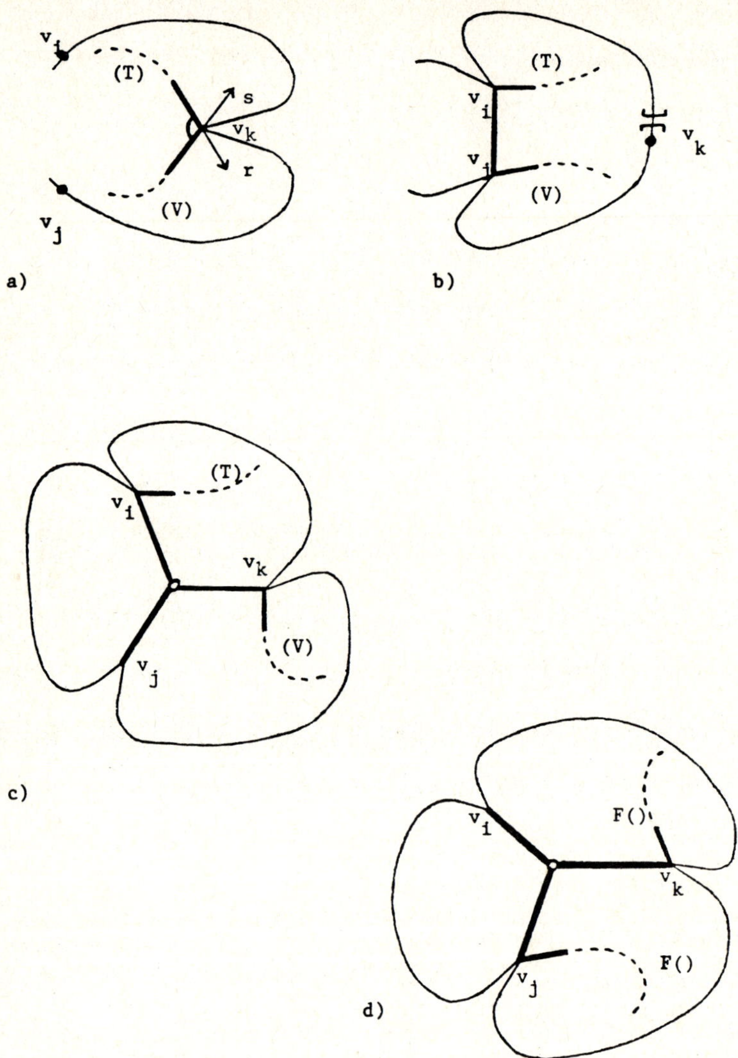

Figure 12 Computing the function <ARG>.

Case 1. $\langle F(i,k), B(k,j)\rangle$ (Fig.12-a).

Let $F(i,k) = T$ and $B(k,j) = V$, with r and s their respective arms. If $\angle(r,s) < 180$ and $T \neq 0$ and $V \neq 0$, then set $\langle F(i,k), B(k,j)\rangle = (|S(i,k)| + |S(k,j)| + 1, Y\text{-pattern: } T \bigcup V)$, else $\langle F(i,k), B(k,j)\rangle = 0$.

Case 2. $\langle B(i,k-1), F(k,j)\rangle$ (Fig.12-b).

Let $B(i,k-1) = T$ and $F(k,j) = V$. If an extended X_2-pattern is possible between v_i and

v_j

then set $\langle B(i,k-1), F(k,j)\rangle = (|S(i,k-1)| + |S(k,j)| + 1, Y\text{-pattern: } \{v_iv_j\} \bigcup T \bigcup V)$,
else $\langle B(i,k-1), F(k,j)\rangle = 0$.

Case 3. $\langle B(i,k-1), B(k,j-1)\rangle$ (Fig.12-c).

Let $B(i,k-1) = T$ and $B(k,j-1) = V$. If an extended X_3-pattern W is possible between v_i, v_j, v_k, then

$\langle B(i,k-1), B(k,j-1)\rangle = (|S(i,k-1)| + |S(k,j-1)| + 1, T \bigcup V \bigcup W)$,
else $\langle B(i,k-1), B(k,j-1)\rangle = 0$.

Case 4. $\langle F(i+1,k), F(k+1,j)\rangle$ (Fig.12-d).

Let $F(i+1,k) = T$ and $F(k+1,j) = V$. If an extended X_3-pattern W is possible between v_i, v_j, v_k, then

$\langle F(i+1,k), F(k+1,j)\rangle = (|S(i+1,k)| + |S(k+1,j)| + 1, T \bigcup V \bigcup W)$,
else $\langle F(i+1,k), F(k+1,j)\rangle = 0$.

Because of Relation (2), it is clear that STEP 2 computes the Y-pattern connecting v_i and v_j (if any is to be found) such that the number of compatible patterns which can be applied in $V(i,j)$ is maximum. All we have to check is that all cases are indeed handled in STEP 2. Consider the path from v_i to v_j in any such Y-pattern. If it contains an N2-node, it will be detected in Q_1. Otherwise one N3-node may appear on this path and all these candidates will be reported in Q_3 and Q_4. The final case, handled by Q_2, assumes that the path from v_i to v_j is free of N2 and N3-nodes.

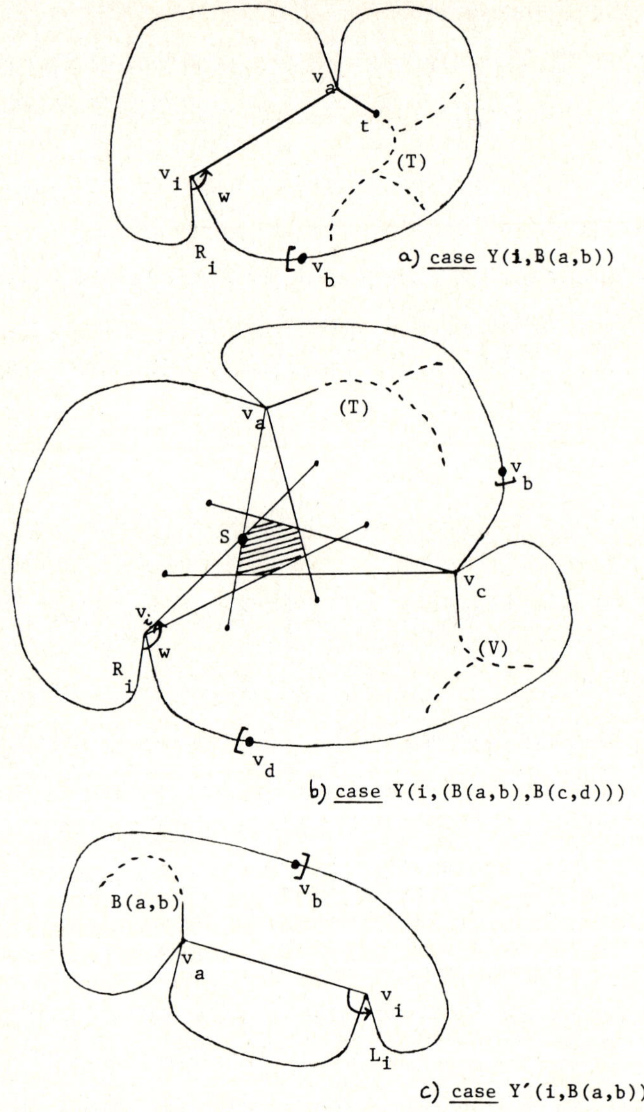

Figure 13 Computing Y(i,ARG) and Y'(i,ARG).

2. *Computing* $S(i,j)$ (STEP 3)

Assume by induction that $S(k,l)$ has been computed for all $v_k, v_l \in V(i,j)$ (except for $S(i,j)$). The algorithm investigates the three following cases in turn:

1. Disallow the presence of any pattern having both v_i and v_j as vertices.
2. Consider the possibility of an X_4-pattern connecting v_i and v_j.
3. Consider the possibility of a Y-pattern connecting v_i and v_j.

Lemma 7 shows that STEP 2 and STEP 3 can be accomplished in polynomial time. Correctness follows directly from previous discussion.

3. *Constructing* Y*-subtrees* (STEP 4)

We compute $B(i,j)$ and $F(i,j)$ by iteratively patching Y-subtrees together via two functions, $Y(i, ARG)$ and $Y'(i, ARG)$. ARG is an argument of the form $B(a,b)$ or $(B(a,b), B(c,d))$ (or the same with F). We describe these functions with respect to B's only, all other cases being similar.

Case 1. $Y(i, B(a,b))$ (Fig.13-a)

The vertices v_a, v_b, v_i occur in clockwise order. Let $T = B(a,b)$. Extend the notch at v_a to take into account the arm of T. Extend the notch at v_i by making its associated wedge be the entire plane. If an extended X_2-pattern is possible between v_a, v_i and if $w = \angle(R_i, v_i v_a) < 180$, set $Y(i, B(a,b)) = (Y\text{-subtree: } \{v_i v_a\} \bigcup T)$, else $Y(i, B(a,b)) = 0$.

Case 2. $Y(i, (B(a,b), B(c,d))$ (Fig.13-b)

The vertices v_a, v_b, v_c, v_d, v_i occur in clockwise order. Let $T = B(a,b)$ and $V = B(c,d)$. Extend the notch at v_a (resp. v_c) to take into account the arm of T (resp. V). Extend the notch at v_i by making its associated wedge be the entire plane. If an extended X_3-pattern is possible between v_a, v_c, v_i, compute the locus of its N3-node. Let S be the point in the locus which maximizes the angle $w = \angle(R_i, v_i S)$.

If $w < 180$, set $Y(i, (B(a,b), B(c,d)) = (Y\text{-subtree:}\{Sv_i\} \bigcup \{Sv_a\} \bigcup \{Sv_c\} \bigcup T \bigcup V)$, else $Y(i, (B(a,b), B(c,d)) = 0$.

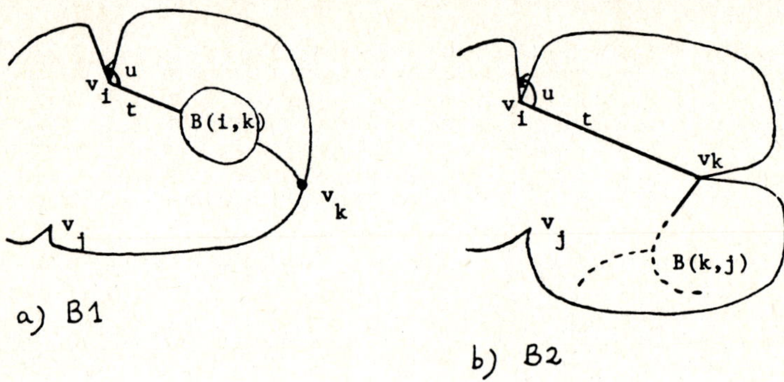

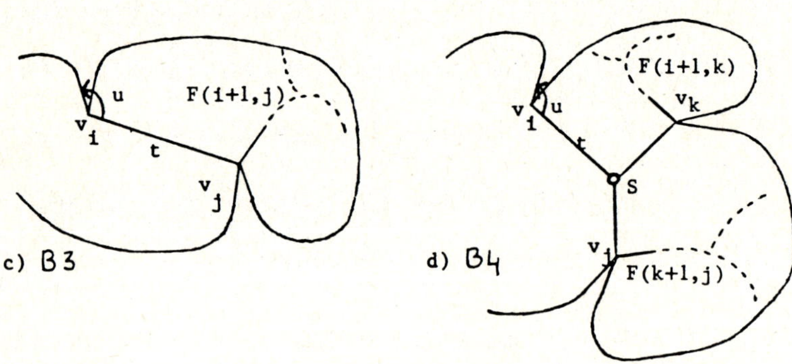

Figure 14 Computing B(i,j).

We define $Y'(i, B(a,b))$ in the same way as $Y(i, B(a,b))$, with $\angle(R_i, v_iv_a)$ replaced by $\angle(v_iv_a, L_i)$ (Fig.13-c). $Y'(i, F(a,b))$ is defined similarly, so we omit the details. We are now ready to implement STEP 4 of the decomposition algorithm. We will only describe the computation of $B(i,j)$, since the case of $F(i,j)$ is strictly similar. We begin by computing the four sets B_1, B_2, B_3, B_4. Let C be the value of $|S(i,j)|$ computed in Step 3.

$B_1 = \{\ Y\text{-subtree of } B(i,k)\}$, for all $v_k \in V(i, j-1)$ such that $|S(i,k)| + |S(k+1,j)| = C$.

/* v_j is not a notch of the Y-subtree */

$B_2 = \{\ Y'(i, B(k,j))\}$, for all $v_k \in V(i+1, j-1)$ such that $|S(i+1, k-1)| + |S(k,j)| = C$.

/* v_i's neighbor is an N2-node */

$B_3 = \{\ Y'(i, F(i+1, j))\}$, if $|S(i+1,j)| = C$.

/* v_i's neighbor is an N2-node */

$B_4 = \{\ Y'(i, (F(i+1,k), F(k+1,j)))\}$, for all $v_k \in V(i+1, j-1)$
such that $|S(i+1,k)| + |S(k+1,j)| = C$.

/* v_i's neighbor is an N3-node */

Let T be the Y-subtree of $B_1 \bigcup B_2 \bigcup B_3 \bigcup B_4$ which maximizes the angle $u = \angle(t, L_i)$, where t is understood here as the arm of T directed outward from v_i (Fig.14). We define $B(i,j)$ as a pointer to the arm of T (now understood to be directed towards v_i).

Computing $B(i,j)$ can be clearly done in polynomial time, according to Lemma 7. Note that Y-subtrees can be merged in constant time by linking their respective arms together. B_1 through B_4 evaluate all candidate Y-subtrees adjacent to v_i and lying in $V(i,j)$, and keep a single candidate, i.e. the subtree which has maximum angle u. We can show by induction that it is sufficient to consider only the Y-subtrees in the B's and F's. B_1 considers all subtrees which do not have both v_i and v_j as notches (Fig.14-a). B_2 and B_3 compute the subtrees whose vertex adjacent to v_i is an N2-node. Note that B_1 and B_2 may share common subtrees. The two possible configurations are illustrated in Fig.14-b,c. Finally B_4 detects all candidate subtrees such that the vertex adjacent to v_i is an N3-node (Fig.14-d).

4. *Completing the OCD* (Step 5)

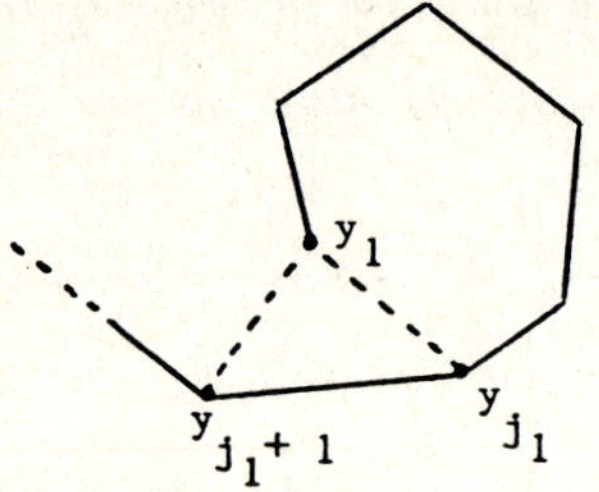

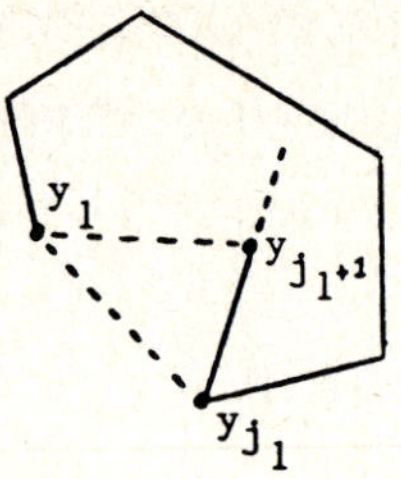

a) The two cases for defining a pseudo-notch.

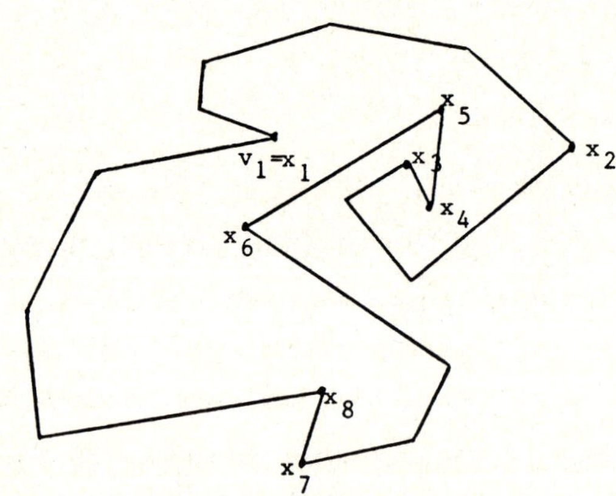

b) The pseudo-notches of a polygon.

Figure 15

The last step of procedure *ConvDec* consists of removing the remaining notches with the naive decomposition. This can be done in polynomial time, leading to the main result of this section.

Theorem: An optimal convex decomposition of a simple polygon can be computed in polynomial time.

4. Towards an Efficient Implementation

We will not make any attempt to evaluate the exponent in the polynomial expression bounding the time complexity of the previous algorithm. Clearly, it is prohibitively high. To produce a substantial savings in the running time of the algorithm, we first identify routines which need to be made more efficient. We essentially have three items to examine:

1. The naive decomposition.
2. Computing (extended) X_2 and X_3-patterns.
3. Computing X_4-patterns.

As we will see, improving the first two routines can be done without otherwise altering procedure *ConvDec*. Unfortunately, following the exact prescriptions of the procedure would lead to computing all possible X_4-patterns. Since we may have on the order of c^4 such patterns, discovering structural facts to limit the number of candidates to examine is in order. To begin, we consider the implementation of the naive decomposition. This allows us to introduce most of the tools used later on.

1. *The Naive Decomposition*

Recall that the naive decomposition involves removing each notch in turn by means of a simple line segment *naively* drawn from the notch. For simplicity we will choose a segment collinear with one of the edges of P adjacent to the notch. The degree requirements we made in the original definition of the naive decomposition can be relaxed here, since their only motivation was the simplification of subsequent proofs. Each line segment extends from a notch to the first intersection with the current decomposition. This can be easily accomplished in $O(n+c)$ time, but this is still too slow for our purposes.

The improvement which we propose involves $O(n)$ time preprocessing at the outset of the computation. Before describing this preprocessing, a few more definitions are in order. A *convex polygonal line* is a sequence of vertices $\{a_1, \ldots, a_p\}$ such that $\angle(a_i a_{i-1}, a_i a_{i+1}) \leq 180$, for each i such that $1 < i < p$. The sequence is called a *convex chain* if $\{a_1, \ldots, a_p, a_1\}$ forms a convex polygon. Note that in this case $a_1, \ldots, a_p$ corresponds to a clockwise traversal of the polygon. It is known from [3,6] that if the vertices of a convex n-gon are stored in a linear array it is possible to compute its intersection with any line in $O(\log n)$ time. Unfortunately, between each pair of notches v_i, v_{i+1}, the boundary of P certainly is a convex polygonal line but not necessarily a convex chain. This motivates the following preprocessing.

Partition the boundary of P between two consecutive notches into contiguous convex chains. Let $L_i = \{y_1, \ldots, y_p\}$ be the convex polygonal line given in clockwise order, with $y_1 = v_i$ and $y_p = v_{i+1}$. If neither the angle $\angle(y_1 y_k, y_1 y_2)$ nor the angle $\angle(y_k y_{k-1}, y_k y_1)$ is reflex for any $2 \leq k \leq p$, L_i is a convex chain and remains unchanged. Otherwise, let j_1 be the smallest k such that $\{y_1, \ldots, y_k, y_{k+1}, y_1\}$ is a non-convex polygon, i.e. such that either $\angle(y_1 y_{k+1}, y_1 y_2)$ or $\angle(y_{k+1} y_k, y_{k+1} y_1)$ is reflex (Fig.15-a). We define C_1 as the convex chain $\{y_1, \ldots, y_{j_1}\}$. Next we apply the same procedure recursively on the remaining part of L_i. This leads to defining C_2 as $\{y_{j_1}, \ldots, y_{j_2}\}$, with j_2 being the smallest $k > j_1$ such that $(y_{j_1} y_{k+1}, y_{j_1} y_{j_1+1})$ or $(y_{k+1} y_k, y_{k+1} y_{j_1})$ is reflex. We iterate on this process until we reach y_p, thus partitioning L_i into t consecutive convex chains $C_1, \ldots, C_t$.

We apply the same treatment to each pair of notches (v_i, v_{i+1}) and renumber the chains accordingly. This leads to a partitioning of the whole boundary of P into m consecutive convex chains, $C_1, \ldots, C_m$, in clockwise order. Letting x_i, x_{i+1} be the endpoints of C_i in clockwise order (with $x_1 = x_{m+1} = v_1$), we call the x_i the *pseudo-notches* of P. Note that all notches are pseudo-notches but the converse is in general not true (Fig.15-b). This preprocessing requires $O(n)$ operations. We next show that the number of convex chains is of the same order of magnitude as the number of notches.

Lemma 8. $m \leq 2(1 + c)$.

Proof: Consider the vertices p_i of P which are *not* notches of P and let U be the sum of all the angles $\angle(p_ip_{i+1}, p_{i-1}p_i)$. Similarly for all vertices p_j which are notches of P, let V be the sum value of all the angles $\angle(p_{j-1}p_j, p_jp_{j+1})$. It is a classical result of geometry that

$$U - V = 360. \tag{3}$$

For each convex chain $C_i = \{a_1, \ldots, a_p\}$ such that a_p is not a notch of P, let a_{p+1} be the vertex of P adjacent to a_p in clockwise order. We define U_i as the sum of all angles $\angle(a_ja_{j+1}, a_{j-1}a_j)$, for all $2 \le j \le p$. Let $U_{i_1}, \ldots, U_{i_t}$ be the values thus obtained. By construction the polygon $\{a_1, \ldots, a_p, a_{p+1}, a_1\}$ has a reflex angle either at a_{p+1} or at a_1. It follows that if c (resp. d) is the angle $\angle(a_{p+1}a_1, a_pa_{p+1})$ (resp. $\angle(a_1a_2, a_{p+1}a_1)$) measured (exceptionally) between -180 and $+180$ degrees, negative if there is a reflex angle at a_{p+1} (resp. a_1), positive otherwise, we have

$$U_i = 360 - (c + d) \ge 180. \tag{4}$$

Since none of the U_i's accounts for the reflex angles of P, we have $\sum_{1 \le j \le t} U_{i_j} \le U$. Also if, between a pair of consecutive notches, P consists of a single convex chain, no U_i is defined on this portion of P, whereas if it consists of p chains, $p - 1$ U_i's are defined. This implies that $t = m - c$. Combining this fact with (4) we derive

$$180(m - c) \le \sum_{1 \le j \le t} U_{i_j} \le U,$$

and from (3)

$$U = 360 + V \le 180(2 + c),$$

which completes the proof. ∎

This leads to the following result.

Lemma 9. In procedure *ConvDec*, STEP 5 can be accomplished in time $O(n + c^2 \log n)$.

Proof: We assume that P has been preprocessed as described above, with each convex chain stored in a linear array. For each remaining notch v, in turn, let t be the ray emanating from v in a direction removing the reflex angle at v (for example the direction of one of the edges

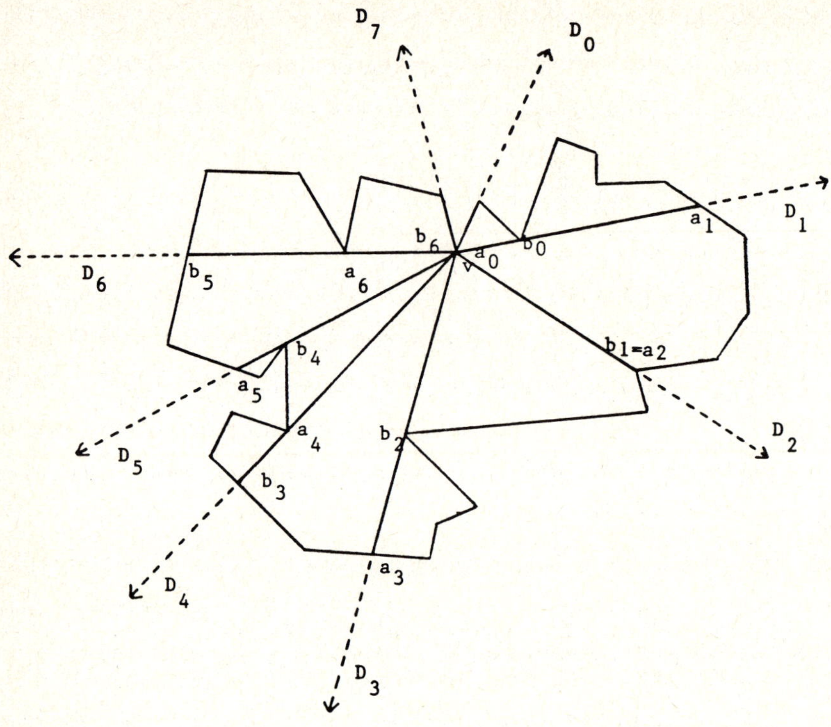

Figure 16 The superrange of a notch v.

adjacent to v). Among the intersection-points of t with the edges of the current decomposition, we must determine the closest to v. To do so, we compute all the intersections between t and the *internal edges* of the decomposition, i.e. the edges not on the boundary of P. It is easy to show by induction that there are $O(c)$ such edges, hence $O(c)$ intersection-points to compute. Next we compute the intersection of t with each convex chain of P in turn, using the fast algorithm of [6]. The running time of this method will be $O(n + c^2 \log n)$, including preprocessing. ∎

2. *Computing Extended Patterns*

We refine the preprocessing described above. Since the computation of X-patterns is intimately based on the notion of ranges, we precompute the visibility-polygon with respect to each notch at an overall cost of $O(c^2 \log n)$ time and space. This may seem a paradox since storing all these polygons may require as much as $\theta(cn)$ storage. The crux is that only significant vertices of the visibility-polygons will be stored. This economical description of the visibility-polygon of a notch v is called the *superrange* of v. In the next paragraphs we describe how to compute superranges and then show how to use these new structures efficiently.

Let v be a notch of P and $t_1, \dots, t_p$ be the list in clockwise order of all the pseudo-notches visible from v. Note that scanning $t_1, \dots, t_p$ corresponds to a clockwise traversal of the boundary of P as well as a clockwise sweep around v. Let D_i be the ray emanating from v with the direction from v to t_i and let D_0 (resp. D_{p+1}) be the ray passing through the edge of P starting from v in clockwise (resp. counterclockwise) order. The set of rays $D_0, \dots, D_{p+1}$ partitions the region of P visible from v into $p+1$ simple polygons, all adjacent to v. Typically a polygon is comprised between D_i, D_{i+1} and a convex polygonal line on the boundary of P. Let a_i and b_i be the endpoints of this convex line (with b_i following a_i in clockwise order – Fig.16). For each notch v, we define the *superrange* of v, denoted SR(v), as the ordered list

$$\mathrm{SR}(v) = \{(a_0, b_0), \dots, (a_p, b_p)\}.$$

The next result states that superranges can be computed very efficiently.

Lemma 10. An n-gon can be preprocessed in $O(n)$ time so that the superrange of any of its notches can be computed in $O(c \log n)$ time.

Proof: The proof of this result is too lengthy to be included here. We refer the reader to [3] for the details of the proof. In [7] El-Gindy and Avis present a linear algorithm for computing the visibility-polygon from any point in P. Although their algorithm cannot be used here since it is too slow for our purposes, we can still use it as a basis to give intuition for our algorithm. El-Gindy and Avis's algorithm can be seen as an extension of a Graham scan. It essentially involves traversing the boundary of P in one direction, occasionally backing up but never more than once per vertex. Our algorithm is conceptually similar to [7]; the only difference comes from the traversing scheme used. Instead of going from one vertex to the next, indeed, we go from one convex chain to the next. In this regard, the data type "edge" in [7] becomes, in our algorithm, the data type "convex chain". An important feature of the former data type which we lose in our algorithm is the property that a ray scanning an edge moves either totally clockwise or totally counterclockwise. A convex chain, instead, can change directions at most twice. However, whenever entering a new chain, we can use the fast algorithm of [6] to compute the changes of direction in $O(\log n)$ time. This means that we can rewrite the entire algorithm of [7], now taking convex chains instead of edges as our basic objects. The price to pay will be a factor $\log n$ in every step of [7]'s algorithm. Since there are only $O(c)$ convex chains, however, the $O(n)$ algorithm of [7] now becomes an $O(c \log n)$ algorithm. ∎

The notion of superrange can be of great use for many geometric problems and is thus interesting in its own right. To appreciate its usefulness to our specific decomposition problem, we need introduce a function of two arguments, $R(v, D)$, where v is a notch of P and D is a ray emanating from v. Let D_i, D_{i+1} be the two rays (introduced earlier in the definition of superranges) between which D lies. If $\angle(vb_i, va_i)$ is reflex, $R(v, D)$ is set to 0, otherwise it is set to the segment vy, where y is the intersection of D with a_ib_i. Clearly, $R(v, D)$ is still well-defined if D is a directed segment emanating from v instead of a ray. The next two results give motivation for these definitions.

Lemma 11. Once the superrange of each notch has been computed, $R(v, D)$ can be computed in $O(c)$ time for any notch v.

Proof: Trivial. ∎

Lemma 12. If v is a notch of P and vx is the edge of an X-pattern, x lies on the segment $R(v, vx)$.

Proof: The result is obvious if x lies on the boundary of P, since it is then a notch of P and $R(v, vx)$ is exactly vx. Suppose that x is an N3-node and vx contains $R(v, vx)$ (Fig.17). Let (a_i, b_i) be the pair of SR(v) such that vx intersects a_ib_i. The segment a_ib_i partitions P into two polygons, P_1 and P_2, with say P_1 containing x. Since the portion of the boundary of P between a_i and b_i is a convex chain, P_1 is a convex polygon, therefore the X-pattern cannot have notches in P_1, which is in contradiction with the fact that all the angles formed by an X-pattern are non-reflex. ∎

We next show how to use these results to compute extended X_2 and X_3-patterns efficiently. For simplicity we will first consider the cases where the patterns are standard, i.e. not extended. It will then be easy to generalize the results obtained to extended patterns.

– *Detecting X_2-patterns*

Lemma 13. With $O(n + c^3 + c^2 \log n)$ preprocessing it is possible to check for the possibility of an X_2-pattern between any two notches in constant time.

Proof: The preprocessing involves computing the superrange of each notch as well as the segments $R(v_i, v_iv_j)$, for all pairs of notches v_i, v_j. Lemmas 10 and 11 show that this can be done in $O(c^3 + c^2 \log n)$ time. From Lemma 12, it then follows that an X_2-pattern between v_i and v_j is possible if and only if $R(v_i, v_iv_j) = v_iv_j$ and the segment v_iv_j removes the reflex angle at both v_i and v_j. ∎

– *Detecting X_3-patterns*

Computing X_3-patterns is somewhat more complicated. We need some additional preprocessing which we next describe. Recall that R_i and L_i are the *directed* segments from v_i to the next vertices of P respectively following and preceding v_i in clockwise order. This notion of direction allows us to define $R(v_i, R_i)$ and $R(v_i, L_i)$ without ambiguity. Similarly to give full meaning to angles of the form $\angle(v_ix, R(v_i, D))$, the segment $R(v_i, D)$ will be understood as assuming the same direction as D.

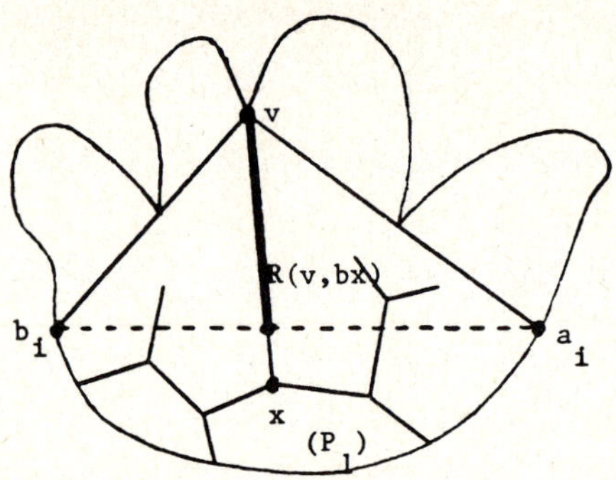

Figure 17   R(v,D) expresses the longest edge vx with direction D of an X-pattern.

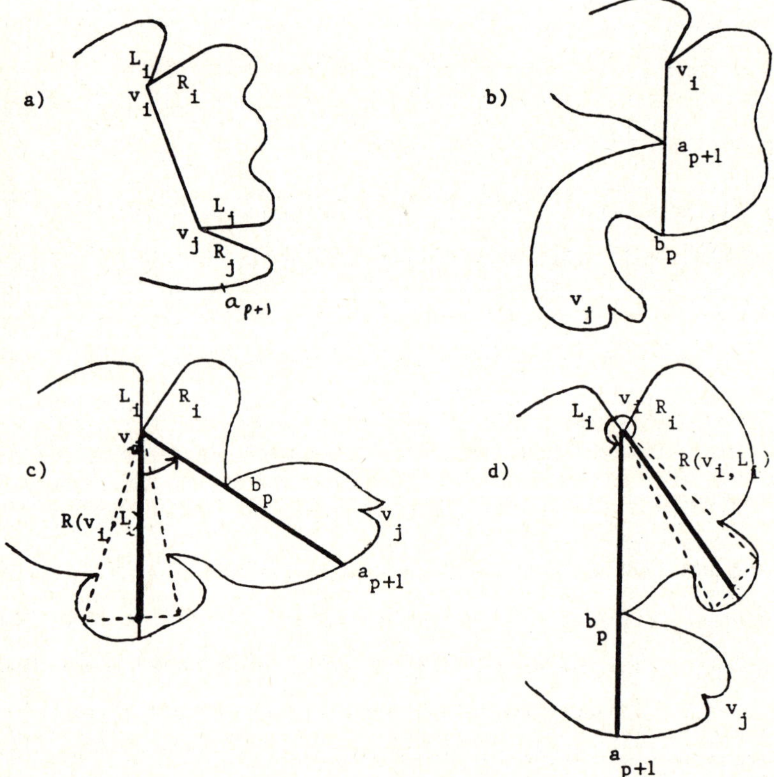

Figure 18   The definition of r_{ij} and l_{ji}.

For each pair of notches v_i, v_j, we define the two points r_{ij} and l_{ij} as follows: $v_i r_{ij}$ is the rightmost segment in the range of v_i, visible from v_j. More precisely, if both R_i and L_j lie on the same side of the line passing through $v_i v_j$ and $\angle(v_i v_j, R_i) < 180$ (Fig.18-a), we determine the pairs (a_p, b_p) and (a_{p+1}, b_{p+1}) of SR(v_i) such that v_j occurs between b_p and a_{p+1} in a clockwise traversal of the boundary of P. We assume that a_{p+1} does not lie on the segment $v_i b_p$ (unlike in Fig.18-b). We may have $v_j = b_p = a_{p+1}$, however. Let t denote the segment $R(v_i, L_i)$ if $\angle(R(v_i, L_i), v_i a_{p+1}) < 180$ (Fig.18-c), and the segment $v_i a_{p+1}$ otherwise (Fig.18-d). If we actually have $\angle(v_i v_j, t) < 180$, we define t as $R(v_i, v_i v_j)$. Finally if $\angle(t, R_i) < 180$, we define r_{ij} as the endpoint of t ($\neq v_i$). If any of the above conditions fails, r_{ij} is 0.

We repeat the same process on v_j with respect to v_i. If R_i and L_j lie on the same side of the line passing through $v_i v_j$, we first determine the pairs (a_p, b_p) and (a_{p+1}, b_{p+1}) from SR(v_j) such that v_i occurs between b_p and a_{p+1} in clockwise order. We will suppose that b_p does not lie strictly between v_j and a_{p+1}. Let t be $R(v_j, R_j)$ if $\angle(v_j b_p, R(v_j, R_j)) < 180$ or $v_j b_p$ otherwise. Similarly, if $\angle(t, v_j v_i) < 180$, t is reset to $R(v_j, v_j v_i)$ so that we can define l_{ji} as the endpoint of t other than v_j if $\angle(L_j, t) < 180$. In all other cases, l_{ji} is set to 0.

With the superrange of each notch at our disposal, we can compute each r_{ij} and l_{ji} in $O(c)$ time, which yields an $O(n + c^3 + c^2 \log n)$ overall preprocessing time. We are now ready to describe the computation of X_3-patterns.

Lemma 14. With $O(n + c^3 + c^2 \log n)$ preprocessing it is possible to check for the possibility of an X_3-pattern between any three notches in constant time.

Proof: Let v_i, v_j, v_k be three notches of P. We wish to give a set of necessary and sufficient conditions for v_i, v_j, v_k to form an X_3-pattern. Let $C_1 = v_i r_{ij} \cap v_j l_{ji}$, $C_2 = v_i r_{ij} \cap v_k r_{ki}$ and $C_3 = v_j l_{ji} \cap v_k l_{kj}$. One of the following is true: 1) C_2 is further from v_i than C_1 is ($\angle(v_k l_{kj}, v_k r_{ki}) + \angle(v_k r_{ki}, v_k C_1) = \angle(v_k l_{kj}, v_k C_1)$); 2) C_3 is further from v_j than C_1 is ($\angle(v_k C_1, v_k l_{kj}) + \angle(v_k l_{kj}, v_k r_{ki}) = \angle(v_k C_1, v_k r_{ki})$); 3) C_1 is further from v_i than C_2 is and further from v_j than C_3 is ($\angle(v_k l_{kj}, v_k C_1) + \angle(v_k C_1, v_k r_{ki}) = \angle(v_k l_{kj}, v_k r_{ki})$). In case 1, set $C = C_2$; in case 2, set $C = C_3$, and in case 3 set $C = C_1$. Similarly we introduce the points A and B, defined like C with respect to (v_j, v_k) and (v_k, v_i) respectively. From now on, we will

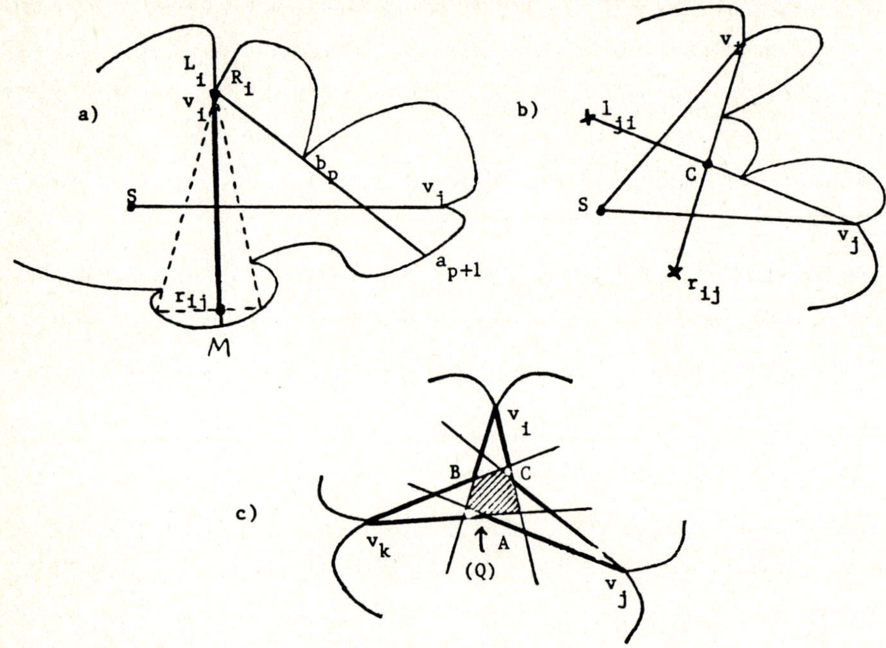

Figure 19 Detecting X3-patterns.

assume that A, B and C fall in case 3. The other cases are treated in a similar way and may thus be omitted. We can show that v_i, v_j, v_k are the notches of an X_3-pattern if and only if:

1. v_i, v_j, v_k occur in clockwise order on the triangle $v_i v_j v_k$.

2. $r_{ij}, r_{jk}, r_{ki}, l_{ik}, l_{kj}, l_{ji}$ are all distinct from 0.

3. The points A, B, C are well-defined.

4. The polygon $Q = v_i C v_j A v_k B v_i$ is simple and has a non-empty kernel (recall that the kernel of a polygon is the region visible from every point in the polygon).

All these conditions can be easily checked in constant time with the preprocessing described above.

We say that a point is *range-visible* from a notch v if it lies in its range, i.e. if the segment lies totally within P and removes the reflex angle at v. We define the *wedge* $W(ax, ay)$ as the region swept by a ray pivoting in clockwise order around a from ax to ay. Let S be the N3-node of an X_3-pattern between v_i, v_j, v_k. To prove that the second condition is also necessary, we show that $r_{ij} \neq 0$, all of the other cases being similar. Since the three edges of the pattern must lie in the triangle v_i, v_j, v_k, the first requirement (illustrated in Fig.18-a) is obvious. Considering the pairs (a_p, b_p) and (a_{p+1}, b_{p+1}) it is equally clear that the configuration of Fig.18-b cannot lead to an X_3-pattern since we must have $\angle(Sv_j, Sv_i) < 180$, where S is the N3-node. Indeed, Sv_j must intersect $b_p a_{p+1}$ with possibly $v_j = b_p$, since S must be visible from both v_i and v_j. This remark shows that not only are the configurations of Fig.18-c,d the only ones possible, but also that S cannot lie in the wedge $W(v_i a_{p+1}, R_i)$. The other conditions to satisfy in order to define r_{ij} express the fact that S lies in the triangle $v_i v_j v_k$ as well as the range of v_i. Also, since we must have $\angle(v_i S, v_i v_j) < 180$, it is legitimate to set t to $R(v_i, v_i v_j)$, if $\angle(v_i v_j, t) < 180$. Finally if $\angle(t, R_i) > 180$ no point visible from v_j can be range-visible from v_i, so we can set r_{ij} to 0. Thus, when an N3-node exists, all these conditions will be satisfied and S cannot lie in the wedge $W(v_i r_{ij}, R_i)$.

As mentioned earlier, the points $a_0, b_0, a_1, b_1, \ldots, a_p, b_p$ occur in clockwise order around the boundary of P, therefore we must have $\angle(v_i l_{ik}, v_i r_{ij}) < 180$ if $l_{ik} \neq 0$ and $r_{ij} \neq 0$, since v_i, v_j, v_k occur in clockwise order. It follows that if A, B, C exist, the polygon Q must be simple.

To prove that these points are well-defined, we first show that Sv_j intersects v_ir_{ij}. As we have seen that Sv_j intersects v_ia_{p+1}, thus implying that v_ir_{ij} is not defined as $R(v_i, v_iv_j)$, we only have to show that Sv_j intersects v_ir_{ij} whenever this segment is defined as $R(v_i, L_i)$. Since S cannot lie in $W(R(v_i, L_i), R_i)$, Sv_j must intersect v_iM (Fig.19-a). Also Sv_j cannot intersect Mr_{ij}, since S would then belong to a convex polygon where no N3-node point can lie (Lemma 12). This proves our claim, and shows that r_{ij} (as well as l_{ji} by a similar reasoning) lies outside the triangle Sv_iv_j (Fig.19-b). Finally, as we know that S cannot lie in the wedges $W(v_ir_{ij}, R_i)$ and $W(L_j, v_jl_{ji})$, we derive $\angle(v_iS, v_ir_{ij}) < 180$ and $\angle(v_jl_{ji}, v_jS) < 180$ which, combined with the previous result, establishes that v_ir_{ij} and v_jl_{ji} intersect. This proves the existence of the point C as well as points A and B, by symmetry. Since S must lie in $W(v_il_{ik}, v_ir_{ij})$, the same reasoning applied to v_j and v_k shows that S lies in the kernel of Q.

The four conditions having proven necessary, we next show that they are sufficient. Assume that they are all satisfied (Fig.19-c). Since v_ir_{ij} is range-visible from v_i, and so is v_jl_{ji} from v_j, Condition 3 shows that C is range-visible from both v_i and v_j. It follows that the boundary of P cannot intersect strictly with v_iC or v_jC, and by symmetry, cannot intersect with the edges of Q. Therefore, any point of its kernel is range-visible from v_i, v_j, v_k and is the N3-node of a possible X_3-pattern. Note that all three angles around the Steiner point are ensured to be < 180 since the kernel of Q lies within the triangle $v_iv_jv_k$. ▮

Our techniques for computing X_2 and X_3 patterns can be used to handle extended patterns as well. Patching together Y-subtrees in STEP 2 can be done along the same lines and no further explanation is necessary. The only remaining question to address concerns the computation of $Y(i, ARG)$ and $Y'(i, ARG)$ in STEP 4. The most general case corresponds to the computation of $Y(i, (B(a, b), B(c, d)))$ (Fig.13) (Y' is handled similarly). We extend the notches v_a, v_c, v_i accordingly, and (referring to Condition 4 in Lemma 14) compute the kernel of Q. In doing so, we drop all requirements involving R_i, since we do not have to remove reflex angles at v_i at this point. Once again, details are tedious but straightforward – see [3]. We conclude

Lemma 15. After $O(n + c^3 + c^2 \log n)$ preprocessing it is possible, in constant time, to

1. check for the possibility of an extended X_2-pattern between any two notches,

2. check for the possibility of an extended X_3-pattern between any three notches,

3. evaluate the functions Y and Y'.

3. *Computing X_4-patterns*

We now have to face the most difficult task in our quest for an efficient implementation. Improving the previous routines simply relied on increasing the amount of preprocessing without otherwise altering the basic nature of the algorithm. The excessive number of possible X_4-patterns require conceptual changes in our algorithm. These changes are based on the observation that of all the $\binom{n}{4}$ potential X_4-patterns, only $O(n^3)$ need be retained for consideration. Implementing the selection requires structural facts about the nature of X_4-patterns.

Definition 4. An X_4-pattern is said to be *loose* if it can be reduced so that each edge adjacent to a notch v_i is made to be collinear with either R_i or L_i (16 configurations should thus be achievable – Fig.20).

The term "reduced" is to be understood here in the sense of Lemma 6. The introduction of loose patterns finds justification in the following.

Lemma 16. Every X_4-pattern which is not reducible to an X_3-pattern or a Y_5-pattern can be reduced to a loose X_4-pattern.

Proof: Let *hull*(T) designate the convex hull of all the points of an X-pattern T. It is clear that every X_4-pattern, T, can be reduced to an X_4-pattern V such that no further reduction of V can lead to another X_4-pattern lying strictly inside *hull*(V), i.e. an X_4-pattern where at least one notch lies strictly in the interior of *hull*(V). We show that if T cannot be reduced to a Y-pattern, V must be loose. Assume that one of the 16 configurations cannot be achieved for V; by applying the reductions shown in Fig.21 we can reach an X_3-pattern, a Y-pattern or an X_4-pattern lying strictly in *hull*(V). Actually, another possibility is to reduce to the alternate case of Fig.21, which can arise only a finite number of times. ∎

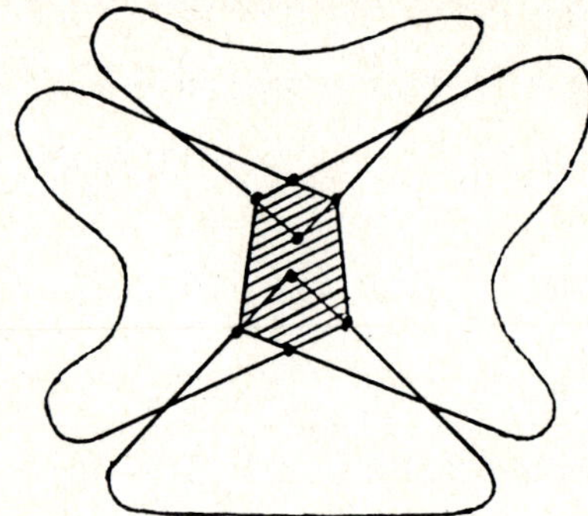

Figure 20 A loose X4-pattern with its 16 extreme configurations.

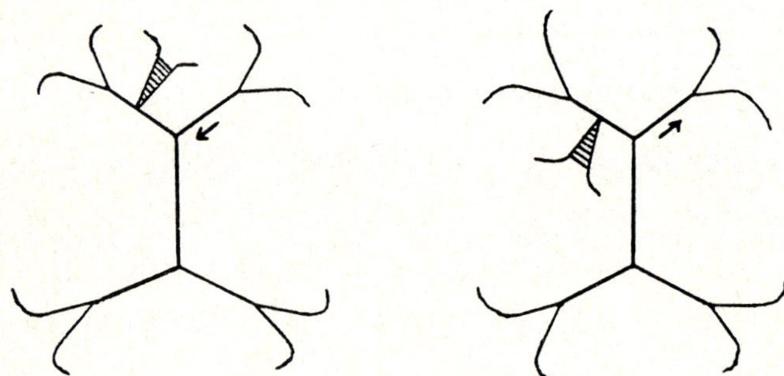

Figure 21 X4-patterns can be assumed to be loose.

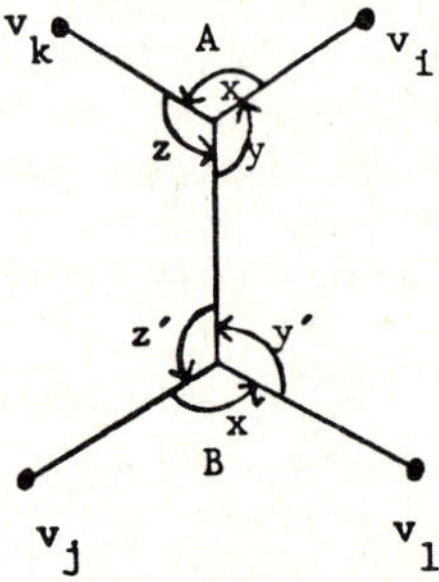

Figure 22 Characteristics of an X4-pattern

To begin with, we determine the points A_{ik} such that $v_k \in V(j+1, i-1)$ (note that if the pair (a_p, b_p) of SR(v_i) used to compute r_i is such that $b_p \in V(j, i)$, all the vertices in r_i will already satisfy this condition). Next we keep only the A_{ik} which lie on the other side of the infinite line passing through R_j than the edge L_j (repeat same operations with respect to r_j and A_{jl}). This ensures that:

Fact 3. v_j, v_k, v_i occur in clockwise order and the angle $\angle z' < 180$ (Fig.22).

Fact 4. v_i, v_l, v_j occur in clockwise order and $\angle y < 180$.

We now retain in r_j only the points A_{jl} for which

$$|S(i,j)| = |S(i+1, l-1)| + |S(l+1, j-1)|.$$

By doing this we keep only the candidates for N3-nodes of an OCD. Similar to the Y-subtrees occurring in $B(i,j)$, candidates must contribute a savings of $|S(i,j)|$ with respect to the removal of reflex angles in $V(i,j)$. Finally we update the lists r_i and r_j with the points just chosen, maintaining the sorted order. We rename the points of r_i (resp. r_j) from v_i (resp. v_j), $A_1, \ldots, A_p$ (resp. $B_1, \ldots, B_q$). This entire step can be done in $O(c)$ time with the preprocessing indicated earlier. We now have a list of all possible N3-nodes for candidate X_4-patterns. It is clear that $E(i,j)$ can be found satisfying the specifications given above. Each A_k will be paired with the point B_l such that A_k and B_l are the N3-nodes of the same optimal X_4-pattern forming the maximum angle $\angle(A_k B_l, A_k v_i)$ (Fig.23). We must now give a precise procedure for accomplishing this task.

II) *Computing a region of safety*

We compute the *region of safety* for added edges in order to ensure Condition 3 of Lemma 17. Note that $A_1 A_p B_1 B_q$ forms a convex quadrilateral (Facts 3,4). The next step is to compute two convex chains $C = \{c_1, \ldots, c_s\}$ and $D = \{d_1, \ldots, d_t\}$ running from r_i to r_j and r_j to r_i, respectively. These chains will have the property that a *middle edge*, (i.e. the edge of an X_4-pattern between the two N3-nodes) from r_i to r_j lies totally in P if and only if it lies totally between the two chains (Fig.27-a). Informally C (resp. D) is the convex hull of the pieces of

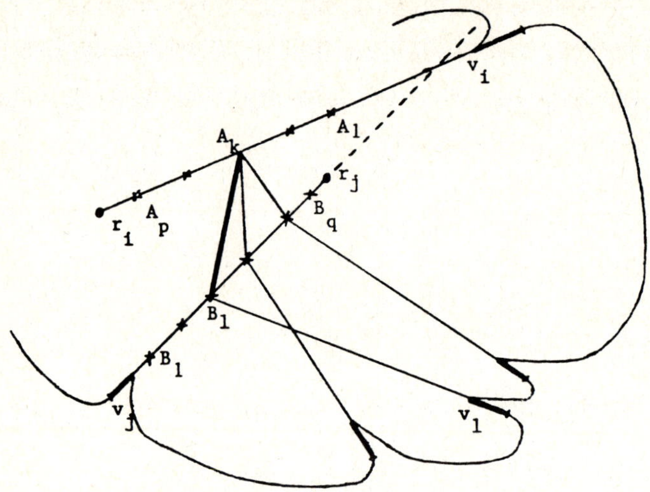

Figure 23 Computing E(i,j).

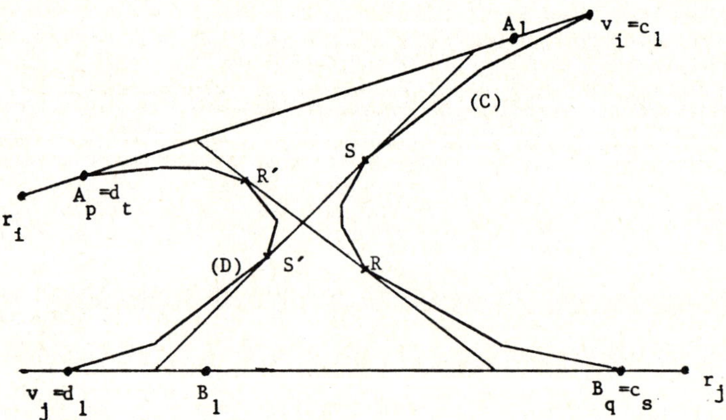

Figure 24 The limits on middle edged RR' and SS'.

the boundary of P which pass through A_1B_q (resp. A_pB_1), and delimits an area of safety for potential middle edges. Actually, as we will see later on, this definition will apply to the *parts* of C and D inside the quadrilateral $A_1A_pB_1B_q$. To preserve the flow of the presentation, we will prove the following result in the next section.

Lemma 18. The convex chains C and D can be computed in $O(c)$ time after $O(n + c^2 \log n)$ time preprocessing.

If the procedure of Lemma 18 determines that any segment drawn between A_1A_p and B_1B_q should intersect the boundary of P, it sets C or D to 0. Otherwise, it effectively returns two convex chains C and D, with the segment c_1d_t (resp. d_1c_s) containing A_1 and A_p (resp. B_1 and B_q). Also, as stated earlier, a segment joining A_1A_p and B_1B_q will intersect the boundary of P if and only if it intersects C or D.

It takes $O(c)$ operations to test whether C and D intersect since both have $O(c)$ vertices. If they do, no middle edge is possible and $E(i,j)$ is set to 0. Otherwise, we can compute RR' and SS' as the limits put upon the middle edges by the polygon P (Fig.24). SS' is computed by beginning at c_1 and d_1 and moving through D until all of D lies above the line passing through d_kc_1, then moving through C until all of C lies below the line passing through d_kc_1. We iterate on this process until termination, which will occur after $O(c)$ steps since no vertex of C or D is visited more than once. Here is a more formal description of the procedure. RR' is obtained in a similar manner.

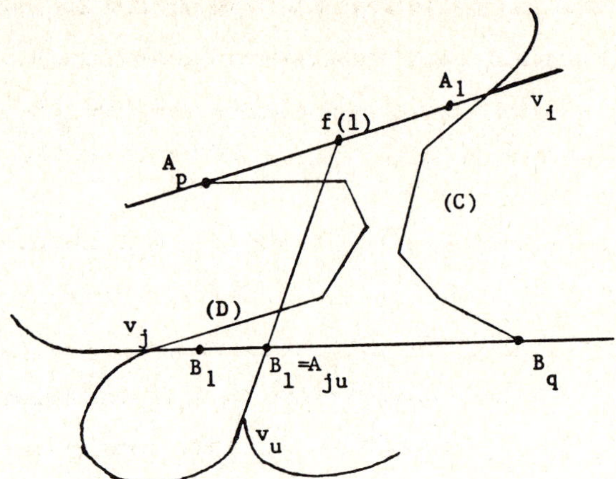

Figure 25 The function f.

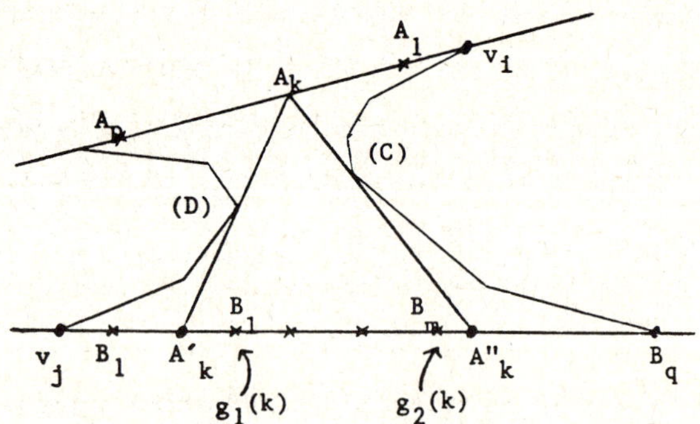

Figure 26 The functions g_1 and g_2.

Computing SS'

```
k:=l:=1
while ∠(d_k d_{k+1}, d_k c_l) < 180 or ∠(c_l c_{l+1}, c_l d_k) < 180
  beg
    while ∠(d_k d_{k+1}, d_k c_l) < 180
      begin k:=k+1 end
    while ∠(c_l c_{l+1}, c_l d_k) < 180
      begin l:=l+1 end
  end
S:= c_l; S':= d_k
```

III) *Computing wedges for possible middle edges*

We are now in a position to compute the wedges where middle edges must lie. We begin by observing that any point in the list r_i not in the wedge formed by RR' and SS' can be discarded, which we can do in $O(c)$ time. For simplicity we still call $A_1, \ldots, A_p$ the elements of the list r_i.

For every $B_l \in r_j$ (recall that $B_l = A_{j_u}$ for some u) we define $f(l)$ as the intersection of the line supporting r_i and the ray emanating from B_l, in the direction of R_u, and not passing through R_u. If this intersection does not exist, we compare the direction of R_u and r_i to decide if the "middle edge" wedge centered at B_l formed by the ray and r_j intersects the line supporting r_i or not. If yes, we set $f(l) = A_p$, else $f(l) = 0$ (Fig.25). Also, for each $A_k \in r_i$, let A'_k (resp. A''_k) denote the point on the segment r_j which is the closest to B_1 (resp. B_q) and such that the segment $A_k A'_k$ (resp. $A_k A''_k$) does not strictly intersect D (resp. C), i.e. the intersection consists of a segment or a single point. We now view r_j as a list of vertices and we find the two vertices B_l $(= g_1(k))$ and B_m $(= g_2(k))$ which lie on the segment $A'_k A''_k$ and are the closest to A'_k and A''_k, respectively (Fig.26). If B_l and B_m do not exist, we define $g_1(k)$ and $g_2(k)$ to be 0. At this stage we need two results whose proofs we postpone till the next section.

Lemma 19. After $O(c)$ preparation, the functions f, g_1, g_2 can be evaluated in constant time.

Lemma 20. Let $V = \{1, \ldots, q\}$ and $W = \{1, \ldots, p+q\}$, $W_1 \bigcup W_2 = W$, $|W_1| = q$, $|W_2| = p$ and let f, g_1, g_2 be defined with: $f \mid V \mapsto W_1$ a bijection and $g_1, g_2 \mid W_2 \mapsto V$ non-decreasing functions, with $g_1(i) \le g_2(i)$ for each $i \in W_2$. Define $x \mid W_2 \mapsto V$ such that for each $i \in W_2$, $x(i)$ is the smallest integer in V with $g_1(i) \le x(i) \le g_2(i)$ and $i \le f(x(i))$ if such an integer exists, and 0 otherwise. If f, g_1, g_2 are computable in constant time, then so is x after $O(p+q)$ preprocessing.

IV) *Computing* $E(i,j)$

Lemma 19 allows us to compute all the values of f, g_1, g_2 in $O(c)$ time. Note that if $f(l) = 0$ no middle edge adjacent to B_l is possible since it must lie in the wedge $W(B_l B_q, B_l f(l))$. Therefore we can eliminate those B_l from r_j. Once again we still represent the resulting list by $B_1, \ldots, B_q$. Similarly, if $g_1(k) = g_2(k) = 0$, A_k cannot be an N3-node and we eliminate all such A_k from r_i. Note that the values of g_1 and g_2 should be computed after the last selection on r_j. We will merge the points $f(l)$ with the remaining vertices A_k, thus forming the set W. Strictly speaking, f maps l not to a point on r_i but to the corresponding index in W. We can always assume that f is injective. Next we define V as the set of vertices left on r_j; instead of mapping to actual vertices of r_j, the functions g_1 and g_2 will map k to the corresponding indices in V. Because of the removals, g_1 and g_2 obey the two conditions of Lemma 20. Finally we define W_1 as the sublist of W corresponding to the $f(l)$ and W_2 as the complement in W, i.e. the indices corresponding to the A_k. It is easy to see that all removals, merges, and settings of functions can be done in $O(c)$ time. Moreover all the conditions of Lemma 20 have been met.

We can thus set up the function x in $O(c)$ time. The last step consists of keeping in $E(i,j)$ all the pairs $(A_k, B_{x(A_k)})$ such that $x(A_k) \neq 0$, with $A_k \in r_i$ and $\angle(A_k v_u, A_k B_{x(A_k)}) < 180$ $(A_k = A_{iu})$. Note that $x(A_k)$ is a shorthand for $x(t)$, with t the element in W_2 corresponding to A_k. Recall that if any of the previous computations fails, we have $E(i,j) = 0$. We can now state our main result:

Lemma 21. After $O(n + c^2 \log n)$ preparation, $E(i, j)$ can be evaluated in $O(c)$ time.

Proof: The preprocessing involves setting up the lists r_i which, as seen earlier, requires $O(n + c^2 \log n)$ time. For any i and j, $E(i, j)$ is then computed in $O(c)$ operations. We review the main phases of the procedure and establish its correctness. In the preprocessing stage, Fact 1 ensures that the angles $\angle x < 180$ and $\angle x' < 180$, and that Condition 1 of Lemma 17 is satisfied. Facts 3 and 4 show that $\angle y < 180$ and $\angle z' < 180$. Then, considering the savings, by an argument now standard, we eliminate the N3-nodes which are not candidates. Next we pair $A_k \in r_i$ with $B_l \in r_j$ $(l = x(A_k))$.

By definition, $x(A_k)$ is the smallest integer in V (i.e. the vertex of r_j that maximizes the angle $\angle y$) lying on the segment $(g_1(k), g_2(k))$ (i.e. ensuring Condition 3 of Lemma 17), and such that A_k lies on the segment A_1F, with F the point on r_i corresponding to $f(x(A_k))$, i.e. ensuring $\angle y' < 180$. Finally, since $x(A_k)$ maximizes the angle $\angle y$ and $\angle(y + z)$ is a constant for all B_l, no vertex of r_j can be paired with A_k if $\angle z > 180$. If $\angle z \leq 180$, all the conditions of Lemma 17 are satisfied, and $A_k B_{x(A_k)}$ can be kept as the middle edge candidate to be adjacent to A_k. Since we know that this edge is indeed the middle edge of a loose X_4-pattern, only savings consideration will later decide whether this edge belongs to an OCD or not. This is, in essence, the only major difference with the Y-subtrees of $B(i, j)$ and $F(i, j)$, where both savings and geometry had to be tested at once in order to determine candidacy. ∎

2. The Proofs of the Lemmas Left Unresolved

We now justify our earlier claims and successively show how to compute the region of safety and set up the functions f, g_1, g_2, x, all in $O(c)$ time.

Lemma 18. The convex chains C and D can be computed in $O(c)$ time after $O(n + c^2 \log n)$ time preprocessing.

Proof: C and D are computed in the same manner, so we may concentrate on C exclusively. We assume that all the superranges have been precomputed, which requires $O(n + c^2 \log n)$ time, as was shown in Lemma 10. Let (a_p, b_p) be the pair of $\mathrm{SR}(v_j)$ such that the segment $a_p b_p$ intersects r_j. Recall that this pair is uniquely defined and must be computed in order to obtain

r_j. If a_p is a notch let w be a_p, otherwise let w be the notch of P next to a_p, counterclockwise. We may always assume that adequate preprocessing allows us to determine w in constant time, given a_p. If v_i lies between w and v_j in clockwise order, we simply define C as $v_i B_q$ (Fig.27-c). Otherwise things are somewhat more complex, and before describing an algorithm formally we will give an overview of the method.

Let L be the line collinear with the edge of C adjacent to v_i (i.e. the first edge of C counterclockwise). L will essentially wrap around the obstacles created by the boundary of P in a counterclockwise motion (Fig.27-d). Let V be the polygonal line on the boundary of P between v_i and w in clockwise order. We will show that all the *obstacles* (which are the vertices of C) are notches of V. Consequently we can expect to wrap around C entirely in $O(c)$ operations if L can pivot around each vertex of C in constant time on the average.

Let x be a vertex of C with L_1 (resp. L_2) designating the line L before (resp. after) pivoting around x (Fig.27-d). We first locate L_1 in the superrange of x, then we scan SR(x) counterclockwise, until we hit a vertex b_k which lies on V. We can show that in general b_k is also the next vertex of C. Recall that locating L in SR(x) involves finding the pair (a_j, b_j) such that L intersects $a_j b_j$. To ensure an $O(c)$ running time, we cannot actually locate L_1 in SR(x). Instead we determine a notch y nearby which will serve the same purpose. This notch is to be determined at the time when L_1 is computed. Thus we define the function NEXT which maps (x, y) to (a_k, b_k). More generally, NEXT maps any pair of notches x, y ($x \in V$) to the pair (a_k, b_k) of SR(x), computed as follows:

1. Find the two pairs (a_j, b_j) and (a_{j+1}, b_{j+1}) of SR(x) such that y lies between b_j and a_{j+1} in clockwise order.
2. Scan the pairs of SR(x) counterclockwise, starting at (a_j, b_j) (i.e. $(a_j, b_j), (a_{j-1}, b_{j-1}), \ldots$) and determine the pair (a_k, b_k) such that b_k is a notch of V and a_{k+1} lies outside of V. If we fail to find such a pair, return (0).
3. When NEXT is evaluated and a pair (a_k, b_k) is actually returned, the function sets a global variable *cnext* to a_{k+1} if it is a notch, else the notch of P next to a_{k+1}, counterclockwise (Fig.27-d).

We first show that the function NEXT is well-defined and can be evaluated in time proportional to the number of pairs scanned in step 2. Recall that in any superrange SR(x), the pairs (a_j, b_j) realize a partition of the set of all notches, and more precisely any notch y lies between b_j and a_{j+1} in clockwise order for some j. Thus, once SR(x) has been computed, we can extend the preprocessing to assign each notch of P to each corresponding pair b_j, a_{j+1}. A simple scan through the notches of P will do it in $O(c)$ time. Finally, noticing that we can test if a notch lies in V in constant time, and that *cnext* is also found in constant time, for the reasons seen above, we achieve our claims. We are now ready to set out the algorithm for computing C. Let $a_l b_l$ be the segment intersecting r_i with (a_l, b_l) in SR(v_i) and let t be the notch next to a_l counterclockwise. Let γ be the intersection $r_i \bigcap r_j$ if it is defined, or the endpoint of r_j ($\neq v_j$) otherwise.

Computing C

```
if v_i lies between w and v_j
   then return (C = {v_i, γ, B_q})
C:={v_i}, d:= v_i, e:= NEXT (v_i, t)
while e ≠ 0
   begin
      if ∠(dB_q, de) < 180
         then return (C:= C ⋃{B_q})
      d:= e
      C:= C ⋃{d}
      e:= NEXT (e,cnext)
   end
return (C:= 0)
```

To see that the algorithm runs in $O(c)$ time, it suffices to note that the notch *cnext* moves counterclockwise on the boundary of P, so $O(c)$ pairs (a_j, b_j) will be examined in all the superranges considered by the function NEXT.

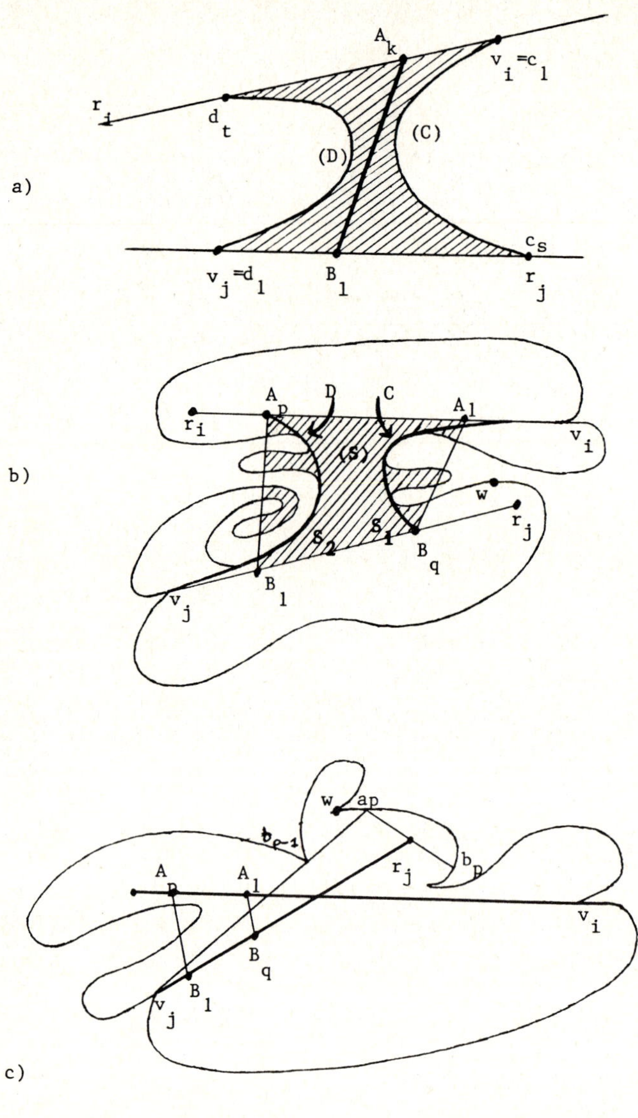
a)
r
i
d
t
A
k
v
i
=c
1
(D)
(C)
v
j
=d
1
B
1
c
s
r
j
b)
A
p
D
C
A
1
r
i
(S)
v
i
w
r
j
S
2
S
1
B
q
B
1
v
j
c)
w
ap
b
r
j
A
p
A
1
v
i
B
q
B
1
v
j

(Figure 27 .../...)

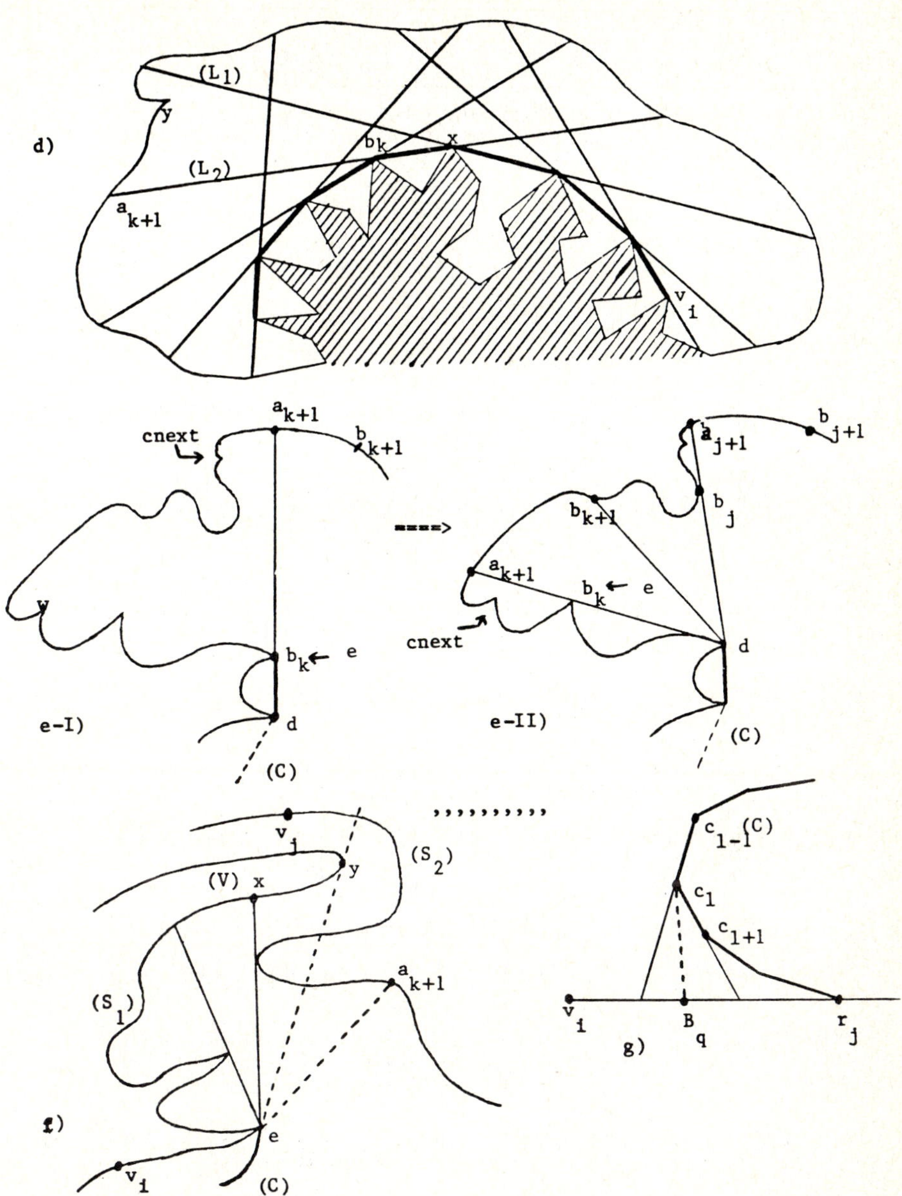

Figure 27 Computing the chains C and D.

We next show how a middle edge intersects the boundary of P if and only if it intersects C or D. In order to connect r_i and r_j via a middle edge, the intersection of P and the quadrilateral $A_1B_qB_1A_q$ must contain a polygon S with edges A_1A_p and B_1B_q. Note that this intersection may actually consist of several polygons. The boundary of S consists of these two segments joined by two polygonal lines, S_1 and S_2. S_1 has all its vertices in V and S_2 in V', where V' is defined as V, switching the roles of v_i and v_j (Fig.27-b).

To avoid computing S_1 and S_2 explicitly (they may have on the order of n vertices), we first notice that no middle edge can intersect the convex hull of S_1 without intersecting S_1. The same observation on S_2 leads to computing the convex hulls C_1 and C_2, respectively. For convenience, we can replace the vertex A_1 in S_1 (resp. B_1 in S_2) by v_i (resp. v_j) and still preserve the initial property that a middle edge lies totally in P if and only if it does not strictly cross C or D. We now turn to the actual computation of C and D.

The first case considered assumes that v_i lies between w and v_j in clockwise order. S is then reduced to the single intersection point of r_i and r_j and S can be set to v_iB_q (Fig.27-c). Note that we can always assume that in this case r_i and r_j intersect, otherwise the lists r_i and r_j would be empty. If v_i does not lie between w and v_j, V is not empty, and we will prove by induction that C is actually the convex hull of S_1 or 0 if no middle edge is possible. Fig.27-e illustrates the computation of the next edge of C. To ensure convexity, all of S_1 must lie on the same side of the line passing through this edge. Therefore the next vertex of C after the vertex labelled e in Fig.27-e-I) must be the point x of V, visible from b_k, which minimizes the angle $\angle(ea_{k+1}, ex)$. b_k and a_{k+1} are the vertices in SR(d) returned by the previous call on NEXT.

Since the endpoints of V are notches of P, x must be one of the vertices listed in SR(e) between which a_{k+1} lies, so we must start scanning SR(e). We only have to perform a counterclockwise scan since, by induction hypothesis, a_{k+1} does not lie in V, therefore the next vertex of C must be some a_l or b_l in SR(e) for $l \leq j$. Once again, the crux is that a counterclockwise scan in SR(e) corresponds both to a counterclockwise scan through the vertices of P and a counterclockwise angular sweep. Note that a_{k+1} is a point of SR(d) which has to be located in SR(e). Since a_{k+1} is not a notch in general this operation seems too complex.

Instead we can find the pair b_j, a_{j+1} of SR(e) between which *cnext* lies (recall that *cnext* is the first notch after a_{k+1} in a counterclockwise scan through the boundary of P). Since *cnext* is a notch, this operation can be done in constant time. Thus the function NEXT will return the next vertex of C. Note that if the point determined by NEXT is not a notch, NEXT returns 0. Indeed, this point could not be the next vertex of C and the actual next vertex y would not be visible from e. Consequently, C would intersect S_2 and no middle edge would then possible (Fig.27-f). We observe that if C is well-defined, there exists l such that B_q lies in the wedge $W(C_{l-1}C_l, C_lC_{l+1})$ (Fig.27-g). Thus the algorithm terminates by substituting B_q for $C_{l+1}, C_{l+2}, \ldots$ and the remaining vertices of C; this is legitimate since this last portion of C cannot have any effect on middle edges. ∎

Lemma 19. In $O(c)$ time, it is possible to precompute the functions f, g_1, g_2 so that any evaluation can be done in constant time.

Proof: $f(l)$ can be evaluated directly in constant time by intersecting the line passing through r_i with the ray emanating from B_l, collinear with R_u, yet not passing through R_u with $B_l = A_{j_u}$. If there is no intersection, $f(l) = 0$. Next we show how to compute all the values of g_1 and g_2 in $O(c)$ time. We start by computing the intersection of r_i with all the lines $(d_k d_{k+1})$ for consecutive values of k. These points partition r_i into segments, and the previous computation provides a sorted list of their endpoints in $O(c)$ time. To each of these segments corresponds a unique vertex of D. Then, for each $A_1, \ldots, A_p$ in turn, we find the segment where A_k lies. Let d_m be the corresponding vertex of D. We compute A'_k by intersecting r_j with the line passing through $A_k d_m$ (Fig.28). This also gives us a sorted list of the points A'_k, since D is convex. Finally we can merge the A'_k and B_l in $O(c)$ time, and in one scan through the list, find for each A'_k the nearest B_l on the same side as B_q. We then set $g_1(k)$ to 1. We iterate on this process with respect to C, defining $g_2(k)$ for each A_k on r_i. Finally, for each A_k, we check that $g_1(k) \leq g_2(k)$. If this is not the case, we set $g_1(k)$ and $g_2(k)$ to 0. All of this work clearly requires $O(c)$ time. ∎

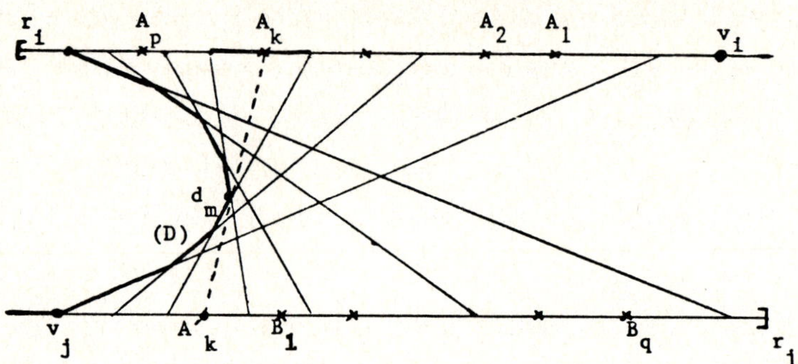

Figure 28 Computing g_1.

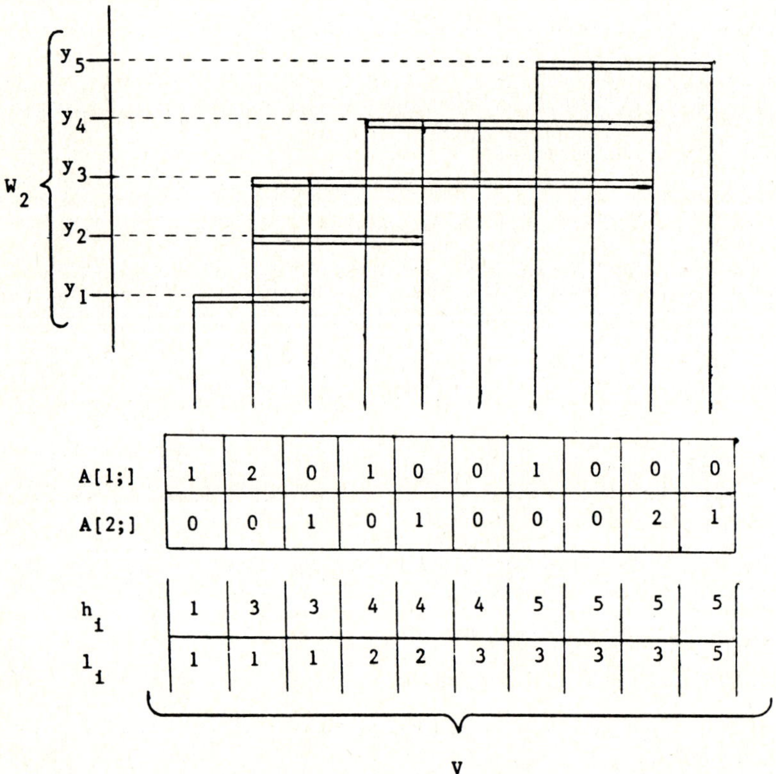

Figure 29 The setting of the function x.

Lemma 20. Let $V = \{1, \ldots, q\}$ and $W = \{1, \ldots, p+q\}$, $W_1 \bigcup W_2 = W$, $|W_1| = q$, $|W_2| = p$ and let f, g_1, g_2 be defined with: $f \mid V \mapsto W_1$ a bijection and $g_1, g_2 \mid W_2 \mapsto V$ non-decreasing functions, with $g_1(i) \leq g_2(i)$ for each $i \in W_2$. Define $x \mid W_2 \mapsto V$ such that for each $i \in W_2$, $x(i)$ is the smallest integer in V with $g_1(i) \leq x(i) \leq g_2(i)$ and $i \leq f(x(i))$ if such an integer exists, and 0 otherwise. If f, g_1, g_2 are computable in constant time, then so is x after $O(p+q)$ preprocessing.

Proof: Note that the naive method for computing all the values of x runs in $O(pq)$ time. We present an $O(p+q)$ time algorithm for achieving all these computations and establish its correctness. Let $y_1, \ldots, y_p$ be the elements of W_2 in increasing order ($1 \leq y_j \leq p+q$). First of all, we consider the set of $y \in W_2$ such that $g_1(y) \leq i \leq g_2(y)$ for a given i between 1 and q, and observe that it is a contiguous (possibly empty) subset of W_2 since g_1 and g_2 are non-decreasing. We compute the largest and smallest y, denoted y_{h_i} and y_{l_i} respectively, as follows (if there is no such y, we set (l_i, h_i) to 0).

```
Initialize an array A (2 × q) to 0.
for i = 1,...,p
  begin
    A[1, g1(yi)]:= A[1, g1(yi)]+1
    A[2, g2(yi)]:= A[2, g2(yi)]+1
  end
l:= 1; h:= 0
for i = 1,...,q
  begin
    h:= h+A[1, i]
    if l ≤ h
      then (li, hi):= (l, h)
      else (li, hi):= 0
    l:= l+A[2, i]
  end
```

The algorithm clearly achieves a time bound of $O(p+q)$. To establish its correctness, we observe that the first loop sets A$[1,j]$ to the number of y in W_2 such that $j = g_1(y)$, i.e. the number of intervals $[g_1(y), g_2(y)]$, starting at j. Similarly, A$[2,j]$ counts the number of intervals finishing at j. Then since, as i increases, $g_1(y_i)$ and $g_2(y_i)$ cannot decrease or pass each other, we can derive (l_i, h_i) from (l_{i-1}, h_{i-1}) by counting the number of intervals which have to be added and removed. More precisely, the difference $h_i - h_{i-1}$ is exactly the number of y in W_2 such that $i-1 < g_1(y) \leq i$, which is equivalent to $g_1(y) = i$ and shows that this number is A$[1,i]$. Likewise, if $i-1 < g_2(y_{l_{i-1}})$ we have $l_i = l_{i-1}$. Else if $g_2(y_{l_{i-1}}) = i-1$, $l_i - l_{i-1}$ is the number of y such that $g_2(y) = i-1$, i.e. A$[2,i]$ (see example in Fig.29).

We are now ready to set up the function x by computing all its values. If $y_{l_1} \leq f(l) \leq y_{h_1}$, all $x(i)$ with i between y_{l_1} and $f(l)$ must be set to 1. Now if $y_{l_1} \leq y_{h_1} \leq f(l)$, all $x(i)$ with i between y_{l_1} and y_{h_1} must be set to 1. In both cases, no other i in W should have $x(i)$ equal to 1. Then we can carry out the same reasoning with 2, assigning this value only to the $x(i)$ which have not been set yet. Since, as i increases from 1 to q, l_i and h_i cannot decrease or pass each other (unless $(l_i, h_i) = 0$) a possible implementation is:

Initialize all $x(i)$ to 0 for all $i = 1, \ldots, q$

M:= 0

for $i = 1, \ldots, q$

  begin

    a:= max(y_{l_i}, M)

    b:= min$(f(i), y_{h_i})$

    if $a \leq b$ and $(l_i, h_i) \neq 0$

      then

        for $j = a, \ldots, b$

        begin $x(j)$:= i **end**

        M:= $b + 1$

  end

Note that if i belongs to W_1, the value assigned to $x(i)$ is meaningless. The algorithm runs in time $O(p+q)$, which completes the proof. ∎

3. The Cubic Algorithm

We are now prepared to use the information contained in $E(i,j)$ to produce an OCD. $E(i,j)$ may be computed in STEP 4 of the algorithm *ConvDec*, with the additional preprocessing described earlier. We can now replace the former computation of M in STEP 3 by the following:

Initialize M as the empty set.

For each $v_k \in V(i+1, j-2)$ such that $E(k,j)$ contains a pair (A_{ki}, B_{jl}), assign to M the maximum (with respect to cardinality) of M and

$X4(v_i, v_k, v_l, v_j, A_{ki}, B_{jl}) \bigcup S(i+1, k-1) \bigcup S(k+1, l-1) \bigcup S(l+1, j-1)$.

For each $v_k \in V(i+2, j-1)$ such that $E(i,k)$ contains a pair (A_{ij}, B_{kl}), assign to M the maximum (with respect to cardinality) of M and

$X4(v_j, v_i, v_l, v_k, A_{ij}, B_{kl}) \bigcup S(i+1, l-1) \bigcup S(l+1, k-1) \bigcup S(k+1, j-1)$.

Note that we investigate the two possible topologies of an X_4-pattern lying in $V(i,j)$ and adjacent to v_i and v_j (Fig.30). The procedure for computing M requires $O(c)$ time and, from Lemma 21, we know that the additional preprocessing requires $O(n + c^2 \log n)$ time. We are now ready to evaluate the complexity of each step of the decomposition algorithm:

- Preprocessing: $O(n + c^3 + c^2 \log n)$
- Step 2: $O(c^3)$
- Step 3: $O(c^3)$
- Step 4: $O(c^3)$
- Step 5: $O(n + c^2 \log n)$

The total running time of the decomposition algorithm is therefore $O(n + c^3 + c^2 \log n)$, which is easily shown to be also $O(n + c^3)$.

Main Theorem. It is possible to compute an optimal convex decomposition of a simple polygon with n vertices and c reflex angles in time $O(n + c^3)$.

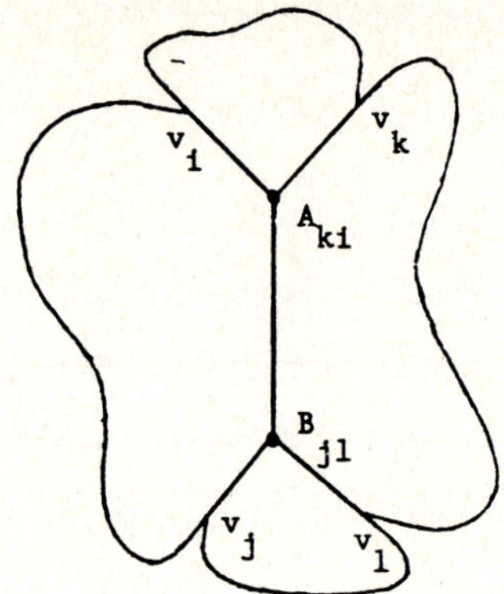

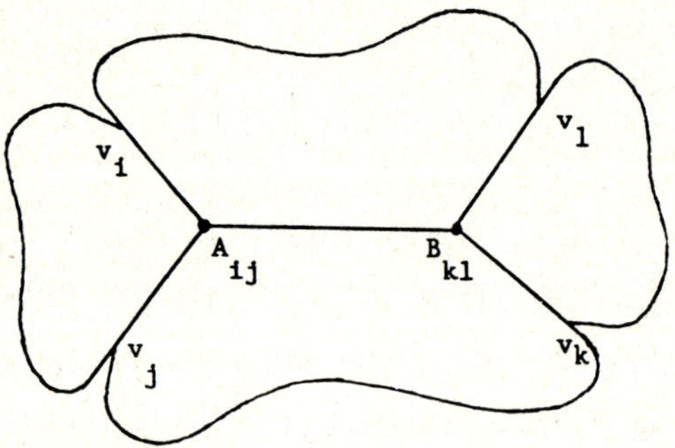

Figure 30 Computing B with the topologies of X4-pattern

We should note that in $S(i, i-1)$ the algorithm provides us with c optimal, yet not all necessarily identical, decompositions.

5. Concluding Remarks

The main result of this paper is an algorithm for decomposing a polygon into a minimum number of convex parts. The algorithm is linear in the number of vertices and cubic in the number of reflex angles. From a theoretical viewpoint, our main achievement has been to show that the decomposition problem was polynomial, even when Steiner points are allowed in the decomposition. Also one merit of our algorithm is to have its complexity expressed in the form $O(n + f(c))$. This is to our knowledge the only decomposition algorithm with this property; one interesting open problem is to decide whether the preprocessing used in our algorithm can be applied to the algorithms known for the case where no Steiner points are allowed in the decomposition. Indeed, Greene's $O(n^2c^2)$ and Keil's $(c^2n \log n)$ methods for this problem are less efficient than our algorithm.

On the practical side, the complexity of our algorithm might be acceptable, given the near-convexity of most polygons in practice. Unfortunately the algorithm seems inherently intricate and implementing it in its most elaborate form is certainly a formidable task. We might be willing, however, to sacrifice a little efficiency in order to achieve greater simplicity. Computing only X_2-patterns and doing away with superranges may often be found an acceptable compromise. Even the naive decomposition, when implemented efficiently, may turn out the best alternative if we can afford to miss an optimal solution by at worst a factor of two in the number of convex parts.

Of course, only the cubic algorithm reveals the genuine "beauty" of the problem. Its long development involves many subproblems, most of which are interesting in their own right (we believe). For example the concept of superranges might provide an effective means of dealing with visibility problems and its fast computation ($O(c \log n)$) makes it very appealing. This can be viewed as a first step towards adapting non-convexity to algorithms for convex designs.

Acknowledgment: We thank Diane Souvaine for reading this (long) manuscript very carefully and giving us many helpful comments and suggestions.

REFERENCES

[1] Asano, T., Asano, T. *Minimum partition of polygonal regions into trapezoids*, Proc. 24th Annuual FOCS Symp., 1983, pp. 233–241.

[2] Aho, A.V., Hopcroft, J.E. and Ullman, J.D. *The design and analysis of computer algorithms*, Addison-Wesley, 1974.

[3] Chazelle, B. *Computational geometry and convexity*, PhD Thesis, Yale University, 1980. Also available as Technical Report CMU–CS–80–150, Carnegie-Mellon University, Pittsburgh PA, 1980.

[4] Chazelle, B. *Convex partitions of polyhedra: a lower bound and worst-case optimal algorithm*, SIAM J. on Computing, August 1984.

[5] Chazelle, B., Dobkin, D.P. *Decomposing a polygon into its convex parts*, Proc. 11th Annual SIGACT Symp., pp. 38–48, 1979.

[6] Chazelle, B., Dobkin, D. *Detection is easier than computation*, Proc. 12th Annual SIGACT Symp., pp. 146-153, 1980.

[7] El Gindy, H., Avis, D. *A linear algorithm for computing the visibility polygon from a point*, Journal of Algorithms, 2(1981), pp. 186-197.

[8] Feng, H., Pavlidis, T. *Decomposition of polygons into simpler components: Feature generation for syntactic pattern recognition*, IEEE Trans. Comp., Vol. C-24, No. 6, June 1975, pp. 636-650.

[9] Ferrari, L., Sankar, P.V. and Sklansky, J. *Minimal rectangular partitions of digitized blobs*, Proc. 5th International Conference on Pattern Recognition, Miami Beach, Dec. 1981, pp. 1040-1043.

[10] Garey, M.R., Johnson, D.S., Preparata, F.P. and Tarjan, R.E. *Triangulating a simple polygon*, Info. Proc. Lett., Vol. 7(4), June 1978.

[11] Graham, R.L. *An efficient algorithm for determining the convex hull of a finite planar set*, Info. Proc. Lett., 2, pp. 18-21, 1973.

[12] Greene, D.H. *The decomposition of polygons into convex parts*, Advances in Computing Research, Jay Press, pp. 235–259, 1983.

[13] Keil, M. *Decomposing polygons into simpler components*, PhD Thesis, University of Toronto, 1983.

[14] Lingas, A. *The power of non-rectilinear holes*, Proc. 9th Colloquium on Automata, Languages and Programming, Aarhus, 1982.

[15] Lingas, A. *Heuristics for minimum edge length rectangular decomposition*, Unpub. Man., MIT, Nov. 1981.

[16] Lingas, A., Pinter, R., Rivest, R. and Shamir, A. *Minimum edge length partitioning of rectilinear polygons*, Proc. 20th Annual Allerton Conference on Communication, Control, and Computing, Monticello, Ill., Oct. 1982.

[17] O'Rourke, J., Supowit, K.J. *Some NP-hard polygon decomposition problems*, IEEE Trans. Information Theory, 29(2), 1983, pp. 181–190.

[18] Schachter, B. *Decomposition of polygons into convex sets*, IEEE Trans. on Computers, C–27, pp. 1078–1082, 1978.

[19] Shamos, M.I. *Computational geometry*, PhD thesis, Yale University, 1978.

[20] Toussaint, G.T. *Pattern recognition and geometrical complexity*, Proc. 5th Int. Conf. on pattern recognition, Miami Beach, 1980, pp. 1324–1347.

[21] Toussaint, G.T. *Decomposing a simple polygon with the relative neighborhood graph*, Proc. 18th Annual Allerton Conf. on Communication, Control, and Computing, Monticello, Ill., Oct. 1980.

COMPUTATIONAL GEOMETRY, G.T. Toussaint (Editor)
© Elsevier Science Publishers B.V. (North Holland), 1985

EXPECTED TIME ANALYSIS OF ALGORITHMS IN COMPUTATIONAL GEOMETRY

Luc Devroye

McGill University
School of Computer Science
805 Sherbrooke Street West
Montreal, Canada H3A 2K6

We give a brief inexhaustive survey of recent results that can be helpful in the expected time analysis of algorithms in computational geometry. Most fast average time algorithms use one of three principles: bucketing, divide-and-conquer (merging), or quick elimination (throw-away). To illustrate the different points, the convex hull problem is taken as our prototype problem. We also discuss searching, sorting, finding the Voronoi diagram and the minimal spanning tree, identifying the set of maximal vectors, and determining the diameter of a set and the minimum covering sphere.

1. INTRODUCTION.

There has been an increasing interest in the study and analysis of algorithms in computational geometry; a recent survey paper by Toussaint (1980) has 168 references. Most of the emphasis has been placed on the study of the worst-case complexity of various algorithms under several computational models. It is well-known that many algorithms perform considerably better on the average than predicted by the worst-case analyses. In this note, we would like to point out a few recent developments in the analysis of the expected complexity of some algorithms. To keep the general discussion simple and yet insightful we make a few convenient assumptions.

The Assumptions.

The input data $X_1,\dots,X_n$ can be considered as a sequence of independent identically distributed R^d-valued random vectors with common density f. (Here the unrealistic assumption is that real numbers can be stored in a computer.)

An algorithm takes time $T=T(X_1,\dots,X_n)$, a Borel measurable function of the input data, and it halts with probability one, i.e. $T < \infty$ almost surely.

The common operations (+,-,/,*,mod,compare,move) take time uniformly bounded over all values of the operands. For example, a*b or a mod b take time bounded by a constant not depending upon a or b. (Once again, this is unrealistic, because the multiplication or comparison of two real numbers takes infinite time.)

We note here that the floor of a number x is x- x mod 1, so that it too is a constant time operation. The assumptions made here are not entirely unreasonable: real numbers can be approximated very accurately, and the common operations take constant time for all operands that can be stored in a word (i.e., whose value does not exceed a large threshold value). In many situations, the benefit of being able to use the rich theory for random variables with a density, far outweighs the disadvantages of our assumptions.

Expected Time.

We are interested in the expected time E(T) taken by certain algorithms. Obviously, E(T) depends upon f and n because the averaging is done over all random samples of size n drawn from the density f.

Fast Expected Time Algorithms.

In many applications, fast expected time algorithms can be obtained by the bucketing principle: find the smallest rectangle C covering $X_1,\ldots,X_n$; we recall here that a rectangle in R^d is a d-fold product of intervals of the type [a,b]. Now, divide C into equal-sized rectangles (buckets), and solve the problem by traveling from bucket to bucket while performing some local operations.

We will also discuss the dramatic savings in expected time that can be obtained by the proper application of the divide-and-conquer principle. Finally, it is sometimes possible to take a superficial look at the data and to eliminate without much work all useless points. The more involved work is then performed on the reduced data sequence: this will be called the quick elimination (or: throw-away) principle.

Throughout the paper, we will use n for the sample size. For behaviour as $n \to \infty$, we use the Landau symbols O, o, $\sim$, and the newer notation Ω,Θ. If a_n and b_n are two positive number sequences, then $a_n = O(b_n)$ $(\Omega(b_n))$ when there exist positive numbers n_0 and K_0 such that for all $n \ge n_0$, $a_n \le K_0 b_n$ $(a_n \ge K_0 b_n)$. Next, $a_n = \Theta(b_n)$ if $a_n = O(b_n)$ and $a_n = \Omega(b_n)$. Finally, $a_n = o(b_n)$ if $\lim_{n\to\infty} a_n/b_n = 0$, and $a_n \sim b_n$ when $\lim_{n\to\infty} a_n/b_n = 1$.

The paper is not meant as a survey of algorithms, or as a state-of-the-art article describing the fastest known expected time algorithm. Its sole purpose is to give selected analytical results that can be used to obtain upper bounds for E(T), or to make inferences about the asymptotic behavior of E(T). But the reader should be aware that even when E(T) is small, it is possible that the distribution of T is such that the algorithm is undesirable. While it is true, by Markov's inequality, that

$$P(T \geq cE(T)) \leq \frac{1}{c} \ , \text{ all } c > 0,$$

this bound is often too crude to be practical. What one usually wants is a property such as: $P(T-E(T) > cE(T)) \to 0$ as $n \to \infty$, all $c > 0$. This requires usually a deeper analysis. For example, if Var(T) is the variance of T, we have

$$P(T-E(T) > cE(T)) \leq \frac{\operatorname{Var}(T)}{c^2 E^2(T)} \ , \ c > 0,$$

and the right-hand-side tends to 0 when $\operatorname{Var}(T)/E^2(T)$ has a zero limit.

We will not discuss <u>all</u> the fast expected time algorithms. For example, to find the convex hull of $X_1,\ldots,X_n$ in R^2 under certain computational models, at least c n log n time is needed (Avis, 1979; Yao 1981). Jarvis' algorithm (Jarvis, 1973) takes expected time O(nE(N)) where N is the number of convex hull points. In the design of this algorithm, no special care is taken to obtain fast expected time behavior. Nevertheless, for certain distributions E(N)=O(1) (see Carnal (1970); this is true for multivariate t-distributions, etc.), so that Jarvis' algorithm runs in linear expected time for a fairly large class of distributions.

2. THE BUCKETING PRINCIPLE.

Let C be the smallest closed rectangle covering $X_1,\ldots,X_n$ and let C be divided into m^d equal-sized rectangles (buckets) where $m=\operatorname{int}(^{1/d})$ the floor of $n^{1/d}$. Bucket memberships can thus be obtained for all data points in time O(n). Often one keeps track of these memberships by using m^d linked lists, one per bucket, so that O(n) space is used.

The obvious application in R^1 involves sorting $X_1,\dots,X_n$. Here one empties the buckets from left to right and performs a subsequent sort within each bucket, if necessary. If this subsequent sort is a comparison-based sort (e.g., heapsort, bubble sort, shell sort, merge sort or quicksort) with average time g(n) (this number is independent of f, since we have a comparison-based sort), then the overall average time for sorting is

$$E(T) = O(n) + E(\sum_{i=1}^{n} g(N_i))$$

where $N_1,\dots,N_n$ are the cardinalities of the n buckets. Devroye and Klincsek (1981) addressed the question of when $E(T)=O(n)$. They showed that when $g(u)/u \uparrow \infty$ as $u \to \infty$, $g(u)/u^2$ is nonincreasing, and g is convex, then $E(T)=O(n)$ if and only if f has compact support (i.e., f=0 outside some finite interval) and

$$\int g(f(x))\, dx < \infty \,. \qquad (1)$$

Notice that they put no continuity or boundedness assumptions on f. In experiments, Akl and Meijer (1982) found that for sufficiently smooth densities, bucket sort (with slight ad hoc improvements) compares favorably with even the best version of quicksort.

These results do not contradict the fact that if $X_1,\dots,X_n$ have to be sorted by using comparisons only, that the average number of comparisons is at least $\Omega(n \log n)$, because our model allows the floor function to be counted as a constant time operation.

Consider now the following generalization of the previous result: travel from bucket to bucket, performing within the j-th bucket operations taking expected time bounded between $ag(N_j)$ and $bg(N_j)$ when N_j is given. Here $0 < a \le b < \infty$. Once again, the expected time of the entire algorithm is

$$E(T) = \Theta(n+E(\sum_{i=1}^{m^d} g(N_i))).$$

Assume that $g(u)/u \uparrow \infty$ and $g(u)/u^K \downarrow 0$ for some positive K as $u \to \infty$. Also, assume that g is convex. Then E(T)=O(n) if and only if f has compact support and (1) holds (Devroye, 1981a). These basic results have several applications. We cite just two examples.

Examples.

1. Searching in constant expected time.

Assume that $X_1,\ldots,X_n$ are stored in the bucket data structure given above, and that we are presented with X_Z (where Z is uniformly distributed over 1,...,n). We have to determine the index i such that $X_i=X_Z$. This is the classical problem of successful search. If the X_i's are stored in the buckets in order of arrival, then the expected search time is

$$\Theta\,(E(\frac{1}{n}\sum_{i=1}^{m^d} N_i^2 + 1)).$$

By a simple extension of the previous result, we see that E(T)=O(1) if and only if f has compact support and $\int f^2(x)\,dx < \infty$. If within each bucket the data are organized into a binary search tree rather than a linked list, by considering one of the coordinates of the X_1's as the key for sorting, then E(T)=O(1) if and only if f has compact support and

$$\int f(x) \log_+ f(x)\,dx < \infty, \text{ where } \log_+ u = \max(0, \log(u)).$$

2. Convex hull algorithms that are based upon sorting.

The convex hull of $X_1,\ldots,X_n$ is a subsequence $X_{i_1},\ldots,X_{i_k}$ of $X_1,\ldots,X_n$ such that for all X_{i_j} there exists a hyperplane through X_{i_j} and all X_q's, $q \neq i_j$, belong to the same closed halfspace determined by this hyperplane. In R^2, it can be obtained from $X_1,\ldots,X_n$ as follows: (i) Find a point x that belongs to the interior of the convex hull of $X_1,\ldots,X_n$. Sort all the X_i's according to the polar angles of $X_i - x$ (by using the bucket sort described above). This yields a polygon P. (ii) Visit all vertices of P in turn by pushing them on a stack. Pop the stack when non-convex-hull points are encountered. In essence, this is Graham's algorithm (1972) with a modification in the sorting method that is used. Step (ii) takes time O(n). The average time taken by (i) is O(n) when the density of the polar angle of $X_i - x$ is square integrable. Since x itself is a random vector, one must be careful before making any inference about f. Nevertheless, it is sufficient that f is bounded and has compact support.

The previous applications have one feature in common: the times taken by the algorithms on individual buckets just depend upon the number and/ or position of the data points within these buckets (and not on, say, the number of data points in neighboring buckets). In more involved problems, we cannot avoid looking at neighboring buckets. For example, consider the class of "closest point problems" in R^d (Shamos and Hoey, 1975) such as: find all nearest neighbor pairs, construct the Voronoi graph, find the minimal spanning tree, etc. (see Bentley and Friedman (1979) for other applications). Shamos (1978) and Weide (1978) discuss many applications of the bucketing principle, and Bentley, Weide and Yao (1980) give a fairly comprehensive treatment of the time analysis of bucketing algorithms for closest point problems. We take the liberty to cite a couple of examples from their study:

Examples.

1. The all-nearest-neighbor problem.

   All nearest neighbor pairs can be found in $O(n \log n)$ time (worst-case) (Shamos and Hoey, 1975). Weide (1978) proposed a bucketing algorithm in which for a given X_i, a "spiral search" is started in the bucket of X_i, and continues in neighboring cells, in a spiraling fashion, until no data point outside the buckets already checked can be closer to X_i than the closest data point already found. Bentley et al. (1980) showed that Weide's algorithm halts in average time $O(n)$ when there exists a bounded open convex region B such that the density f of X_1 is 0 outside B and satisfies $0 < \inf_B f(x) \le \sup_B f(x) < \infty$.

2. The Voronoi diagram.

   The Voronoi diagram in R^2 can be found in time $O(n \log n)$ (worst-case) (Shamos (1978), Horspool (1979), Brown (1979)). Bentley et al. (1980) have a bucketing algorithm that uses spiral search and has some additional features. The Voronoi diagram can be found in average time $O(n)$ when d=2 and the density f of X_1 satisfies the condition of Example 1. From the Voronoi diagram, the convex hull can be obtained in linear time (Shamos, 1978).

3. The minimal spanning tree.

For a graph (V,E), Yao (1975) and Cheriton and Tarjan (1976) give algorithms for finding the minimal spanning tree (MST) in worst-case time $O(|E|\log\log|V|)$. The Euclidean minimal spanning tree (EMST) of n points in R^d can therefore be obtained in $O(n \log\log n)$ time if we can find a supergraph of the EMST with $O(n)$ edges in $O(n \log\log n)$ time. Yao (1982) suggested to find the nearest neighbor of each point in a critical number of directions; the resulting graph has $O(n)$ edges and contains the MST. This nearest neighbor search can be done by a slight modification of the algorithm in Example 1. Hence, the EMST can be found in expected time $O(n \log\log n)$ for any d and for all distributions given in Example 1. The situation is a bit better in R^2. We can find a planar supergraph of the EMST in expected time $O(n)$ (such as the Delaunay triangulation (the dual of the Voronoi diagram), the Gabriel graph, etc.) and then apply Cheriton and Tarjan's (1976) $O(n)$ algorithm for finding the MST of a planar graph. Thus, in R^2 and for the class of distributions given in Example 1, we can find the EMST in linear expected time.

Finally, we should mention a third group of bucketing algorithms, where special buckets are selected based upon a global evaluation of the contents of the bucket. For example, assume that not more than a_n buckets are selected according to some criterion (from the approximately n original buckets) in time $O(n)$, and that only the data points within the selected buckets are considered for further processing. If N is the number of selected points, then we assume that "further processing" takes time $O(g(N))$ for a given function g. Because the global evaluation procedure is not specified, we should assume the worst-case, and this leads to the study of the order statistics of the cardinalities of the buckets. The following results can be found in Devroye (1981b). When M is the maximum of n i.i.d. Poisson (1) random variables, then $E(M) \sim \log n/ \log\log n$. The same is true if M is the maximum of $N_1,\dots,N_n$, where N_i is the cardinality of $[\frac{i-1}{n}, \frac{i}{n}]$ and the data is $U_1,\dots,U_n$, a sequence of i.i.d. uniform (0,1) random variables. Using tight bounds on the upper and lower tails of M, one can show that

$$E(g(N)) = O(g(a_n \frac{\log n}{\log\log n}))$$

where $a_n \geq 1$, g is nondecreasing, $g(x)=O(1+x^{\beta})$ (some $\beta > 0$), $\sup_{x>0} \frac{g(cx)}{g(x)} < \infty$ (all $c > 1$), and the X_i's have a bounded f with compact support.

Example.

The convex hull in R^2.

Shamos (1979) suggested to construct the convex hull in R^2 in the following fashion: mark all the nonempty extremal buckets in each row and column (the extremes are taken in the north and south directions for a column, and east and west directions for a row); mark all the adjacent buckets in the same rows and columns; apply Graham's $O(n \log n)$ convex hull algorithm to all the points in the marked buckets. It is clear that $a_n=O(\sqrt{n})$ and that the expected time of the algorithm is $O(n)+O(E(g(N)))$ = where $g(n)=n \log n$. This is

$$O(n) + O(\sqrt{n} \frac{(\log n)^2}{\log \log n}) = O(n).$$

Furthermore, the expected time spent on determining the bucket membership divided by the total expected time tends to 1.

3. THE DIVIDE-AND-CONQUER PRINCIPLE.

A problem of size n can often be split into two similar subproblems of size approximately equal to n/2, and so forth, until subproblems are obtained of constant size for which the solutions are trivially known. For example, quicksort (Sedgewick (1977, 1978)) is based on this principle. The average time here is $\Theta(n \log n)$, but, unfortunately enough, since the sizes of the subproblems in quicksort can take values 0,1,2,... with equal probabilities, the worst-case complexity is $\Theta(n^2)$. One can start in the other direction with about n equal-sized small problems, and marry subsolutions in a pairwise manner as in mergesort. Because of the controlled subproblem size, the worst-case complexity becomes $\Theta(n \log n)$ (Knuth, 1975). Both principles will be referred to as divide-and-conquer principles. They have numerous applications in computational geometry with often considerable savings in expected time. The first general discussion of their value in the design of fast expected time algorithms can be found in Bentley and Shamos (1978).

Let us analyze the divide-and-conquer algorithms more formally. Assume that $X_1,\dots,X_n$ are R^d-valued independent random vectors with common distribution, and that we are asked to find $A_n=A(X_1,\dots,X_n)$, a subset of $X_1,\dots,X_n$, where $A(.)$ satisfies:

1) $A(x_1,\dots,x_n)=A(x_{\sigma(1)},\dots,x_{\sigma(n)})$, for all $x_1,\dots,x_n \in R^d$, and all permutations $\sigma(1),\dots,\sigma(n)$ of $1,\dots,n$.
2) $x_1 \in A(x_1,\dots,x_n) \Rightarrow x_1 \in A(x_1,\dots,x_i)$ for all $x_1,\dots,x_n \in R^d$, and all $i \le n$.

The convex hull satisfies these requirements. If $Q_1(x),\dots,Q_{2^d}(x)$ are the open quadrants centered at $x \in R^d$, then we say that X_1 is a maximal vector of $X_1,\dots,X_n$ if some quadrant centered at X_i is empty (i.e., contains no X_j, $j\neq i$, $j\le n$). The set of maximal vectors also satisfies the given requirements. Let N=cardinality(A_n). For $p \ge 1$, we known by Jensen's inequality that

$$E(N^p) \ge (E(N))^p.$$

In the present context, we would like an inequality in the opposite direction. For random sets A_n satisfying 1) and 2), and under very weak conditions on the behavior of $E(N)$, we have

$$E(N^p) = 0((E(N))^p)$$

(Devroye, 1983). For example, it suffices that $E(N)$ is nondecreasing, or that $E(N)$ is regularly varying at infinity. Also, if $E(N) \le a_n \uparrow$, then $E(N^p)=0(a_n^p)$. In essence, the results of Devroye (1983) imply that under weak conditions on the distribution of X_1, $E(N^p)$ and $(E(N))^p$ are comparable. The same is true for other nonlinear functions of N. For example, if $E(N) \le a_n \uparrow$, then $E(N \log(N+e))=0(a_n \log a_n)$. Thus, the knowledge of $E(N)$ allows us to make statements about other moments of N. Here are some known results about $E(N)$.

Examples.

1. A_n is the convex hull. X_1 has a density f.

(i) $E(N)=o(n)$ (Devroye, 1981b).

(ii) If f is normal, then $E(N)=0((\log n)^{(d-1)/2})$ (Raynaud, 1970). For $d=2$, $E(N) \sim 2\sqrt{2\pi \log n}$ (Renyi and Sulanke, 1963, 1964).

(iii) If f is the uniform density in the unit hypersphere of R^d, then $E(N)=0(n^{(d-1/d+1)})$ (Raynaud, 1970).

(iv) If f is the uniform density on a polygon of R^2 with k vertices, then $E(N) \sim \frac{2k}{3} \log n$ (Renyi and Sulanke, 1963, 1968).

(v) If f is a radial density, see Carnal (1970). For example, if f is radial, and $P(||X_1||>u)=L(u)/u^r$ where $r \geq 0$ and L is slowly varying (i.e., $L(cx)/L(x) \to 1$ as $x \to \infty$, all $c > 0$), then $E(N) \to c(r) > 0$. If $P(||X_1||>u) \sim c(1-u)^r$ for some $c,r >$ as $u \uparrow 1$, and $P(||X_1||>1) = 0$, then $E(N) \sim c(r)\ n^{1/(2r+1)}$ for some $c(r) > 0$.

2. $\underline{A_n \text{ is the set of maximal vectors.}}$

When X_1 has a density and the components of X_1 are independent, then E(N) is nondecreasing (Devroye, 1980) and $E(N) \sim 2^d(\log n)^{d-1}/(d-1)!$ (Barndorff-Nielsen, 1966; Devroye, 1980).

A_n can be found by the following merging method. Assume for the sake of simplicity that $n=2^k$ for some integer $k > 1$.

1. Let $A_{1i}=A(X_i)$, $1 \leq i \leq n$. Set $j \leftarrow 1$.
2. Merge consecutive A_{ji}'s in a pairwise manner (A_{j1} and A_{j2}; A_{j3} and A_{j4}; etc.)
3. Set $j \leftarrow j+1$. If $j > k$, terminate the algorithm ($A_n=A_{k1}$). Otherwise, go to 2.

We assume that merging and editing of A_{ji} and $A_{j\ i+1}$ with cardinalities k_1 and k_2 can be done in time bounded from above by $g(k_1)+g(k_2)$ for some nondecreasing positive-valued function g, and assume that $E(g(N)) \leq b_n \uparrow$ where, as before, N=cardinality(A_n). Then the given algorithm finds A_n in expected time

$$0(n \sum_{j=1}^{2n} b_j/j^2).$$

If the merging and editing take time bounded from below by $a(g(k_1)+(k_2))$ and $E(g(N)) \geq sb_n$ where g and b_n are defined above, and $a,s > 0$ are constants, then we take at least

$$\gamma n \sum_{j=1}^{n} b_j/j^2$$

expected time for some $\gamma > 0$ and all n large enough (Devroye, 1983).

Thus, the divide-and-conquer method finds A_n in linear expected time if and only if

$$\sum_{j=1}^{\infty} b_j/j^2 < \infty .$$

Examples.

1. The set of maximal vectors.

Merging of two sets of maximal vectors can be achieved in quadratic time by pairwise comparisons, for any dimension d. We can thus take $g(u)= u^2$ in the previous analysis if we merge in this way. If $E(N) \sim a_n \uparrow \infty$, then we can check that the divide-and-conquer algorithm runs in linear expected time if and only if

$$\sum_{j=1}^{\infty} a_j^2/j^2 < \infty .$$

2. Convex hulls in R^2.

Two convex hulls with ordered vertices can be merged in linear time into a convex hull with ordered vertices (Shamos, 1978). Thus if $E(N)=0(a_n)$ and $a_n \uparrow \infty$, then

$$\sum_{j=1}^{\infty} a_j/j^2 < \infty \qquad (2)$$

is sufficient for the linear expected time behavior of the divide-and-conquer algorithm given here. When $\liminf E(N)/a_n > 0$, then (2) is also necessary for linear expected time behavior. Notice here that (2) is satisfied when, say, $a_n=n/\log^{1+\delta} n$ or $a_n=n/(\log n.\log^{1+\delta}\log n)$ for some $\delta > 0$. This improves the sufficient condition $a_n=n^{1-\delta}$, $\delta > 0$, given in Bentley and Shamos (1978).

3. Convex hulls in R^d.

Merging can trivially be achieved in polynomial (n^{d+1}) time for two convex hulls with total number of vertices equal to n. When $E(N) \le a_n \uparrow$ and

$$\sum_{j=1}^{\infty} a_j^{d+1}/j^2 < \infty,$$

we can achieve linear expected time. This condition is fulfilled for the normal density in R^d and the uniform density on any hypercube of R^d.

4. THE QUICK ELIMINATION (THROW-AWAY) PRINCIPLE.

In extremal problems (e.g., find the convex hull, find the minimal covering ellipse, etc.) many of the data points can be eliminated from further considerations without much work. The remaining data points then enter the more involved portion of the algorithm. Often the worst-case time of these elimination algorithms is equal to the worst-case time of the second part of the algorithm used on all n data points. The expected time is sometimes considerably smaller than the worst-case time. We illustrate this once again on our prototype problem of finding the convex hull.

Examples.

1. The convex hull.

Assume that we seek the extrema $e_1,\ldots,e_m$ in m carefully chosen directions of R^d, form the polyhedron P formed by these extrema, and eliminate all X_i's that belong to the interior of the extremal polyhedron P. The remaining X_i's are then processed by a simple worst-case $O(g(n))$ convex hull algorithm. What can we say about the expected time of these algorithms? An expected time of $o(g(n))$ would indicate that the elimination procedure is worthwhile on large data sequences. We could also say that the elimination procedure achieves 100% asymptotical efficiency. In Devroye (1981c) it is shown that this happens when (i) the open half-spaces defined by the hyperplanes through the origin perpendicular to the e_i's cover R^d except possibly the origin; and (ii) X_1 has a radial density f, where

$$\alpha(u) = \inf\{t:P(||X_1||>t)=u\}$$

is slowly varying at 0 ($\lim_{t\downarrow 0} \alpha(tu)/\alpha(t) = 1$, all $u>0$), and $\alpha(u) \to \infty$ as $u\downarrow 0$. Condition (i) holds when the e_i's are determined by the d+1 vertices of the regular (d+1)-vertex simplex in R^d centered at the origin. One could also take 2d directions defined by $(0,0,\ldots,0,\pm 1)$, etc. Condition (ii) is satisfied by the normal density and a class of radial exponential densities (Johnson and Kotz, pp. 298).

The previous result can be sharpened in specific instances. For example, if f is normal, $g(n)=n \log n$, and (i) holds, then the expected time is $O(n)$. Furthermore, the expected time spent by the algorithm excluding the elimination is $o(n)$ (Devroye, 1981c).

When $d=2$ and f is bounded away from 0 and infinity on a non-degenerate rectangle of R^2 ($f=0$ elsewhere), and $e_1,\ldots,e_8$ are equi-spaced directions, then the expected time of the elimination algorithm is $O(n)$ even when $g(n)=n^2$ (Devroye and Toussaint, 1981).

Eddy (1977) has given a slightly different elimination algorithm in which the number of directions and the directions themselves depend upon the data. Akl and Toussaint (1978) report that in R^2, for certain distributions, almost all the elimination algorithms achieve extremely fast average times provided that $e_1,\ldots,e_m$ are easily computed (e.g., they are axial or diagonal directions).

2. <u>Finding a simple superset</u>.

Assume that we wish to find $A_n=A(X_1,\ldots,X_n)$ in the following manner: (i) Find a set $B_n=B(X_1,\ldots,X_n)$ where B_n is guaranteed to contain A_n, in expected time T_n; (ii) Given that cardinality (B_n) is equal to N, find A_n from B_n in worst-case time bounded by $g(N)$. Note that the expected time of the entire elimination algorithm is bounded by

$$T_n + E(g(N)).$$

2.1. <u>A_n is the convex hull, B_n is the set of maximal vectors</u>.

We discussed some distributions for which $T_n = O(n)$. In R^2, step (ii) can be executed with $g(n)=n^2$ (Jarvis' algorithm, 1973) or $g(n)=n \log n$ (Graham's algorithm, 1972). Thus, the entire algorithm takes expected time $O(n)$ when $E(N \log_+ N)=O(n)$ or $E(N^2)=O(n)$, according to the algorithm selected in step (ii). When the components of X_1 are independent and X_1 has a density, then these conditions are satisfied by the results of Devroye (1980, 1983) given in section 3. The linearity is not lost in this case in R^d even when $g(n)=n^{d+1}$ in step (ii).

2.2. A_n is the diameter of $X_1,\ldots,X_n$; B_n is the convex hull.

$A_n=\{X_i,X_j\}$ is called a diameter of $X_1,\ldots,X_n$ when $||X_k-X_m|| \le ||X_i-X_j||$ for all $1 \le k \le m \le n$, $(k,m)\neq(i,j)$. Given B_n, some A_n can be found by comparing all $\binom{N}{2}$ distances between points in B_n (see Bhattacharya (1980) for an in-depth treatment of the diameter problem, and a survey of earlier results). But B_n can be found in linear expected time for many distributions. In such cases, our trivial diameter algorithm runs in linear expected time provided that $E(N)=0(\sqrt{n})$. Assume for example that f is the uniform density in the unit hypersphere of R^d, then the trivial diameter algorithm runs in linear expected time if and only if (i) the convex hull can be found in linear expected time, and (ii) $d \le 3$ (section 3, example 1 (iii)).

2.3 A_n is the minimum covering circle, B_n is the convex hull.

The minimum area circle in R^2 covering $X_1,\ldots,X_n$ has either three convex hull points on its perimeter, or has a diameter determined by two convex hull points. Again, it can be found (trivially) from the convex hull in worst-case time $0(n^4)$ (see Elzinga and Hearn (1972, 1974), Francis (1974) and Shamos (1978) for $0(n^2)$ algorithms and subsequent discussions). Thus, A_n can be identified in linear expected time if the convex hull B_n can be found in linear expected time, and if $E(N)=0(\sqrt{n})$ (or $0(n^{1/4})$, if the trivial algorithm is used). In fact, by using the $0(n \log n)$ algorithms of Preparata (1977) or Shamos (1978), the condition on E(N) can be relaxed even further.

5. REFERENCES.

[1] S.G. Akl and H. Meijer, Recent advances in hybrid sorting algorithms, Utilitas Mathematica, 21 (1982) 325-343.

[2] S.G. Akl and G.T. Toussaint, A fast convex hull algorithm, Information Processing Letters, 7 (1978) 219-222.

[3] Avis, D., On the complexity of finding the convex hull of a set of points, Technical Report SOCS 79.2, School of Computer Science, McGill University (1979).

[4] O. Barndorff-Nielsen and M. Sobel, On the distribution of the number of admissible points in a vector random sample, Theory of Probability and its Applications, 11 (1966) 249-269.

[5] J.L. Bentley and J.H. Friedman, Data Structures for range searching, Computing Surveys, 11 (1979) 398-409.

[6] J.L. Bentley and M.I. Shamos, Divide and conquer for linear expected time, Information Processing Letters, 7 (1978) 87-91.

[7] J.L. Bentley, B.W. Weide and A.C. Yao, Optimal expected-time algorithms for closest point problems, ACM Transactions of Mathematical Software, 6 (1980) 563-580.

[8] Bhattacharya, B., Applications of computational geometry to pattern recognition problems, Ph.D. Dissertation, School of Computer Science, McGill University (1980).

[9] K.Q. Brown, Voronoi diagrams from convex hulls, Information Processing Letters, 9 (1979) 227-228.

[10] H. Carnal, Die konvexe Hülle von n rotationssymmetrische verteilten Punkten, Zeitschrift für Wahrscheinlichkeitstheorie und verwandte Gebiete 15 (1970) 168-176.

[11] D. Cheriton and R.E. Tarjan, Finding minimum spanning trees, SIAM Journal of Computing, 5 (1976) 724-742.

[12] L. Devroye, A note on finding convex hulls via maximal vectors, Information Processing Letters, 11 (1980) 53-56.

[13] Devroye, L., Some results on the average time for sorting and searching in R^d, Manuscript, School of Computer Science, McGill University (1981a).

[14] L. Devroye, On the average complexity of some bucketing algorithms, Computers and Mathematics with Applications, 7 (1981b) 407-412.

[15] L. Devroye, Moment inequalities for random variables in computational geometry, Computing, 30 (1983) 111-119.

[16] L. Devroye, How to reduce the average complexity of convex hull finding algorithms, Computers and Mathematics with Applications, 7 (1981c) 299-308.

[17] L. Devroye and T. Klincsek, Average time behavior of distributive sorting algorithms, Computing, 26 (1981) 1-7.

[18] L. Devroye and G.T. Toussaint, A note on linear expected time algorithms for finding convex hulls, Computing, 26 (1981) 361-366.

[19] W.F. Eddy, A new convex hull algorithm for planar sets, ACM Transactions of Mathematical Software, 3 (1977) 398-403.

[20] J. Elzinga and D. Hearn, The minimum covering sphere problem, Management Science, 19 (1972) 96-104.

[21] J. Elzinga and D. Hearn, The minimum sphere covering a convex polyhedron, Naval Research Logistics Quarterly, 21 (1974) 715-718.

[22] Francis, R.L. and White, J.A., Facility Layout and Location: An Analytical Approach (Prentice-Hall, 1974).

[23] R. Graham, An efficient algorithm for determining the convex hull of a finite planar set, Information Processing Letters, 1 (1972) 132-133.

[24] Horspool, R.N., Constructing the Voronoi diagram in the plane, Technical Report SOCS 79.12, School of Computer Science, McGill University (1979).

[25] R.A. Jarvis, On the identification of the convex hull of a finite set of points in the plane, Information Processing Letters, 2 (1973) 18-21.

[26] Johnson, N.L. and Kotz, S., Distributions in Statistics: Continuous Multivariate Distributions (John Wiley, New York, 1972).

[27] Knuth, D., The Art of Computer Programming, vol. 3: Sorting and Searching (Addison-Wesley, Reading, Mass., 2nd Ed., 1975).

[28] Lipton, R.J. and Tarjan, R.E., Applications of a planar separator theorem, 18th Annual IEEE Symposium on the Foundation of Computer Science (1977) 162-170.

[29] Preparata, F., Steps into computational geometry, Report R-760, Coordinated Sciences Laboratory, University of Illinois (1977).

[30] H. Raynaud, Sur le comportement asymptotique de l'enveloppe convexe d'un nuage de points tirés au hasard dans R^n, Comptes Rendus de l'Académie des Science de Paris, 261 (1965) 627-629.

[31] A. Renyi and R. Sulanke, Uber die konvexe Hülle von n zufällig gewählten Punkten I, Zeitschrift für Wahrscheinlichkeitstheorie und verwandte Gebiete, 2 (1963) 75-84.

[32] A. Renyi and R. Sulanke, Uber die konvexe Hülle von n zufällig gewählten Punkten II, Zeitschrift für Wahrscheinlichkeitstheorie und verwandte Gebiete 3 (1964) 138-147.

[33] A. Renyi and R. Sulanke, Zufällige konvexe Polygone in einem Ringgebiet, Zeitschrift für Wahrscheinlichkeitstheorie und verwandte Gebiete 9 (1968) 146-157.

[34] R. Sedgewick, The analysis of quicksort programs, Acta Informatica, 7 (1977) 327-355.

[35] R. Sedgewick, Implementing quicksort programs, Communications of the ACM, 21 (1978) 847-857.

[36] Shamos, M.I., Computational geometry, Ph.D. Dissertation, Dept. of Computer Science, Yale University (1978).

[37] Shamos, M.I., Seminar given at McGill University (1979).

[38] Shamos, M.I. and Hoey, D., Closest-point problems, Proceedings of the 16th IEEE Symposium on the Foundations of Computer Science (1975) 151-162.

[39] Toussaint, G.T., Pattern recognition and geometrical complexity, Proceedings of the 5th International Conference on Patter Recognition and Image Processing (Miami, Florida, 1980).

[40] Weide, B.W., Statistical methods in algorithm design and analysis, Ph.D. Dissertation, Department of Computer Science, Carnegie-Mellon University (1978).

[41] A.C. Yao, An $O(|E|\text{loglog}|V|)$ algorithm for finding minimum spanning trees, Information Processing Letters, 4 (1975) 21-23.

[42] A.C. Yao, On constructing minimum spanning trees in k-dimensional space and related problems, SIAM Journal on Computing, 11 (1982) 721-736.

[43] A.C. Yao, A lower bound to finding convex hulls, Journal of the ACM, 28 (1981) 780-787.

© Elsevier Science Publishers B.V. (North Holland), 1985

Practical Use of Bucketing Techniques in Computational Geometry

Takao ASANO*, Masato EDAHIRO*, Hiroshi IMAI*, Masao IRI*

and

Kazuo MUROTA†

Techniques for using "buckets" to improve the efficiency of several computational-geometrical algorithms are described, together with examples illustrating the practical importance of the bucketing techniques. Specifically, they are applied to the problems of minimum-weight perfect matchings in the plane, two-dimensional Voronoi diagrams, point location and range search in the plane, and shortest paths in networks.

1. Introduction

In the last decade we have seen an explosive growth in the field of geometrical algorithms [Toussaint 1980, 1982]. Although much of the work is theoretical in nature, some newly discovered algorithms are proved to be quite practical. However, it has also become evident that the "theoretically best" algorithm is apt to be outperformed by more naive methods in some practical situations. Therefore, if we seriously wish the results of our research to be of some use in practice, we are confronted with the problem of the balance between theory and practice: to investigate the practicality of a "theoretical" algorithm and to analyse theoretically the performance of a "practical" algorithm --- both in the worst case and in the average case ("average" being defined appropriately). It should be noted in this connection also that there are very large problems for which only linear-time algorithms are practical, and that, in such a case, the magnitude of the coefficient of proportionality before the problem size in the complexity function affects considerably the practicality of an algorithm.

From such a theoretical-practical standpoint, the authors and their former colleagues have developed algorithms for a number of fundamental computational-geometrical problems using "buckets", and investigated their applicability to various large-scale problems connected with geographical data-bases and regional sciences [Iri et al. 1983] and to VLSI CAD.

In the most general terms, "buckets" mean the subregions into which the entire region under consideration is partitioned. There have been several proposals for making use of some kinds of buckets in geometrical algorithms. L. Devroye [1981a, b] analysed theoreti-

*Department of Mathematical Engineering and Instrumentation Physics, Faculty of Engineering, University of Tokyo, Bunkyo-ku, Tokyo 113, Japan.

†Institute of Socio-Economic Planning, University of Tsukuba, Sakura, Ibaraki 305, Japan.

cally the role played by buckets mainly for the problem of convex hulls and sorting. J. L. Bentley, B. W. Weide and A. C. Yao [1980] proposed a kind of bucket algorithm for the Voronoi diagrams. The hierarchical triangular subdivision used in the point-location algorithm by D. Kirkpatrick [1983] as well as the J-way division in the range-search algorithm by D. E. Willard [1982] might be regarded as highly sophisticated variants of buckets. Needless to say, we have long used the technique of partitioning a vast region into smaller subregions in order to efficiently store a large amount of data for the entire region in, and retrieve them from, a rather slow secondary memory of a computer system.

The basic properties of buckets which the authors have taken advantage of in developing their algorithms may be summarized as follows.

(1) If buckets have a simple shape, i.e., if they are square or rectangular with the sides parallel to the coordinate axes, it is computationally quite easy to partition the region under consideration into the buckets and to determine to which bucket a point belongs and which buckets a line segment intersects. (Highly sophisticated partitions of the region, or of the set of objects, lack this property. At the end of this paper, the possibility of using "triangular" buckets will briefly be discussed.)

(2) Unless the distribution of geometrical objects (points and/or line segments) is extremely skewed (which is seldom the case in practical situations), the number of objects belonging to a bucket is usually bounded by a constant if the entire region is partitioned into a number of buckets equal in order to the number of objects.

(3) In most geometrical problems, the mutual influence between two objects is likely to decrease with their distance, so that it is usually a good strategy to first construct a local or myopic solution within each bucket and then to combine local solutions to make a global solution.

If we can take advantage of all these properties of buckets, we shall be led to an algorithm which runs in linear time on the average. However, a simple-minded use of buckets will not give a satisfactory result in many problems. It is crucial for the success of an algorithm using bucketing techniques whether or not we can devise a well-tuned algorithm for processing the information within each bucket as well as that for combining the local results into a global solution. This last part is highly problem-dependent.

Specifically, this paper surveys the algorithms using bucketing techniques for five computational-geometrical problems which the authors and their former colleagues have developed and the efficiencies of which have been tested against problems of practical size.

In Section 2, we summarize algorithms for the minimum-weight perfect matching in the plane [Iri-Murota-Matsui 1981a, 1983] and their application to the optimization of plotter pen movement [Iri-Taguchi 1980], [Reingold-Tarjan 1981], [Iri-Murota-Matsui 1982], where sinuous traversing of buckets plays an essential role.

In Section 3, a kind of incremental algorithm for constructing the Voronoi diagram in the plane [Ohya-Iri-Murota 1984a,b] is shown to be more efficient than the theoretically optimal algorithm of the divide-and-conquer type for most problems arising in practical situations, where buckets are provided with a sophisticated but easy-to-handle structure of a quaternary tree. The algorithm is so fast that it can be used as a subroutine to be called repeatedly in a geometrical optimization problem [Iri-Murota-Ohya 1984], [Suzuki-Iri-Murota 1983].

In Section 4, buckets are used for the point-location problem [Edahiro-Kokubo-Asano 1983a,b] in order to get an algorithm the performance of which compares favourably with existing algorithms [Kirkpatrick 1983], [Lee-Preparata 1977], [Preparata 1981], where a considerably elaborate algorithm is used for manipulating the information within each bucket. In Section 5, a range-search algorithm, based upon the idea of introducing an artificial topology among originally isolated given points, is proposed and its efficiency tested by computational experiments [Edahiro-Iri-Asano 1984a].

In Section 6, it is shown that the idea of bucketing can be used also for the shortest-path problem [Imai-Iri 1982, 1984], where buckets are used not for partitioning the geometrical region itself but for a preliminary classification of the set of vertices with respect to the current values of their potentials, thus relaxing the geometrical-topological interference among the vertices belonging to one and the same bucket.

While the buckets used in the previous examples are either square or rectangular, there is theoretically the possibility of using buckets of other shapes. Probably the next simplest shape is a triangle. In Section 7, a brief discussion on the use of "triangular buckets" is given with respect to Voronoi diagrams and minimum-weight perfect matchings in the plane [Sanae 1984].

2. Minimum-Weight Perfect Matching in the Plane

For a given set of n (even) points in the plane, the minimum-weight perfect matching problem is to determine how to match the n points in pairs so as to minimize the sum of the distances between the matched points. This problem is not only of some theoretical interest in itself, but it has also important practical applications. For example, it can be applied to the problem of optimizing the sequence of drawing lines by a mechanical plotter [Iri-Taguchi 1980], [Reingold-Tarjan 1981]. The best known exact algorithm for this problem requires $O(n^3)$ time [Lawler 1976], which seems to be too expensive from the practical point of view. Even the approximation algorithms with running times of $O(n^2)$ or $O(n \log n)$ [Supowit-Plaisted-Reingold 1980] would not be satisfactory or need some improvement for the application to real-world problems of a size, say, n greater than 10^4.

Based upon theoretical analysis of the worst-case as well as the average-case performance of linear-time approximation algorithms which make use of the technique of traversing buckets, it has been found [Iri-Murota-Matsui 1981a, 1983] that a well-tuned algorithm can practically compete with $O(n \log n)$ algorithms (see also [Avis 1983] for a general survey), and that a remarkable improvement in the plotting speed of a mechanical plotter can be achieved by means of an approximate matching algorithm [Iri-Murota-Matsui 1982].

2.1. Bucketing techniques for Euclidean matching

Suppose we are given a set of points with even cardinality $\boldsymbol{P}=\{P_1,P_2,...,P_n\}$ in the unit square. A (perfect) matching M is a set of $n/2$ edges connecting points of $\boldsymbol{P}$ such that each point of $\boldsymbol{P}$ is an endpoint of exactly one edge. The cost of M is the sum of the lengths of edges in M, i.e., the total length of the matching. The "length" may be in the Euclidean L_2-metric, the L_1-metric, or the L_∞-metric; the L_∞-metric being appropriate for the plotter applications [Iri-Murota-Matsui 1981a, 1982, 1983].

The class of bucket-traversing algorithms, called "straightforward algorithms with preprocess and tour" in [Iri-Murota-Matsui 1983], is described as follows:

1° Partition the unit square into k^2 square *buckets* by dividing each side into k equal parts, and, for each of the n points, determine the bucket to which it belongs.

2° Order the buckets according to some order, forming the sequence $(C_0, C_1, ..., C_{k^2-1})$. Let $(D_0, D_1, ..., D_m)$ be the subsequence of $(C_0, C_1, ..., C_{k^2-1})$ consisting of all the buckets containing at least one point.

3° For each i $(=0, ..., m)$, match arbitrarily as many pairs of points as possible within D_i. Then, we have the subsequence $(E_0, ..., E_j)$ of $(D_0, ..., D_m)$ consisting of all those buckets which contain an unmatched point. Then, link the j points in E_i's consecutively to get a tour.

4° Choose the smaller in total length of the two matchings which are constructed by selecting every second edge of the tour.

Note that each point belongs to one of the k^2 buckets, and that we can determine the bucket to which a point belongs by multiplying the coordinates of the point by k and then truncating off the fractional parts. The quality of an approximate solution may depend on the bucket order in 2°. Here, the quality, or the cost, of a matching is measured in terms of the total length of the matching itself, and not by its ratio to the total length of the optimal solution. If we define the "bucket distance" between two buckets with respect to a bucket order by the difference of their bucket numbers in that order, it is seen that a good bucket order is an order such that, as the bucket distance increases, the geometrical distances between points in one of the two buckets and those in the other increase as slowly as possible. Among others, the performances of the two types of bucket order, called *serpentine* and *spiral rack*, respectively, shown in Fig.2.1, are analysed in detail in [Iri-Murota-Matsui 1981a, 1983] as follows. (Another bucket order called *serpentine rack* is also useful, especially for a graph drawn in a rectangular region [Matsui-Murota-Iri 1981]. It has quite a similar property to the spiral-rack order, provided the rectangular region is divided into square buckets proportional in number to n.)

Let α be a parameter independent of n and let $k=\lceil \alpha\sqrt{n}\,\rceil$. Then, it is easy to see that the complexity of the algorithm based on each of these orders is $O(n)$ both in time and in space. The supremum $\hat{M}_n(\alpha\sqrt{n})$ of the costs of the matchings obtained by the algorithm, taken over all possible configurations of n points in the unit square, is asymptotically proportional to $\sqrt{n}$, so that

$$\hat{\mu}(\alpha) = \lim_{n\to\infty} \hat{M}_n(\alpha\sqrt{n})/\sqrt{n}, \tag{2.1}$$

or its minimum with the optimum parameter value $\hat{\alpha}_0$

$$\hat{\mu}_0 = \hat{\mu}(\hat{\alpha}_0) = \min_{\alpha} \hat{\mu}(\alpha), \tag{2.2}$$

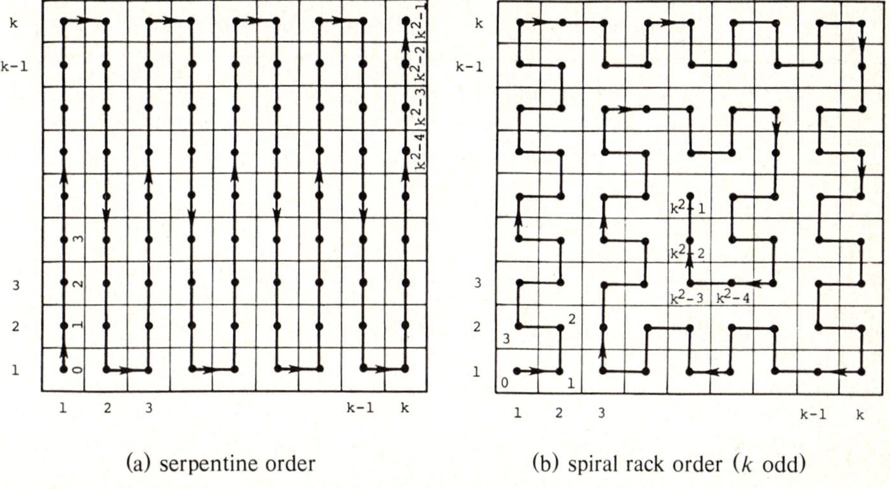

(a) serpentine order (b) spiral rack order (k odd)

Fig. 2.1. Typical bucket orders

is a measure characterizing the quality of the algorithm for the worst case. Another characteristic measure may be defined by supposing that n points are randomly distributed uniformly in the unit square. The cost of the matching obtained by the algorithm is then a random variable M_n, and the asymptotic expectation $\mu=\mu(\alpha)$ of $M_n/\sqrt{n}$, or its minimum $\mu_0=\mu(\alpha_0)=\min_{\alpha}\mu(\alpha)$ with the optimum parameter value α_0, is a characteristic measure for the average case. $\hat{\mu}(\alpha)$, and consequently $\hat{\mu}_0$ and $\hat{\alpha}_0$, for the bucket algorithms either with the serpentine order or with the spiral-rack order can be determined, to within a small interval at worst, by means of a linear-programming approach, while $\mu(\alpha)$, and consequently μ_0 and α_0, can be determined by elementary probability calculation. They are shown in Table 2.1, which is extracted from [Iri-Murota-Matsui 1983], where it is seen that the performance of the spiral-rack algorithm is more than satisfactory when it is compared with the "strip" algorithm in [Supowit-Plaisted-Reingold 1980], the best known $O(n \log n)$ approximation algorithm, with respect to cost and computing time. It is recommended in [Iri-Murota-Matsui 1982,1983] to use, for the actual application to the plotter problem, the spiral-rack algorithm with $\alpha=1.29$ for the L_2-metric and $\alpha=1.26$ for the L_∞-metric as a compromise of the worst case and the average case.

2.2. Drawing connected graphs by a mechanical plotter

When drawing a connected graph by a mechanical plotter, the wasted plotter-pen movement can be minimized by finding a minimum-weight perfect matching of the odd-degree vertices, augmenting the graph with the additional edges between the matched pairs of those vertices, and traversing an Eulerian path on the augmented graph, where the additional edges are to be traversed with the pen off the paper [Iri-Taguchi 1980], [Iri-Murota-Matsui 1981a, 1982, 1983], [Reingold-Tarjan 1981]. Here, the distance between two points

Table 2.1. Asymptotic supremum $\hat{\mu}_O$ and expectation μ_O of the cost/$n^{1/2}$ for the optimal value $\hat{\alpha}_O$ and α_O of the parameter $\alpha=k/n^{1/2}$

Algorithm	Complexity	L_2-Distance				L_∞-Distance			
		$\hat{\mu}_O$	$\hat{\alpha}_O$	μ_O	α_O	$\hat{\mu}_O$	$\hat{\alpha}_O$	μ_O	α_O
Serpentine	O(n)	1.189	1.189	0.637[a]	0.79[a]	1	1	0.603[a]	0.75[a]
Spiral-rack	O(n)	⩽1.014 ⩾0.932	[1.712][b]	0.484[a]	1.12[a]	0.866	1.732	0.443[a]	1.07[a]
Strip[c]	O(n log n)	0.707[c]		0.474[a,d]		?		0.436[a,d]	
Exact	O(n^3)	⩾0.537[c]		0.35[e]		⩾0.5		?	

[a]It is assumed that $M_n/n^{1/2}$ converges in probability to a constant for the serpentine, the spiral-rack and the strip algorithms.
[b]The value corresponding to an upper bound for $\hat{\mu}_O$, which is not asymptotically tight.
[c]From [Supowit-Plaisted-Reingold 1980]. (α is fixed to $1/\sqrt{2}$ for the strip algorithm.)
[d]By elementary probability calculation. See [Iri-Murota-Matsui 1981b] for details.
[e]This is the value conjectured in [Papadimitriou 1977]. However, according to [Iri-Murota-Matsui 1981b], this value is more likely to lie between 0.32 and 0.33.

is measured in the L_∞-metric. Thus, if a linear-time approximation algorithm for the minimum-weight perfect matching in the plane is used, the total time required to find an approximately optimum plotter-pen movement is linear in the size of the graph to be drawn. Fig.2.2 shows an example of application to the drawing of a Voronoi diagram with 4000 points, the diagram itself being a connected graph with about 8000 vertices, all of odd degree, and about 12000 edges (see also Section 3). Fig.2.2(a) is the diagram to be drawn, Fig.2.2(b) and Fig.2.2(c) show the wasted pen movement when the plotter is instructed to draw the edges according to the edge numbers produced by the divide-and-conquer algorithm and to those produced by the quaternary incremental algorithm, respectively. Fig.2.2(d) shows the wasted pen movement when the edges are renumbered by means of the serpentine-rack approximation algorithm.

2.3. Drawing unconnected graphs

In case the graph is unconnected, we may first make it connected by adding new blank edges and, possibly, new vertices, in such a way that the total length of the added edges is as small as possible, and then apply the method in Section 2.2. If the original graph has c connected components, $c-1$ additional edges are necessary and sufficient to make it connected. It would be possible to introduce new vertices as well as new edges to make the graph connected, but, the problem of minimizing the total length of the additional edges with new vertices is a kind of Steiner problem and therefore too hard to solve in practical time for a large graph. Under the restriction that new vertices are not allowed, i.e., that new edges should connect only vertices of the original graph, the problem of finding the additional edges with the exact minimum total length is an extension of the minimum spanning tree problem, and could be solved in a polynomial time, but not in linear time, with the aid of the Voronoi

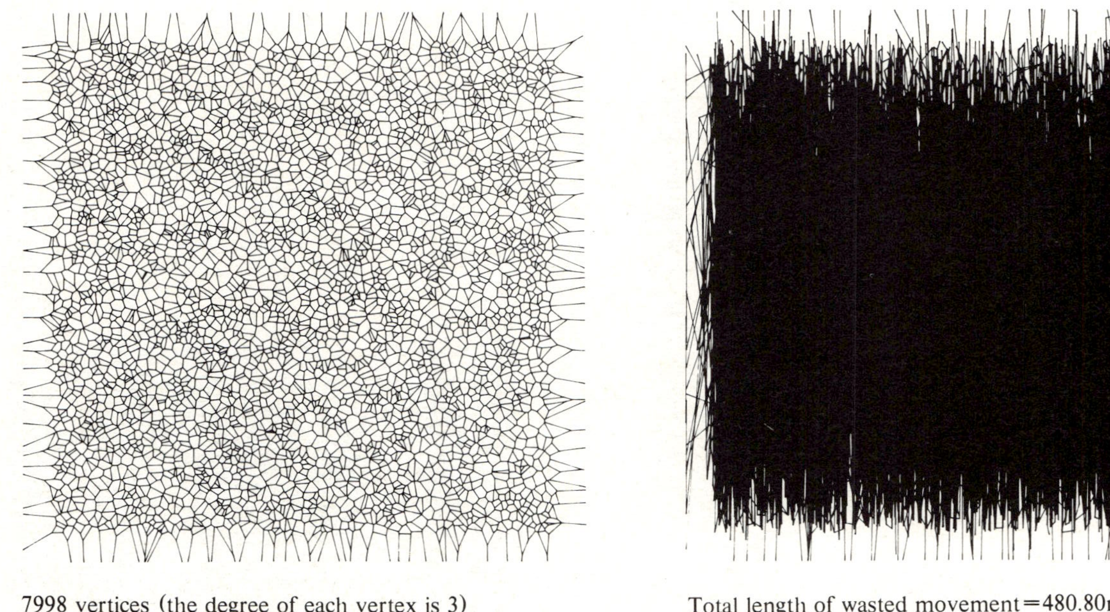

7998 vertices (the degree of each vertex is 3)
11870 edges
1 connected component
Total L_∞-length of lines (on 22cm×22cm sheet): 24.14m

(a) Voronoi diagram

Total length of wasted movement=480.80m
Total plotting time = 1h 52min

CPU time of the host computer (HITAC M-280H)
to input data : 1.75s
to output plotter file: 2.28s

(b) Pen movement wasted --- According to the numbers produced by the divide-and-conquer algorithm

Fig.2.2. Drawing a Voronoi diagram (to be continued)

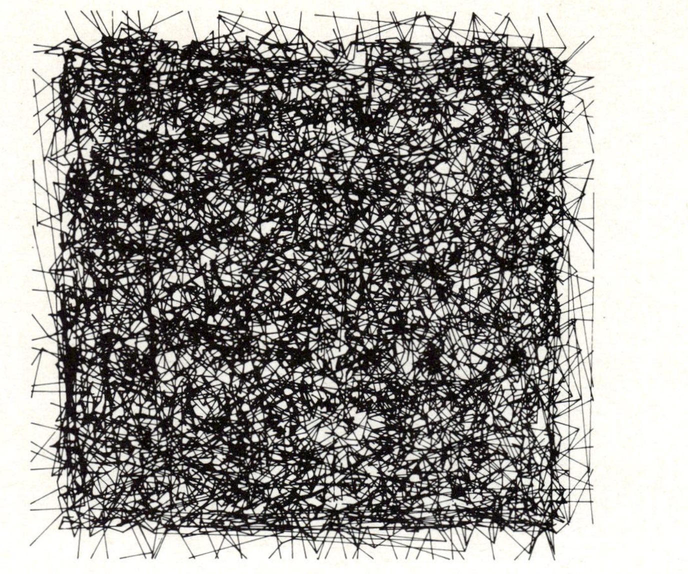

Total length of wasted movement=71.86m
Total plotting time = 50min 50s

CPU time of the host computer (HITAC M-280H)
to input data : 1.75s
to output plotter file: 2.28s

(c) Pen movement wasted --- According to the edge numbers produced by the quaternary incremental algorithm

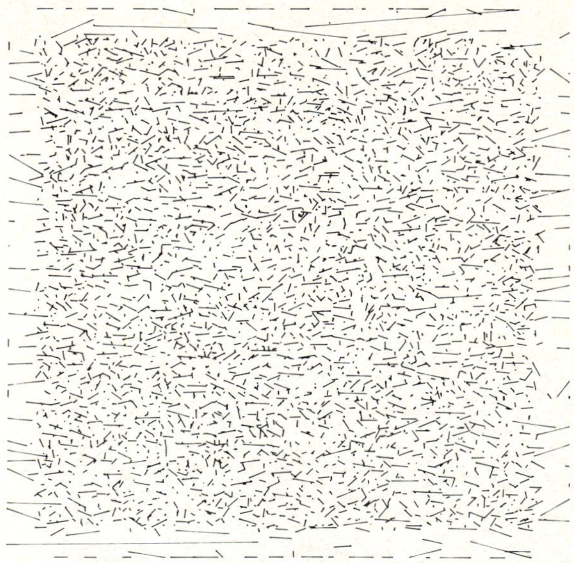

Total length of wasted movement=9.36m
Total plotting time = 24min 20s

CPU time of the host computer (HITAC M-280H)
to input data : 1.75s
to output plotter file : 1.54s
to renumber the edges: 0.37s

(d) Pen movement wasted --- According to the edge numbers renumbered by the serpentine-rack algorithm

Fig.2.2. Drawing a Voronoi diagram (continued)

20726 nodes (including degree-2 nodes)
25336 links (straight line segments)
25 connected components
Total L_∞-length of lines (on 60cm×60cm sheet): 44.47m

(a) Road map of the Kanto district of Japan

Total length of wasted movement = 4702.53m
Total plotting time = 9h 50min

CPU time of the host computer (HITAC M-200H)
to input data : 7.9s
to output plotter file: 14.0s

(b) Pen movement wasted --- Input data shuffled

Fig.2.3. Drawing the road map of the Kanto district of Japan (to be continued)

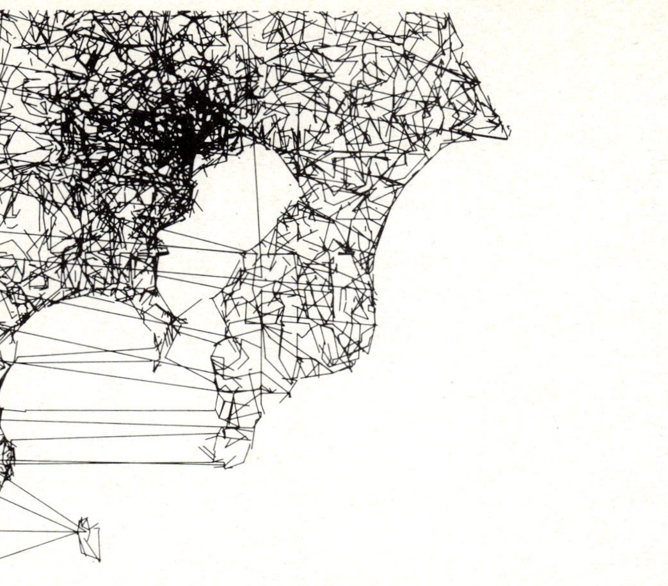

Total length of wasted movement=57.80m
Total plotting time = 51min

CPU time of the host computer (HITAC M-200H)
to input data : 7.9s
to output plotter file: 5.1s

(c) Pen movement wasted --- Input data prepared by an intelligent person

Total length of wasted movement = 13.62m
Total plotting time = 36min 45s

CPU time of the host computer (HITAC M-200H)
to input data : 7.9s
to output plotter file: 3.9s
to preprocess data : 0.85s

(d) Pen movement wasted --- Input data preprocessed by the serpentine-rack algorithm

Fig.2.3. Drawing the road map of the Kanto district of Japan (continued)

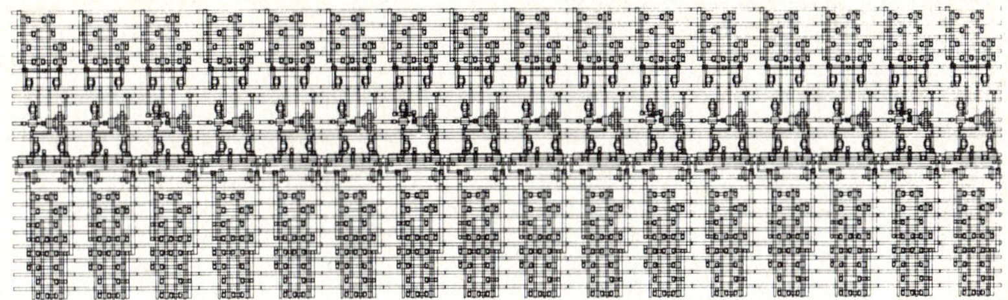

4860 rectangles (19440 vertices, 19440 edges)
4860 connected components
Total L_∞-length of lines (on 83.2cm×23.5cm sheet): 133.8m

(a) VLSI pattern

Total length of wasted movement = 1515m
Total plotting time = 3h 45min

CPU time of the host computer (HITAC M-280H)
to input data : 1.05s
to output plotter file: 2.05s

(b) Pen movement wasted --- Input data produced by a CAD system

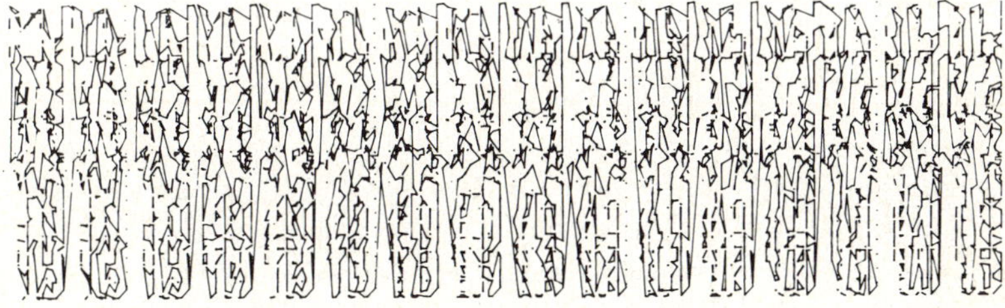

Total length of wasted movement = 33.3m
Total plotting time = 56min 45s

CPU time of the host computer (HITAC M-280H)
to input data : 1.05s
to output plotter file: 2.39s
to preprocess data : 0.72s

(c) Pen movement wasted --- Input data preprocessed by the serpentine-rack algorithm

Fig.2.4. Drawing a VLSI pattern

diagram [Shamos-Hoey 1975], [Lee-Wong 1980a], [Iri et al. 1983].

However, a linear-time approximation algorithm using buckets works well enough also for this problem [Iri-Murota-Matsui 1982]. The algorithm is basically in the same vein as the bucket algorithm for the matching. The given graph (generally unconnected) with its vertices lying in the unit square is first decomposed into connected components of which the number is assumed to be c. (This can be done in linear time by a standard graph algorithm.) The outline of the algorithm is as follows:

1° Divide the unit square into $k \times k$ square buckets and order them.

2° Determine to which bucket each vertex belongs.

3° Scan all vertices, bucket by bucket, according to the prescribed bucket order.

4° If two consecutive vertices belong to distinct connected components, connect them with a new edge, and merge the two connected components into one.

Evidently this algorithm adds $c-1$ new edges and can be implemented to run in $O(n)$ time if $k^2=\alpha^2 n$, where n is the number of vertices of the graph and α is a parameter.

The average-case performance of the algorithm is analysed theoretically under the assumption that the vertices are uniformly distributed in the unit square, independently of the connectedness among them. The average length in the L_∞-measure of an additional edge is estimated theoretically and confirmed experimentally; for the spiral-rack bucket order, it asymptotically attains the minimum of $0.9/\sqrt{n}$ for $k \simeq 1.2\sqrt{n}$ [Iri-Murota-Matsui 1982].

A FORTRAN program containing all the procedures explained in this Section 2, which has been used to draw quite a number of "big" graphs, is found in [Imai 82].

An example of the road map of the Kanto district of Japan is given in Fig.2.3, and that of a VLSI mask pattern in Fig.2.4.

3. Voronoi Diagrams

For a set of n points P_i ($i=1,\ldots,n$) (to be called *generators* in the following) in the Euclidean plane $\mathbf{R}^2$, the Voronoi region of P_i is defined by

$$V_n(P_i)=\bigcap_{j\neq i}\{P\in\mathbf{R}^2 \mid d(P,P_i)<d(P,P_j)\}, \tag{3.1}$$

where d denotes the Euclidean distance. The planar skeleton V_n formed by the boundaries of $V_n(P_i)$ ($i=1,\ldots,n$) is a planar straight-line graph and is called the *Voronoi diagram*, which plays a fundamental role in computational geometry [Shamos-Hoey 1975]. The vertices of the Voronoi diagram are called the *Voronoi points* and the edges the *Voronoi edges*. Two generators P_i and P_j are called *contiguous* in V_n if their Voronoi regions $V_n(P_i)$ and $V_n(P_j)$ have a boundary edge in common.

The divide-and-conquer algorithm, first proposed by D. T. Lee [1976] and described by M. I. Shamos and D. Hoey [1975], and improved subsequently, constructs the Voronoi diagram in $O(n \log n)$ time, which is known to be optimal, at least in the asymptotic order, with respect to the theoretical worst-case performance. However, this still seems to be too expensive for some applications such as the geometrical optimization problems treated in

[Iri-Murota-Ohya 1984], [Suzuki-Iri-Murota 1983]. It was recently found that a more classical and more primitive incremental method [Green-Sibson 1978] can be polished up to make a practically faster algorithm [Ohya-Iri-Murota 1984a,b]. The incremental method is the method which augments the diagram step by step by adding generators one by one. What is new in the algorithm in [Ohya-Iri-Murota 1984a,b] is the special preprocessing procedure for determining the order in which the generators are to be added, where an elaborate bucketing technique is used. The algorithm is called the "quaternary incremental" algorithm since it makes use of a quaternary tree for the structure of buckets.

3.1. Idea of the incremental Voronoi-diagram algorithm

A primitive incremental algorithm [Green-Sibson 1978] with worst-case complexity $O(n^2)$ may be described as follows. Suppose that the n generators are given in the order P_1 through P_n and let V_m denote the Voronoi diagram for the first m generators $P_1,...,P_m$. Starting from a trivial diagram, say V_3, the incremental algorithm constructs V_m from V_{m-1} $(m \leqslant n)$ by adding a new generator P_m to the current diagram V_{m-1} and modifying V_{m-1} in two phases as follows.

Phase 1 (Nearest neighbour search): Among the generators $P_1,...,P_{m-1}$ of V_{m-1}, find the nearest generator, $P_{N(m)}$, to P_m.

Phase 2 (Local modification): Starting with the perpendicular bisector of line segment $P_mP_{N(m)}$, find the Voronoi edge on the boundary of $V_{m-1}(P_{N(m)})$ which the bisector intersects as well as the generator $P_{m'}$ corresponding to the other Voronoi region incident to that edge; then draw the perpendicular bisector of $P_mP_{m'}$, and continue similar procedures until the region of P_m is created in V_{m-1} to obtain V_m (see Fig.3.1).

Phase 1 is equivalent to finding the generator $P_{N(m)}$ $(1 \leqslant N(m) \leqslant m-1)$ such that $V_{m-1}(P_{N(m)})$ contains P_m. The following simple algorithm works pretty well from the prac-

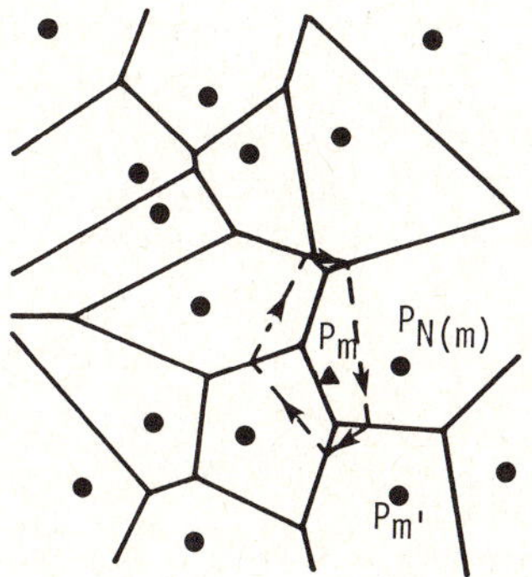

Fig.3.1. Phase 2 of a stage of the incremental algorithm

tical point of view.

Start with an initial guess $P_{i(0)}$ $(1 \leqslant i(0) \leqslant m-1)$ and set $i{:=}i(0)$. If

$$d(P_m, P_i) \leqslant d(P_m, P_j) \tag{3.2}$$

for each P_j contiguous to P_i in V_{m-1}, then finish with $N(m)=i$; Otherwise set $i{:=}j$ for any j such that (3.2) fails and repeat this process.

The initial guess, not specified above, has a substantial influence on the efficiency of the whole algorithm. Desirably, it should be chosen in such a way that the search may be finished in at most a constant number of steps in the neighbourhood of P_m.

As for Phase 2, the situation is more subtle. Although the number of edges of a Voronoi region is mostly six for randomly distributed generators, this does not imply that all the regions relevant to the incremental procedure in the intermediate diagrams necessarily have the same property, but it is possible that an intermediate diagram has a region with very many edges. However, it is intuitively obvious that, if the generators are ordered in such a way that $P_1, \ldots, P_{m-1}$ are distributed approximately uniformly around P_m for each m, the modification in Phase 2 for each m can be performed in constant time on the average.

3.2. Quaternary incremental algorithm --- A fast algorithm with quaternary-tree bucketing [Ohya-Iri-Murota 1984a,b]

Let the generators lie in the unit square $S=\{(x,y) \mid 0<x<1, 0<y<1\}$, and partition S into k^2 square buckets by dividing each side of S into k equal parts, where k is taken as $k=2^M$ $(M=\lfloor \log_4(\alpha^2 n) \rfloor)$ with parameter α to be properly chosen. The order of the generators is determined by means of a quaternary tree T of height M; T has $4^M (=k^2)$ leaves corresponding to the k^2 buckets, and a node at depth h (counted from the root at the top) represents a subsquare of a side of length 2^{-h}. (In particular the root represents the entire unit square.) The leaves of T are numbered from 0 to 4^M-1 in the natural manner *from left to right*; four consecutive nodes at the same depth have a common father. Each bucket is given a two-dimensional address (I,J) $(0 \leqslant I \leqslant k-1,\ 0 \leqslant J \leqslant k-1)$ in the usual way. The leaf r is identified with the bucket $(I(r), J(r))$ using the correspondence defined between "bucket number" r and "bucket address" (I,J) by the following procedure (a "folding-over" algorithm), whose complexity is obviously $O(n)$:

```
c:=⌊2^M/3⌋;  I(0):=c;  J(0):=c;  u:=(−1)^(M+1);  m:=1;  r:=0;  s:=2×c+u;
for t:=1 to M do
  | for p:=1 to m do [r:=r+1;  I(r):=s−I(r−m);  J(r):=J(r−m)];
  | m:=2×m;
  | for p:=1 to m do [r:=r+1;  I(r):=I(r−m);  J(r):=s−J(r−m)];
  | m:=2×m;  u:=−2×u;  s:=s+u
```

An example of this bucket order is shown in Fig.3.2 for $M=3$ $(k^2=64)$.

To each leaf of T, now being identified with a bucket, the generators in that bucket are assigned, whereas the non-leaf nodes of T are still empty. Then, by scanning the leaves *from left to right*, pick up one of the generators, if any, contained in a leaf and put it in all

J \ I	0	1	2	3	4	5	6	7
7	[47] 24	[46] 59	[42] 60	[43] 25	[59]	[58]	[62] 61 14	[63] 37
6	[45]	[44] 64 16	[40] 57	[41] 52	[57]	[56]	[60] 15 5	[61]
5	[37] 23	[36]	[32] 17	[33] 18 7	[49] 33 8	[48] 29 9 1	[52] 36 32	[53] 53 51
4	[39]	[38] 22 13	[34]	[35] 19	[51] 31	[50] 54	[54]	[55] 50 49
3	[7] 45 43 3	[6]	[2] 30 27	[3] 2	[19] 63	[18] 62	[22] 20 4	[23]
2	[5] 46	[4]	[0] 11	[1] 10	[17] 55	[16] 58 28	[20]	[21] 48
1	[13] 6	[12] 26	[8] 44 38	[9] 21 12	[25]	[24] 39	[28] 35	[29] 65
0	[15] 47	[14]	[10] 34	[11] 66	[27] 42 41	[26]	[30]	[31] 56 40

Fig.3.2. The quaternary bucket order for $k^2=64$ buckets (bucket numbers indicated in brackets) and an example of distribution of 66 generators (represented by numbers without brackets) in them

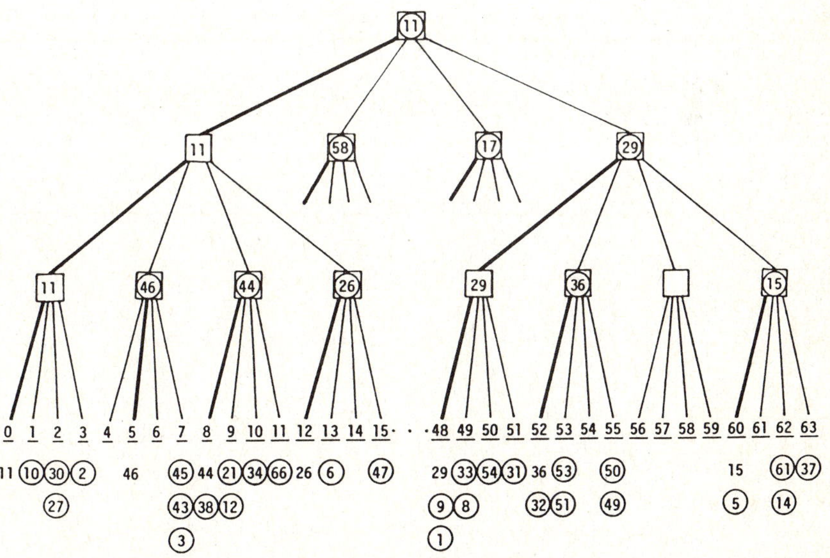

Fig.3.3. Quaternary-tree structure of buckets (underlined numbers) and generators (numbers without underline) (cf. Fig.3.2)

the empty ancestor nodes of the leaf. Thus a quaternary tree is obtained, whose nodes, except for leaves, are filled with at most one generator (see the example in Fig.3.3). Note that a nonempty node has a nonempty father.

The ordering of the generators in the incremental process is determined by traversing the tree from the root in a breadth-first manner, *from right to left* at the same depth. Every time a node with a "new" generator is encountered (see the encircled numbers in Fig.3.3 for example), the generator is added to the Voronoi diagram. Since the father of this node must contain a generator which has already been processed, the generator is adopted as the initial guess $P_{i(0)}$ in Phase 1. When a leaf contains more than one generator left unadded, they are added in an arbitrary order among themselves, where the generator in the same bucket which was added previously is adopted as the initial guess $P_{i(0)}$ in Phase 1. Using this order, the uniformity of distribution of generators $P_1,...,P_m$ at an intermediate stage is achieved, at least approximately, and a good initial guess can be chosen.

3.3. Experimental results

Fig.3.4 shows the CPU times needed (by HITAC M-280H, VOS3/JSS4, Optimizing FORTRAN77, OPT=2, Computer Centre of the University of Tokyo) for constructing the Voronoi diagrams by three algorithms, where the generators are randomly uniformly distributed in the unit square. The ordinate represents the "CPU time/n" (the time per generator), averaged for 10 independent instances, for each n, where the number of generators n, the abscissa, ranges from 2^7 (=128) to 2^{15} (=32,768).

The divide-and-conquer algorithm [Lee 1976], [Shamos-Hoey 1975] of the worst-case complexity $O(n \log n)$ did not run in $O(n)$ time even on the average. (There is a theoretical basis for this fact; see [Ohya-Iri-Murota 1984b].) The primitive incremental method, i.e., the method without any particular device for ordering generators and choosing initial guesses, took about $O(n^{3/2})$ time. The quaternary incremental algorithm constructed a diagram for n generators in $O(n)$ time on the average. The optimal value of parameter α was found to be approximately equal to unity, and, with α=1, the constant time of about 0.25 millisecond was needed per generator, independently of the total number n of the generators involved. The quaternary incremental algorithm was proved to be quite robust against the variation of α and against the non-uniformity of the distribution of generators [Ohya-Iri-Murota 1984b].

Detailed theoretical and experimental results concerning the quaternary incremental method are reported in [Ohya-Iri-Murota 1984b]. The applicability of the incremental method to more general types of Voronoi diagrams, such as Voronoi diagrams for line segments, polygons and circles, those in non-Euclidean metrics [Lee 1980], [Lee-Drysdale 1981], [Lee-Wong 1980a], those in the Laguerre geometry [Imai-Iri-Murota 1984], etc., are now being investigated.

3.4. Application to the geographical optimization problems

Since the quaternary incremental method for the Voronoi diagram is fast enough, we

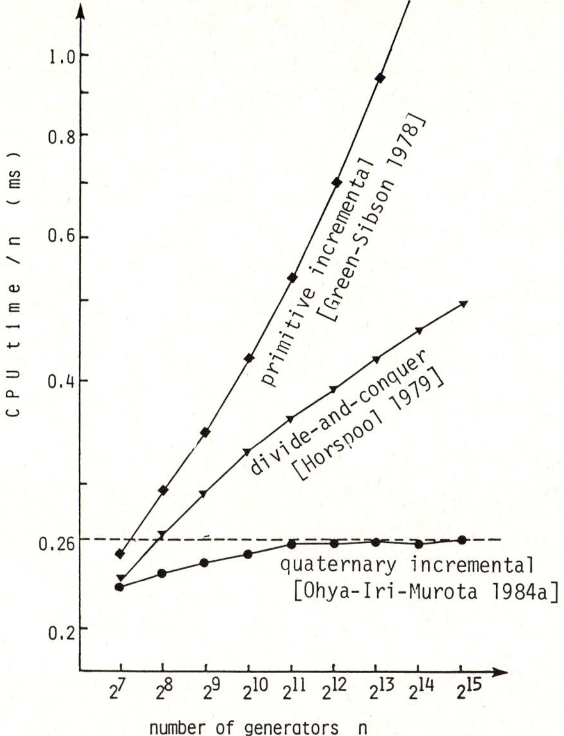

Fig.3.4. Comparison of the speeds of three different Voronoi diagram algorithms

can use it as a fundamental subroutine to be called repeatedly in an optimization problem. For example, the problem of determining the locations $\boldsymbol{x}_1,\ldots,\boldsymbol{x}_n$ of n public facilities in such a way that the total cost for inhabitants to gain access to their nearest facility be minimized is mathematically to find the minimum of the integral

$$F(\boldsymbol{x}_1,\ldots,\boldsymbol{x}_n)=\int_V f(\min\{\|\boldsymbol{x}-\boldsymbol{x}_i\| \mid i=1,\ldots,n\})\mu(\boldsymbol{x})\mathrm{d}\boldsymbol{x}$$
$$=\sum_{i=1}^{n}\int_{V_i} f(\|\boldsymbol{x}-\boldsymbol{x}_i\|)\mu(\boldsymbol{x})\mathrm{d}\boldsymbol{x}, \tag{3.3}$$

where V is the entire region under consideration, $\mu(\boldsymbol{x})$ is the density of inhabitants at point $\boldsymbol{x}$, V_i is the Voronoi region of the i-th facility, and $f(t)$ is the cost paid by a person visiting a facility located at the distance t from his home. If we try to solve this geographical optimization problem by some iterative method, the construction of the Voronoi diagram for the set of n moving points $\boldsymbol{x}_1,\ldots,\boldsymbol{x}_n$ is required for computing the value of the objective function F as well as its derivatives with respect to $\boldsymbol{x}_i$'s at each iteration step.

Fig.3.5 shows an example of a solution for this kind of problem where $n=128$, $f(t)=t^2$

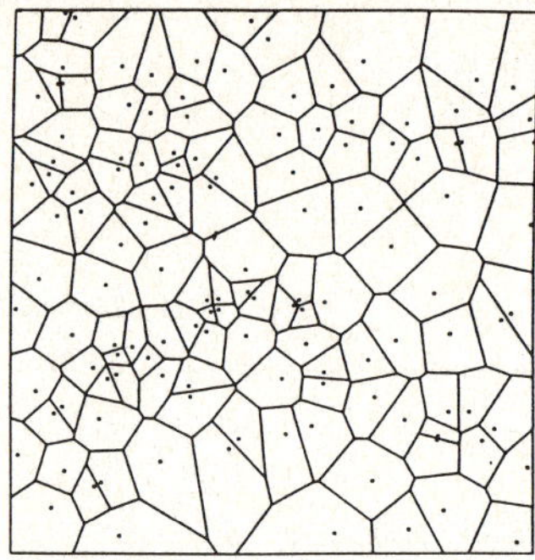

(a) Initial random distribution of facility locations

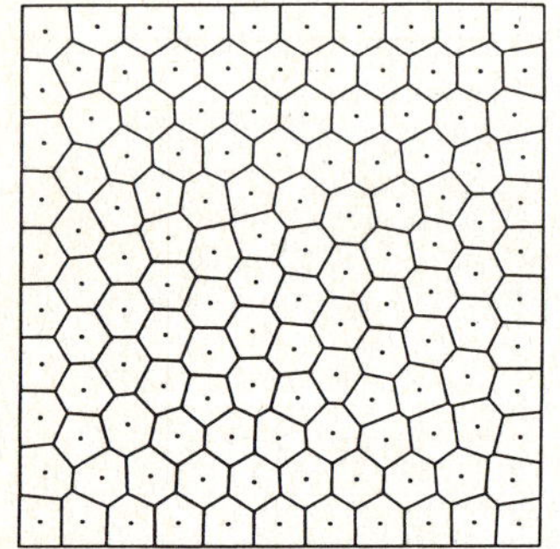

(b) Uniformly-distributed population
--- after 90 iterations (11.9s)

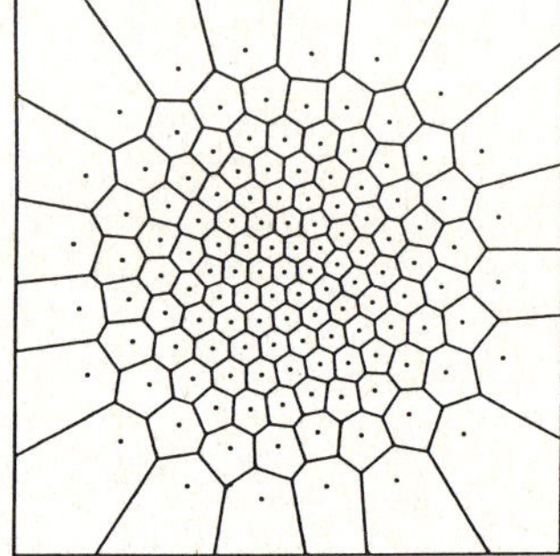

(c) Normally-distributed population
--- after 112 iterations (35.6s)

Fig. 3.5. Example of solution of a geographical optimization problem

and $\mu(\boldsymbol{x})$ is taken as a uniform distribution and a two-dimensional normal distribution in the unit square. Starting with n facilities randomly distributed (Fig.3.5(a)), we resorted to a primitive descent method using the exact gradient of F and a rough approximation to the Hessian of F, and obtained the improved facility locations shown in Fig.3.5(b) after 90 iterations in 11.9 seconds for the uniformly-distributed population $\mu(\boldsymbol{x})$ and those shown in Fig.3.5(c) after 112 iterations in 35.6 seconds for the normally-distributed population $\mu(\boldsymbol{x})$, respectively, in total. The time spent to construct the Voronoi diagrams was 6.4 seconds (for 209 diagrams) for the former case and 11.8 seconds (for 398 diagrams) for the latter. (Each iteration involves a line search, which requires several function evaluations containing the construction of the Voronoi diagrams. The convergence was checked in terms of the variation of F at each step of iteration.) The computation was done on the same machine that was mentioned in Section 3.3. Extensive computational experiments revealed that the computation time needed is roughly proportional to $n^{3/2}$ for a fixed $\mu(\boldsymbol{x})$. (See [Iri-Murota-Ohya 1984] and [Suzuki-Iri-Murota 1983] for more details.)

4. Point-Location Problem

The problem of locating a point in a planar subdivision, which is called the *point-location problem* and has been investigated extensively in recent years, is formally stated as follows: Given a straight-line planar graph G with n vertices and a query point Q, determine which region of the planar subdivision contains Q (Fig.4.1). In many applications, numerous location queries are raised to a fixed planar subdivision. For this reason, algorithms for the point-location problem are generally characterized in terms of the following three attributes: 1) *Preprocessing time* --- the time required to provide G with a structure suitable for search; 2) *Space* --- the storage used for constructing and representing the search structure of G; and 3) *Search time* --- the time required to locate a query point, using the search structure.

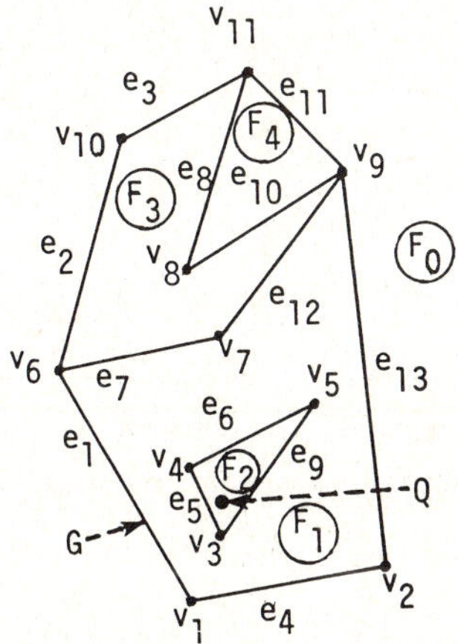

Fig.4.1. A straight-line planar graph G and a query point Q

From the theoretical point of view, optimal algorithms with respect to the worst-case complexity are known [Lipton-Tarjan 1977], [Kirkpatrick 1983]. Table 4.1 shows the theoretical characteristics of some representative algorithms. From the practical point of view, however, it does not seem that much effort has been made to estimate the practical efficiencies of these point-location algorithms, although, in many practical areas such as geographical information processing, computer graphics and VLSI CAD, simple and practically efficient algorithms are needed. Most existing algorithms are too complicated to implement, because of highly sophisticated data structures devised to suit only the theoretical efficiency. Furthermore, theoretically optimal algorithms are not necessarily efficient in practice. For example, the algorithm proposed in [Lipton-Tarjan 1977], which is based on

Table 4.1. Theoretical efficiencies of the typical point-location algorithms

Algorithm		Search Time	Space	Preprocessing Time
[Dobkin-Lipton 1976]		$O(\log n)$	$O(n^2)$	$O(n^2 \log n)$
[Shamos 1975]		$O(\log n)$	$O(n^2)$	$O(n^2)$
[Lee-Preparata 1977]		$O((\log n)^2)$	$O(n)$	$O(n \log n)$
[Lipton-Tarjan 1977]		$O(\log n)$	$O(n)$	$O(n \log n)$
[Preparata 1981]		$O(\log n)$	$O(n \log n)$	$O(n \log n)$
[Kirkpatrick 1983]		$O(\log n)$	$O(n)$	$O(n \log n)$
Bucket Method	(average)	$O(1)$	$O(n)$	$O(n)$
	(worst)	$O(n)$	$O(n\sqrt{n})$	$O(n\sqrt{n})$

the "planar separator theorem" [Lipton-Tarjan 1980], is theoretically optimal, but it is never practical because it is algorithmically extremely complicated, as the authors themselves describe in [Lipton-Tarjan 1977]. On the other hand, Kirkpatrick's algorithm is not only theoretically optimal but also widely believed to be practical for its *conceptual* simplicity [Kirkpatrick 1983]. However, the constant factor in the leading term of its complexity function seems to be so large that it cannot be said to be actually practical.

In this section, an algorithm using buckets which has $O(1)$ search time, $O(n)$ space, and $O(n)$ preprocessing time for almost all graphs is described according to [Edahiro-Kokubo-Asano 1983b].

In the following we assume that a straight-line planar graph G, with n vertices and e edges, is represented by the data structure in $O(n)$ space which is now widely accepted as standard (see, e.g., [Preparata 1981], [Iri et al. 1983], etc.).

4.1. Construction of buckets

Suppose a given graph G is drawn within a rectangular frame R with sides parallel to the axes. The frame R is partitioned into $n_x \times n_y$ rectangular buckets of equal size (Fig.4.2). A bucket which is the i-th from left and the j-th from bottom is denoted by B_{ij}, and the subgraph of G lying in B_{ij} is denoted by G'_{ij} (Fig.4.3). The *bucket coordinate system* is the coordinate system obtained by rescaling the x- and y-axis such that the left-bottom corner of bucket B_{ij} has coordinates (i,j). The bucket to which a query point belongs is easily found by taking the integer parts of the bucket coordinates of the point. n_x and n_y are determined by the formulas:

$$n_x = \lfloor \alpha_x \sqrt{n} \rfloor,\quad n_y = \lfloor \alpha_y \sqrt{n} \rfloor \quad \text{and} \quad \alpha = \sqrt{\alpha_x \alpha_y}, \tag{4.1}$$

with appropriately chosen parameters α_x and α_y, where α is called the *partition parameter* (see Section 4.4). The number of buckets thus constructed is obviously $O(n)$, and consequently, the number of vertices of G in a bucket is $O(1)$ on the average. Moreover, if each bucket contains $O(1)$ edges of G, we have only to look into a planar subgraph G'_{ij} of size $O(1)$ in the bucket containing the query point in order to find the region of G in which it

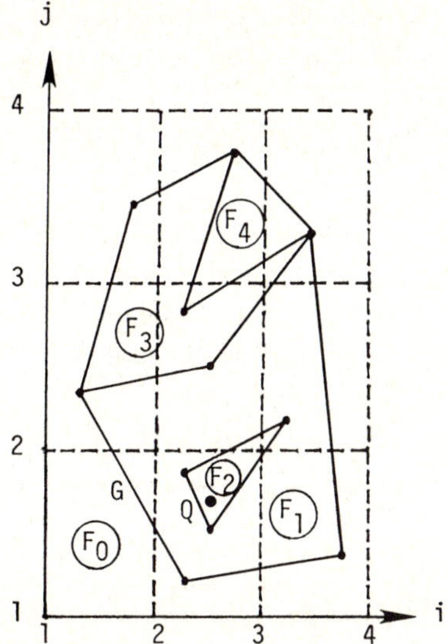

Fig.4.2. Partition into buckets and the bucket coordinate system

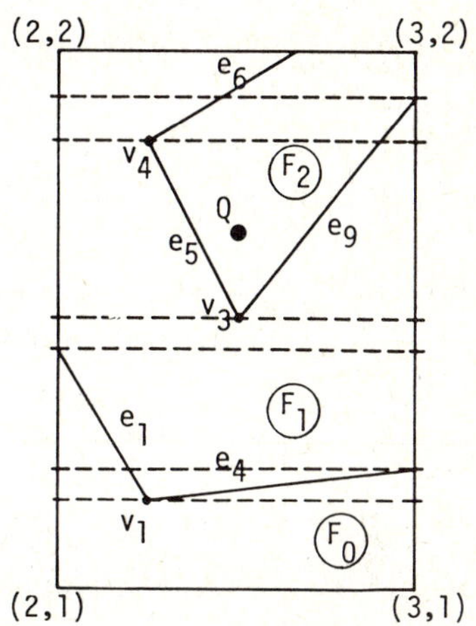

Fig.4.3. G'_{21} and the associated division of bucket B_{21} into slabs

lies. This may be done in O(1) time.

4.2. Preprocessing

At the preprocessing stage, the following data are prepared for each bucket B_{ij}.

N_{ij} : the list of vertices of G'_{ij} (sorted by their y-coordinates).

H_{ij} : the set of edges of G'_{ij} intersecting the upper horizontal side of B_{ij} (not sorted).

V_{ij} : the list of edges of G'_{ij} which intersect the left vertical side of B_{ij} (sorted by the y-coordinates of the intersection points).

F_{ij} : the region of the subdivision by G containing the upper-left corner of B_{ij}.

For example, for bucket B_{21} (Fig.4.2) for the graph of Fig.4.1, they are $N_{21} \to v_4 \to v_3 \to v_1$, $H_{21}=\{e_6\}$, $V_{21} \to e_1$, $V_{31} \to e_9 \to e_4$, and $F_{21}=F_1$ (see Fig.4.3).

These data for the buckets can be constructed in $O(n)$ time on the average and $O(n\sqrt{n})$ time in the worst case. The space required for these data is $O(n)$ in the average case and $O(n\sqrt{n})$ in the worst case.

4.3. Search

The algorithm incorporates the slab technique initiated in [Dobkin-Lipton 1976] and [Shamos 1975]. Note, however, that the slabs are not constructed at the preprocessing stage but at the search stage as they become necessary. Suppose it is known that a query point lies in bucket B_{ij}. Draw a horizontal line through each of the vertices of subgraph G'_{ij} of G in B_{ij}, as shown in Fig.4.3. The bucket B_{ij} is divided into several slabs by those lines.

A slab is further divided into trapezoids (some of which are triangles) by edges of G'_{ij}. It is evident from the construction of the slabs that no two edges of G'_{ij} intersect within a slab, so that the edges intersecting a slab are totally ordered from left to right within the slab.

The location of the query point may therefore be determined as follows.

Step 1. (To form the set S of edges crossing the slab that contains the query point Q, and find the region f containing the left vertical boundary of the slab.) In the bucket B_{ij} containing the query point Q, let $V_{ij}(\mathrm{Q})$ be the set of those edges in V_{ij} which intersect the left side of B_{ij} at a point with the y-coordinate greater than that of Q, and let $N_{ij}(\mathrm{Q})$ be the set of edges having exactly one of the endpoints in N_{ij} that lies above Q. Then

$$S := H_{ij} \oplus V_{ij}(\mathrm{Q}) \oplus N_{ij}(\mathrm{Q}) \oplus V_{i+1,j}(\mathrm{Q}), \tag{4.2}$$

where $A \oplus B = (A-B) \cup (B-A)$ denotes the symmetric difference of two sets A and B. Furthermore, if $V_{ij}(\mathrm{Q}) \neq \varnothing$, f is the lower region of the lowest edge of $V_{ij}(\mathrm{Q})$, and otherwise f is F_{ij}.

For the example of Fig.4.1, $H_{21}=\{e_6\}$, $V_{21}(\mathrm{Q})=\varnothing$, $V_{31}(\mathrm{Q})=\{e_9\}$, $N_{21}(\mathrm{Q})=\{e_6, e_5\}$, $S=\{e_5, e_9\}$, and f is F_{21} because $V_{21}(\mathrm{Q})$ is empty (Fig.4.4).

Step 2. (To determine the region in which the query point Q lies.) If Q lies to the left of the leftmost edge of S, Q is in the region f. Otherwise Q is in the region to the right of the rightmost among those edges of S lying to the left of Q.

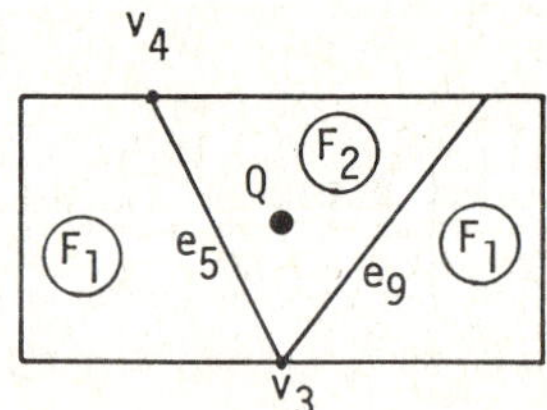

Fig.4.4. Edges in the slab that contains Q

For example, Q in Fig.4.1 is located in F_2, which lies to the right of e_5 (Fig.4.4).

This algorithm runs in time of the order same as the size of the planar subgraph G'_{ij}, so that the search time is O(1) in the average case and O(n) in the worst case.

4.4. Computational experiments

Practical efficiencies of existing algorithms and the bucket algorithm explained in the above were examined by computational experiments on HITAC M-280H (VOS3/JSS4, Optimizing FORTRAN77, OPT=2) at the Computer Centre of the University of Tokyo. The test graphs used are as follows, where n, e and f are, respectively, the number of vertices, that of edges and that of regions of a graph.

1) Voronoi diagrams (n=258~4098, $n=2f$, $e=3f-2$; n points scattered randomly in the unit square in the plane).

Fig.4.5. Map of the administrative sectors of the Kanto district

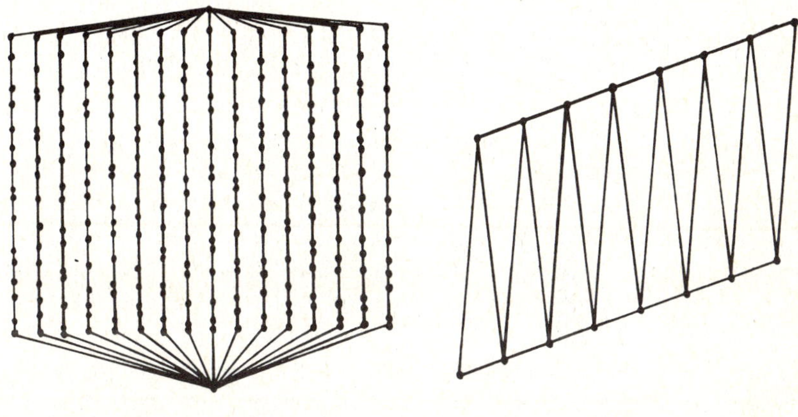

(a) special-shaped graph A (n=258) (b) special-shaped graph B (n=16)

Fig.4.6. Examples of special-shaped graphs

2) Map of the administrative sectors of the Kanto district of Japan (n=2422, e=2711, f=473; Fig.4.5).
3) Special-shaped graphs (Fig.4.6):
   a) graphs of type A (n=258~4098, $n=(f-1)^2+2$, $e=f(f-1)$),
   b) graphs of type B (n=256~4096, $n=f$, $e=2f-3$).

All the graphs have the unit horizontal width and the unit vertical height and query

Table 4.2. Experimental results for the map of the administrative sectors of the Kanto district

Algorithm		Search Time (μs)	Space/n (words)	Preprocessing Time (μs)
[Lee-Preparata 1977]		54.77	10.95	524.5
[Preparata 1981]		43.55	48.36	577.3
[Kirkpatrick 1983]		107.53	61.04	721.5
Bucket Method	$\alpha=1.0$	20.22	10.18	22.9
	$\alpha=2.0$	14.86	24.86	33.3
	$\alpha=3.0$	13.29	47.51	51.0
	$\alpha=\sqrt{e/n}$	19.66	11.23	23.3

points were generated by the uniform distribution in the unit square. The partition parameter α was chosen from $\{1,2,3,\sqrt{e/n}\}$. The average of the search times was taken for ten times as many query points as vertices for each graph and for each parameter value.

The results for the map of the administrative sectors of the Kanto district are shown in Table 4.2. The results for the other graphs are given in Figs.4.7-4.9, where Fig.4.7 shows the search time, Fig.4.8 the trade-off between the search time and the preprocessing time, and Fig.4.9 the trade-off between the search time and the space. With respect to search, the bucket algorithm ran faster than the other algorithms for all graphs, and the theoretical asymptotic complexities in Table 4.1 are experimentally realized for all the algorithms (Fig.4.7). In particular, the bucket algorithm was found to run in O(1) time even for special-shaped graphs of type B. This illustrates the robustness of the bucket algorithm. In preprocessing, the bucket algorithm was the fastest (Fig.4.8). For space, Lee and Preparata's algorithm and the bucket algorithm with $\alpha=1$ are the best (Fig.4.9). The effect of the choice of α is more remarkable for space than for preprocessing time.

From the computational experiments, it may be said that the bucket algorithm is the fastest for almost all graphs, and robust against variations of the types of graphs. The robustness of the bucket algorithm is realized by an appropriate choice of the parameters. Parameters α_x and α_y are to be chosen in such a way that they satisfy $\alpha_x/\alpha_y=S_y/S_x$, where S_x (resp., S_y) is the sum of lengths of edges projected to the x-axis (resp., y-axis), and $\alpha=\sqrt{\alpha_x\alpha_y}$ is to be set equal to $\sqrt{e/n}$.

More details on the bucket algorithm and its properties are found in [Edahiro-Kokubo-Asano 1983b].

5. Range-Search Problems

For a given set of n points in the plane and an arbitrary given query polygon, the range-search problem in the plane is to find the subset of the points lying inside the polygon. It is also a problem of fundamental importance in geographical information processing and in VLSI CAD. Here, it is assumed that, numerous queries are raised to a fixed

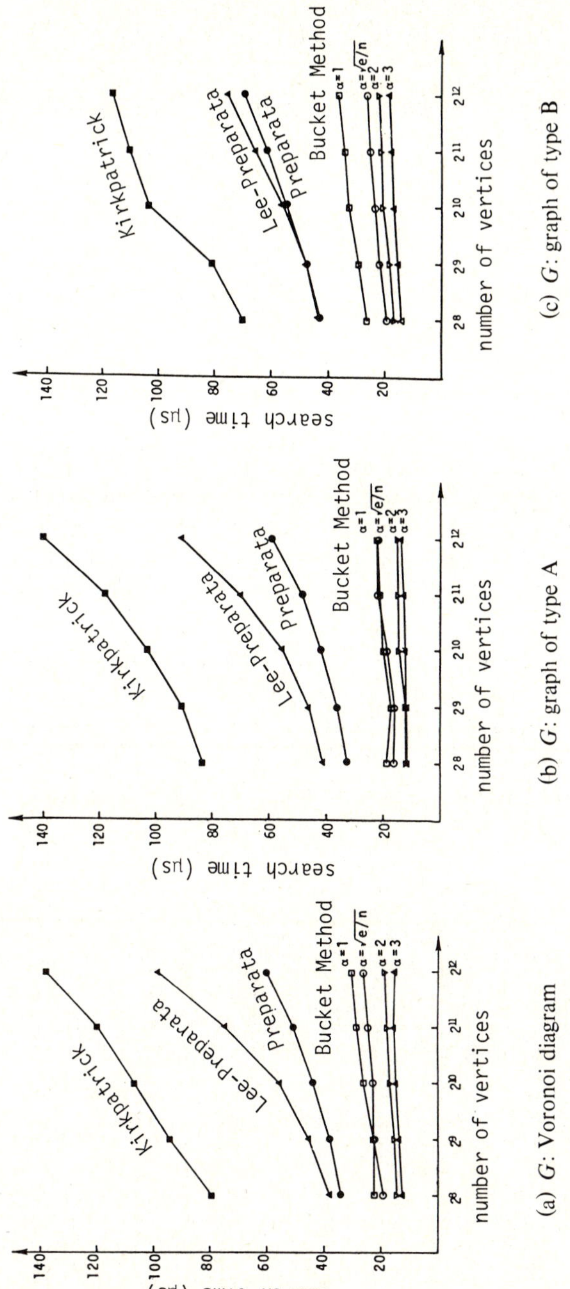

(a) G: Voronoi diagram

(b) G: graph of type A

(c) G: graph of type B

Fig.4.7. Comparison of the search times of various point-location algorithms

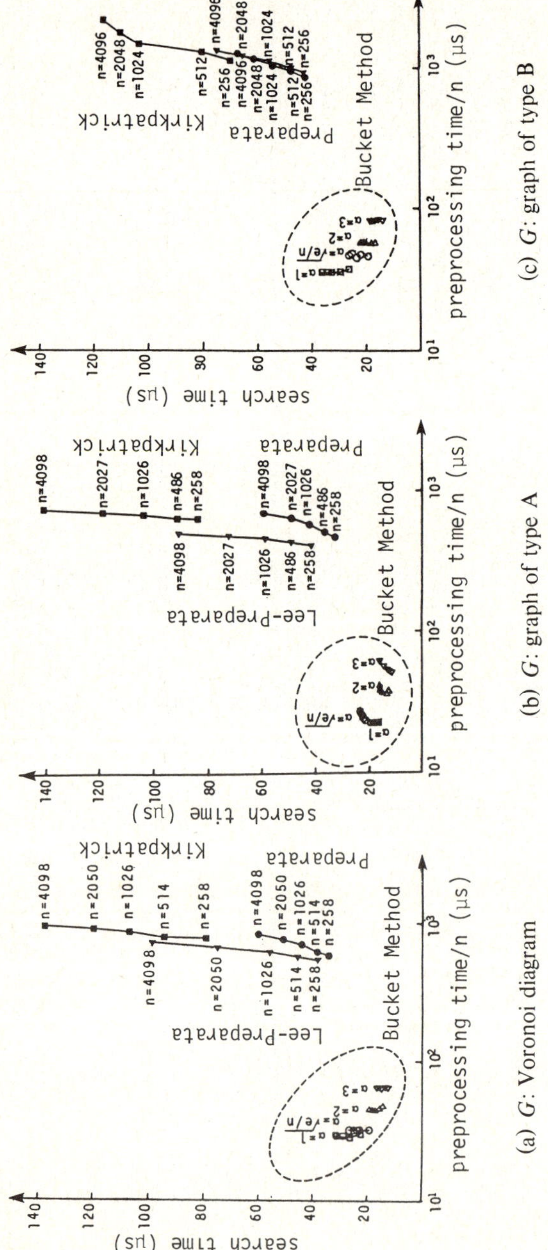

(a) G: Voronoi diagram

(b) G: graph of type A

(c) G: graph of type B

Fig.4.8. Trade-off between search time and preprocessing time in point-location algorithms

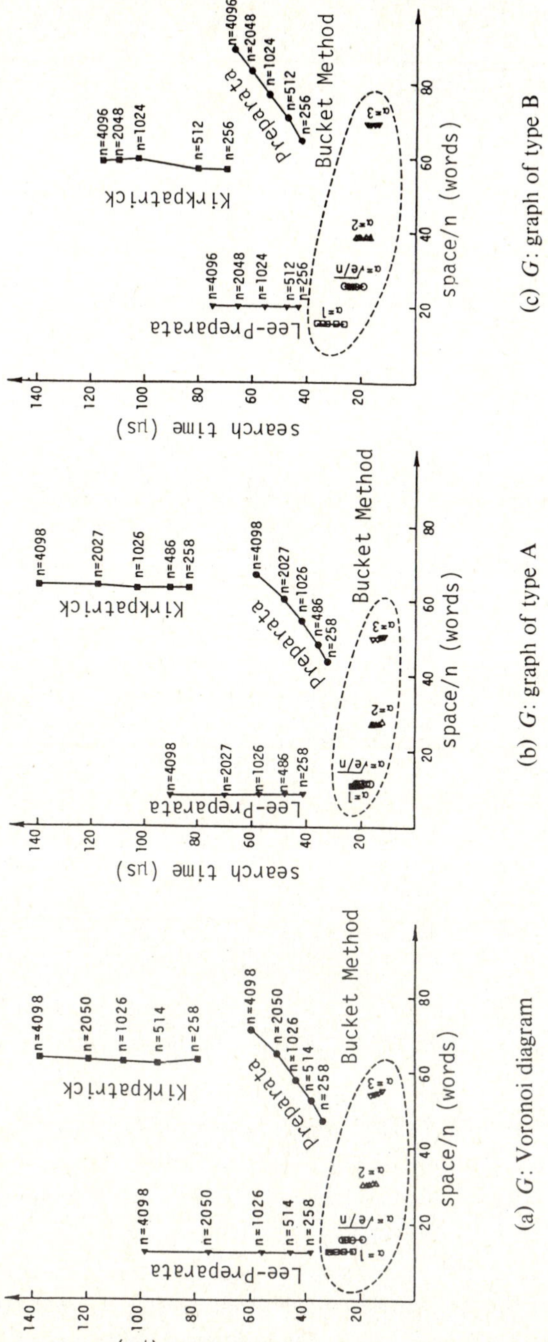

Fig.4.9. Trade-off between search time and space in point-location algorithms

Table 5.1. Efficiencies of range-search algorithms

Algorithm		Search Time	Space	Preprocessing Time
[Willard 1982]	(average)	$O(mn^{\log_6 4}+Q)$	$O(n)$	$O(n(\log n)^2)$
	(worst)	$O(mn^{\log_6 4}+Q)$	$O(n)$	$O(n^2)$
Bucket Method	(average)	$O(m+L\sqrt{n}+Q)$*	$O(n)$	$O(n)$
	(worst)	$O(mn+Q)$	$O(n)$	$O(n \log n)$

* See Section 5.3 for the definition of L

set of points. As in the point-location problem, a range-search algorithm is characterized in terms of the following three attributes: 1) *Preprocessing time* --- the time required to provide the set of points with a structure suitable for search; 2) *Space* --- the storage used for constructing and representing the search structure of the points; and 3) *Search time* --- the time required to enumerate the points lying in the query polygon, using the search structure.

Many algorithms have been proposed for rectilinearly oriented rectangle or hyper-rectangle queries in the plane or in higher-dimensional space [Bentley 1980], [Bentley-Friedman 1979], [Bentley-Maurer 1980], [Finkel-Bentley 1974], [Lee-Wong 1980b], [Lueker 1978], [Willard 1982]. However, only a few results have been found for polygon or polyhedron queries [Willard 1982]. The algorithm proposed by D. E. Willard [1982] uses the structure called the "J-way division" of the plane, which is the division of the plane with straight lines into n regions such that each region has exactly one of the given points. It solves the problem, in $O(n(\log n)^2)$ preprocessing time on the average, in $O(n)$ space and in $O(mn^{\log_6 4}+Q)$ search time, where m is the number of vertices of the query polygon, Q is the number of points inside the polygon and $\log_6 4 \simeq 0.7737$. However, it seems practically too complicated to implement.

In this section, we outline a new simple algorithm to be called the bucket algorithm. Although its worst-case performance in search is a little worse than that of Willard (see Table 5.1), it will be more useful in practice in view of its simplicity and the better average-case performance. The idea of this bucket algorithm was first proposed in [Edahiro-Iri-Asano 1984a]. More details will be reported elsewhere in the near future [Edahiro-Iri-Asano 1984b].

5.1. Basic idea

Let $V=\{v_1,v_2,\ldots,v_n\}$ be a given set of n points and R be a polygonal region in the plane. The region R is represented by the list of m directed arcs $B=e_1 \rightarrow e_2 \rightarrow \ldots \rightarrow e_m$ which trace the boundary of R in the positive direction (Fig.5.1(a)). We construct arbitrarily an auxiliary graph $G(V,E)$ with V as the vertex set and with an arbitrarily defined edge set E (Fig.5.1(b)). Then we execute the following procedures using the graph G.

1° *Marking*: For each arc e_k of B, we search for the edges in E which intersect e_k. Suppose that an edge (v_i, v_j) in E intersects e_k at a point c and v_i lies to the left of e_k. Then we mark v_i "inside" if and only if no intersection with other arcs of B has yet

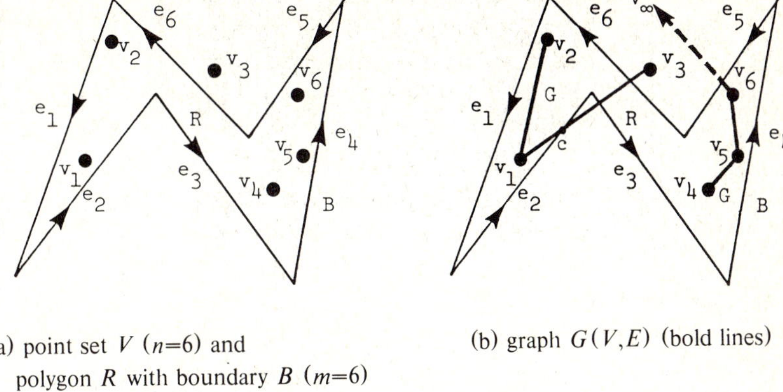

(a) point set V $(n=6)$ and polygon R with boundary B $(m=6)$

(b) graph $G(V,E)$ (bold lines)

Fig.5.1. Range search and the artificial auxiliary graph introduced on the points (Conceptual illustration)

been found on the segment (v_i,c) of the edge (v_i,v_j). Similarly, mark v_j "outside" if and only if no intersection has been found on the segment (c,v_j) yet.

2° *Enumerating*: Let E_1 (resp., E_2) be the subset of those edges in E which intersect (resp., do not intersect) B. We enumerate all the vertices reachable from points marked "inside" by tracing edges of E_2.

Since, having traced all the arcs of B, we can obtain the correct inside-outside answer for both of the two endpoints of every edge in E_1, the above procedure gives the correct solution to the problem if at least one edge in every connected component of G intersects B.

In Fig.5.1(b), no marking takes place when we have checked e_1; v_1 is marked "inside" and v_3 is marked "outside" when we have checked e_2; v_3 is marked "inside" when we have checked e_3; and then v_3 is again marked "outside" when we have checked all the arcs of B. Point v_2 is located "inside" at the enumeration stage because it is reachable from v_1 through edge (v_1,v_2) which belongs to E_2. Points v_4, v_5 and v_6 are not marked because the component of G containing them does not intersect B. However, if we augmented G with an edge extending to a point at infinity, v_∞, as shown by a broken line in Fig.5.1(b), we should be able to have the correct answer.

What is important in this method is how to efficiently find the intersection of arcs of B and edges of E. A bucketing technique is used for that purpose, as will be described in the following section.

5.2. Preprocessing

Let T be a rectangle (desirably, the smallest), with horizontal and vertical sides, containing all the n points of $V=\{v_1,v_2,...,v_n\}$. First, we divide T into αn rectangles, called

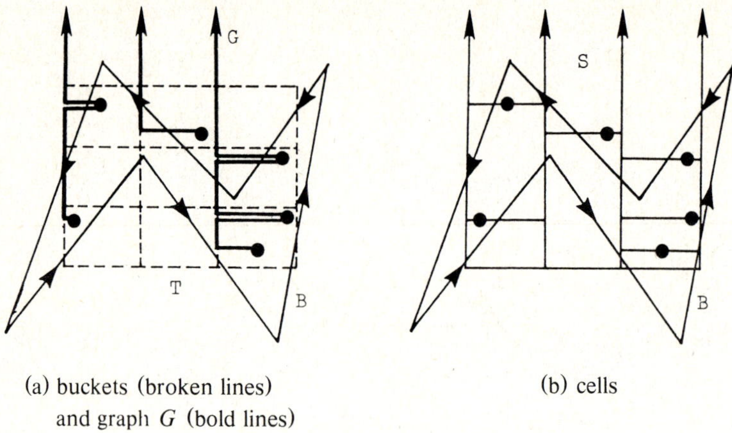

(a) buckets (broken lines) and graph G (bold lines)

(b) cells

Fig.5.2. Buckets, cells and the auxiliary graph

buckets (Fig.5.2(a)), where α is a parameter independent of n. Then, we construct *cells* which are the regions of the subdivision of the plane defined by (i) the lower horizontal side of T, (ii) the upwards semi-infinite vertical lines containing the vertical sides of the buckets, and (iii) the horizontal line segments which pass through points of V and each of which extends between the two vertical sides of the bucket containing the point of V (Fig.5.2(b)). Furthermore, we define S as the region obtained from T by extending it upwards (i.e., the union of all cells), and construct a graph G with $O(\sqrt{n})$ connected components by connecting the points of V lying in the buckets of the same x-coordinate by polygonal-line edges in the order of increasing y-coordinates, and extending the edge starting from the uppermost point infinitely upwards, where each edge is composed of straight-line segments which lie on the vertical boundaries of the buckets and those which horizontally connect a point of V with the left vertical side of its bucket (Fig.5.2(a)). G is represented by a doubly-linked list. Extra memory space of $(2+\alpha)n$ words is needed for G, and it takes $O(n)$ time in the average case, and $O(n \log n)$ in the worst case, to construct the cells and the graph G.

5.3. Search

If G is a specially structured graph defined in Section 5.2, every intersection point of B with G lies on the boundaries of cells. Therefore, in order to find the intersection of each arc e_k of B with G, we may look for the intersection of e_k with cell boundaries, which is simple enough since we may first find the cell to which the initial point of e_k belongs (this can be done in $O(1)$ time on the average using buckets) and then traverse along e_k the cells it intersects. If, in the current cell, e_k intersects the lower or upper horizontal side of the cell, we can find the next cell in a constant time by means of the doubly-linked list representing G. If e_k intersects a vertical side, we can find the next cell in $O(1)$ time on the

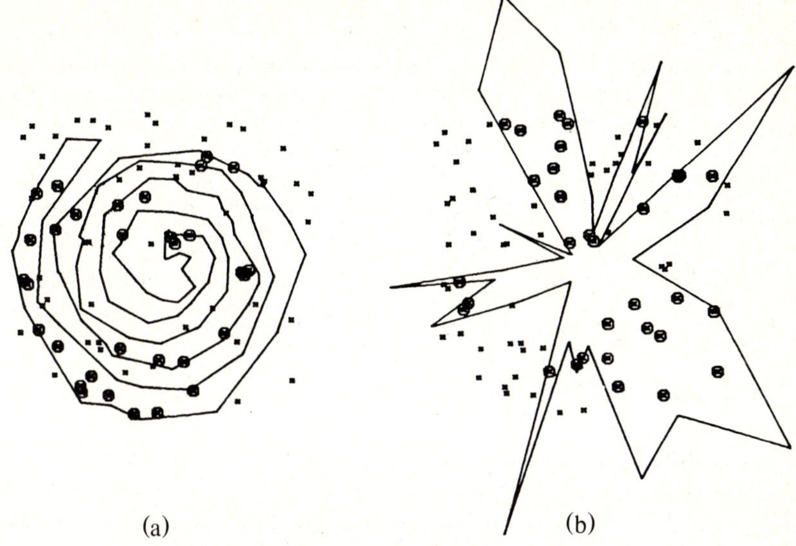

Fig.5.3. Examples of (a) a spiral polygon (m=100) and (b) a star-shaped polygon (m=40) (The point set is the same in (a) and in (b).)

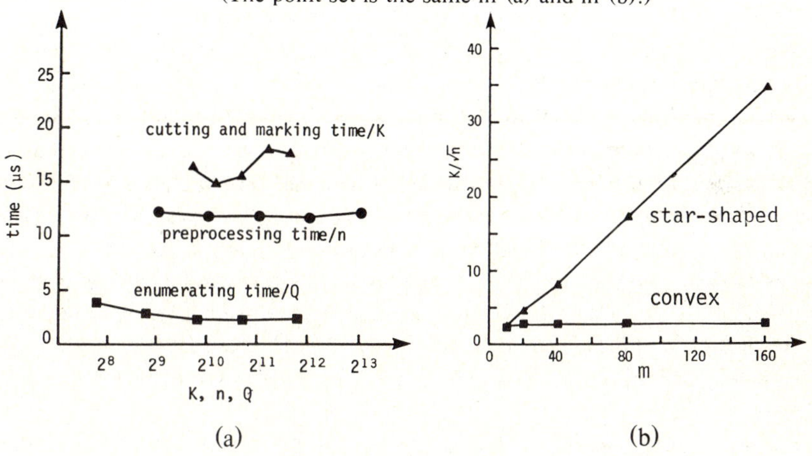

Fig.5.4. Experimental results for the bucket method
(a) Preprocessing time, time for cutting and marking in Steps 1° and 2°, and time for enumerating in Step 3° for a star-shaped polygon (m=160)
(b) Number of points of intersection of star-shaped polygons and convex polygons with the auxiliary graph G for n=8192 points

average again by means of buckets.

The coordinates of a point of intersection of e_k and G can be calculated with at most one multiplication if the slopes of arcs are calculated in advance. In particular, when every arc of B is horizontal or vertical, as is usually the case in VLSI CAD, only comparisons are required.

In summary, the stage of search is performed in the following three steps.

1° Calculate the slopes of arcs of B, cut off that part of each arc which is outside of S, and determine the cell containing the initial point of the remaining part of each arc.

2° Find all the intersections of B and G, and perform "marking" as described in Section 5.1.

3° Enumerate all the points inside R as described in Section 5.1.

Step 1° can be performed in $O(m)$ time in the average case, and in $O(mn)$ time in the worst case. If we define

$$L=\sum_i (l_x(e_i \cap S)+l_y(e_i \cap T)), \tag{5.1}$$

where $l_x(\cdot)$ (resp., $l_y(\cdot)$) is the length in the x- (resp., y-)direction when T is normalized into the unit square by rescaling the x- and y-axis, Step 2° can be performed in $O(L\sqrt{n})$ time in the average case, and in $O(mn)$ time in the worst case. Step 3° can be performed in time proportional to Q, the number of points inside R. Therefore, the search requires $O(m+L\sqrt{n}+Q)$ in the average case, and $O(mn+Q)$ in the worst case.

5.4. Computational experiments

Computational experiments were performed in a manner similar to those in Section 4. Test points were randomly uniformly distributed in the unit square, whereas queries were convex polygons, star-shaped polygons, spiral polygons and rectilinear polygons (Fig.5.3). The results for star-shaped polygons are shown in Fig.5.4. The results for other polygons were quite similar to those for the star-shaped polygons. As is seen from these results, the preprocessing time, the marking time and the time for enumerating the required points are all proportional to the number of relevant objects, i.e., the number of given points (denoted by n), that of intersection points of B and G (denoted by K) and that of the required points (denoted by Q), respectively, where the constants of proportionality are all quite small.

We conclude this section with a few comments. This algorithm can easily be modified so as to be capable of treating also queries of polygonal regions with holes and those consisting of several connected components. Furthermore, it can be extended to the "dynamic" problem where points are added and deleted from time to time.

6. The Shortest-Path Problem

The shortest-path problem is probably the most practical, important, and useful problem in many applications concerning networks. The algorithm which have been developed so far can be classified roughly into two approaches: the label-correcting method (the so-called potential method) and the label-setting method (Dijkstra's method [Dijkstra 1959]). Both methods have many variants.

6.1. The label-correcting method

Let $G(V,A)$ be a directed graph with vertex set V and arc set A, and let $N=(G,d,s)$ be a network with the graph G, a length function $d: A \rightarrow \mathbf{R}_+$ (positive reals) and a vertex $s(\in V)$ prescribed as the *entrance*. We assume all the vertices of G are reachable from s through directed paths. The shortest-path problem on N considered here is to find the shortest paths from the entrance s to all the other vertices in network N.

Most algorithms for the shortest-path problem make use of a *label* $(a(v),p(v))$ attached to a vertex v to indicate that a path has been found from s to v such that its length is $p(v)$ and the last arc on this path is $a(v)$ (of which v is the terminal vertex), where $p(v)$ is called the *potential* of vertex v. The vertices with "infinite" label $(*,\infty)$ are said to be "*unlabelled*" whereas the other vertices are said to be "*labelled*".

The label-correcting method may generally be stated as follows.

1° Attach the special label $(*,0)$ to s, and the "infinite" label $(*,\infty)$ to all the other vertices. All the vertices are defined to be "unsearched".

2° If there is no "labelled and unsearched" vertex, then halts (then the shortest paths have been found); otherwise, take such a vertex v and go to 3°.

3° Search vertex v; that is, for each arc a in $\delta^+ v$ (the set of arcs going out of v), if $p(\partial^- a) > p(v)+d(a)$ (where $\partial^- a$ is the terminal vertex of a), then give the vertex $\partial^- a$ the label $(a,p(v)+d(a))$, and make the vertex $\partial^- a$ "unsearched". Having searched v, make v "searched", and return to 2°.

The efficiency of the label-correcting method depends upon the order of searching "labelled and unsearched" vertices in 2°. Among variants of the label-correcting method, it is said [Dial-Glover-Karney-Klingman 1979] that the two-way sequence method [Pape 1974] is very efficient.

6.2. The label-setting method

The label-setting method may generally be stated as follows.

1° Attach the special "tentative" label $(*,0)$ to s, and give all the other vertices "infinite" label $(*,\infty)$.

2° If all the vertices already have permanent labels, then halt (then the shortest paths have been found); otherwise, find the vertex v which has the smallest potential among all the vertices with tentative labels, and go to 3°.

3° Make the label of vertex v "permanent"; for each arc a going out of v, if $p(\partial^- a) > p(v)+d(a)$, give the vertex $\partial^- a$ the tentative label $(a,p(v)+d(a))$; return to 2°.

The efficiency of the label-setting method depends substantially upon the implementation of that part of algorithm by which we find the smallest-potential vertex to be searched next in 2°. The method now widely believed to be efficient is to maintain the set of vertices with tentative labels using a heap or some other kind of priority queue [Aho-Hopcroft-Ullman 1974]. A method using a (1-level or 2-level) bucket system in [Denardo-Fox 1979] also belongs to this class.

6.3. 1-level bucket method

The 1-level bucket method proposed in [Denardo-Fox 1979] is a variant of the label-setting method. We set $m=\min\{d(a)|\,a\in A\}$ and $M=\max\{d(a)|\,a\in A\}$ (note that $m>0$), and $x=\lceil M/m\rceil+1$, where $\lceil z\rceil$ denotes the smallest integer that is not smaller than z. Noting that the length of each shortest path from s is at most $M(|V|-1)$, we divide the interval from 0 to $M(|V|-1)$ into $x(|V|-1)$ subintervals (called "buckets") of width m, and maintain the set of vertices with tentative labels in the x buckets according to their potentials as follows.

A vertex v with a tentative label $(a(v),p(v))$ is put in the k-th bucket such that $km\leqslant p(v)<(k+1)m$. A set of vertices in each bucket is linked with one another by a two-way list (in an arbitrary order) having a head pointer. In order to find the vertex of the minimum-potential tentative label, we have only to find the smallest-numbered nonempty bucket (a bucket is nonempty if it contains a vertex). Since the bucket width is m, all the tentative labels of the vertices in the smallest-numbered nonempty bucket can be made permanent simultaneously. Moreover, when the potential of a vertex is updated so that it is to be moved to another bucket, we can perform the necessary modification of the two-way lists of the relevant buckets in a constant time. The 1-level bucket method thus can find the shortest paths in $O(|A|+|V|x)$ time. The number of buckets necessary for the computation is $x(|V|-1)$, which is ordinarily prohibitively large. However, noting that the difference of the maximum and the minimum of potentials of tentative labels at any stage of computation is at most M, we can do everything with only x buckets of width m by putting vertex v with potential $p(v)$ in the $(\lceil p(v)/m\rceil \bmod x)$-th bucket, which reduces the total space requirement to $O(|A|+x)$.

In [Denardo-Fox 1979], sophistication of the 1-level bucket method to the 2-level is also proposed.

6.4. Variable bucket method

The 1-level bucket method has the disadvantage that the necessary computation time and memory space depend not only on the topology of the network but substantially on the lengths of arcs in the network. For example, the 1-level bucket method is quite inefficient in the case where there is a very short arc and, at the same time, a very long shortest path. To mitigate this disadvantage, the variable bucket method proposed in [Imai-Iri 1982,1984] uses buckets whose widths vary according to the circumstances but whose number is fixed a priori by the topology of the network.

Suppose that there are k vertices with tentative labels at a stage and that their potentials range from t_l to t_u. Among those k vertices, we maintain the vertices whose potentials are between t_l and t (an appropriately chosen threshold value) by buckets of width w, where t and w are defined, with given parameters α and β, by

$$t=(1-\alpha)t_l+\alpha t_u \quad (\alpha>0), \qquad w=\max\{m, \beta(t-t_l)/k\}. \tag{6.1}$$

If the width w of buckets is greater than m, we cannot change the tentative labels of all the vertices in the smallest-numbered nonempty bucket into permanent ones simultaneously, but we have to solve a *subproblem* for those vertices (this subproblem can ordinarily be solved quite efficiently), as will be explained below. During the course of computation, there appear vertices with tentative labels whose potentials are greater than t. They are kept in a list called the "*overflow-list*". When all the labels with potentials less than t have become permanent, i.e., when all the vertices with tentative labels are in the overflow-list, we reconstruct the buckets with respect to the number k' of vertices in the overflow-list, and the minimum t_l' and the maximum t_u' of the potentials of those vertices, and iterate the above procedure until all the vertices come to have permanent labels. Concerning parameters α and β, the computational experiments [Imai-Iri 1982,1984] show that the choice $\alpha=\beta=1.0$ seems good and that the efficiency of the algorithm is not so sensitive to this choice. In the case of $\alpha\geqslant 1$, the total number of buckets which the variable bucket method requires is at most $|V|/\beta$, independent of the lengths of arcs, which is not the case with the 1-level bucket method.

For the subproblem within a bucket, we may use either of the label-correcting method and the label-setting method. If the label-correcting method is employed for that purpose, the variable bucket method can be regarded as a method which, so to speak, suppresses the instability of the label-correcting method with slight extra work (note that the defect of the label-correcting method is its non-robustness compared with the label-setting method). If the label-setting method is employed, the variable bucket method speeds up finding the minimum-potential vertex by means of buckets.

6.5. Computational experiments

In [Imai-Iri 1982,1984], typical algorithms for the shortest-path problems were tested by computational experiments (on HITAC M-280H, FORTRAN 77), where artificial networks, such as random networks and grid networks, and three real road maps in Japan were used as test networks. The subproblems in the variable bucket method were solved by the label-correcting method using a two-way sequence. Here, we shall pick up from among them the results for the grid networks and the road network of the Kanto district in Japan.

A grid network is a regular $k\times k$ square grid graph with arc distances randomly distributed from 1 to 100. The results are as shown in Fig.6.1.

The road network of the Kanto district as a graph has the same topology as that of Fig.2.3(a), but has 6922 vertices and 10112 arcs with the vertices of degree 2 suppressed. The arc lengths range from 37m to 31506m, where there are 296 arcs longer than 5km. On this network, the following four types of problems, which are different from one another in arc lengths, were considered.

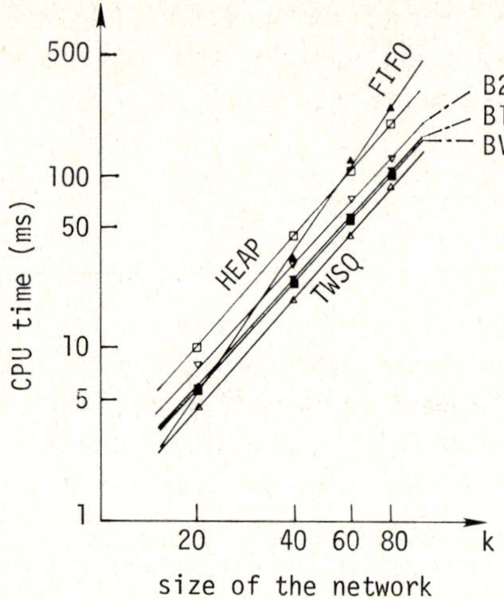

Fig.6.1. Running times for $k \times k$ grid networks (k^2 vertices)

HEAP: label-setting method using a heap
TWSQ: label-correcting method using a two-way sequence
FIFO : label-correcting method using a first-in-first-out queue
B1 : 1-level bucket method
B2 : 2-level bucket method
BV : variable bucket method

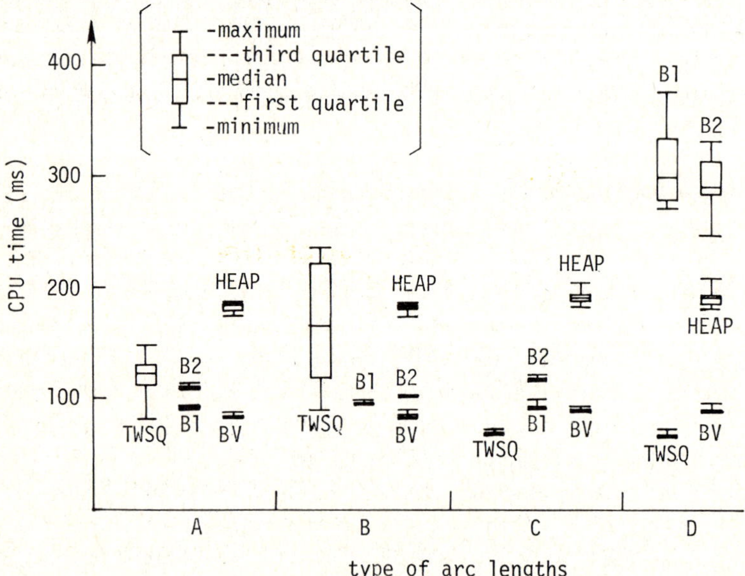

Fig.6.2. Running times for the road network of the Kanto district

(A) The real road lengths are adopted as they are.

(B) The lengths of the above-mentioned 296 long arcs are multiplied by 5.

(C) Random integers between 1 to 100 are adopted for the arc lengths.

(D) Random integers between 1 to 10000 are adopted for the arc lengths.

For each case, 20 different problems with randomly chosen entrances are solved. The distribution of the running times are shown in Fig.6.2.

The variable bucket method is thus seen to be not only nearly as efficient as the best of the widely used methods in each case but it is also robust for a large variety of networks.

7. Possibility of the Use of Nonrectangular Buckets

We have so far considered only *rectangular* buckets. However, there is theoretically the possibility of using buckets of other shapes such as triangles and hexagons. In this section, we shall briefly report some preliminary results on the use of triangular buckets. Certainly, squares or rectangles are the simplest shape for tiling the plane, and, probably, triangles are the next simplest. In fact, the unit square, say, can be divided by a diagonal into two congruent isosceles right triangles, and an isosceles right triangle can be halved into similar triangles recursively. The M-stage subdivision of the unit square of this kind will give rise to $N=2^M$ small isosceles right triangles, which we may use for buckets. To determine the bucket to which a point belongs is a little more complicated than in the case of rectangular buckets, but is still practically feasible. (We need not only the integer parts of the products of the coordinates of a point by the number of subdivisions of a side of the unit square but also have to examine the sign of either the sum or the difference of their fractional parts.)

A natural order to be defined on the set of buckets would be the one illustrated in Fig.7.1 for the case M=7, which might be called the Sierpiński order because the curve obtained by connecting the centers of gravity of consecutive buckets tends to the space

Fig.7.1. Subdivision of the square into triangular buckets with the Sierpiński order

filling curve famous after the name of W. Sierpiński as $M \to \infty$. The Sierpiński order can be determined in $O(N)$ time by slightly modifying the "folding-over" algorithm in Section 3.2 for numbering rectangular buckets. Of course, constructing and numbering triangular buckets takes a few times as much time as constructing and numbering rectangular buckets. Quite recently, the applicability of triangular buckets was examined, mainly experimentally, for the minimum-weight perfect matchings (cf. Section 2) and for the Voronoi diagrams (cf. Section 3) in [Sanae 1984].

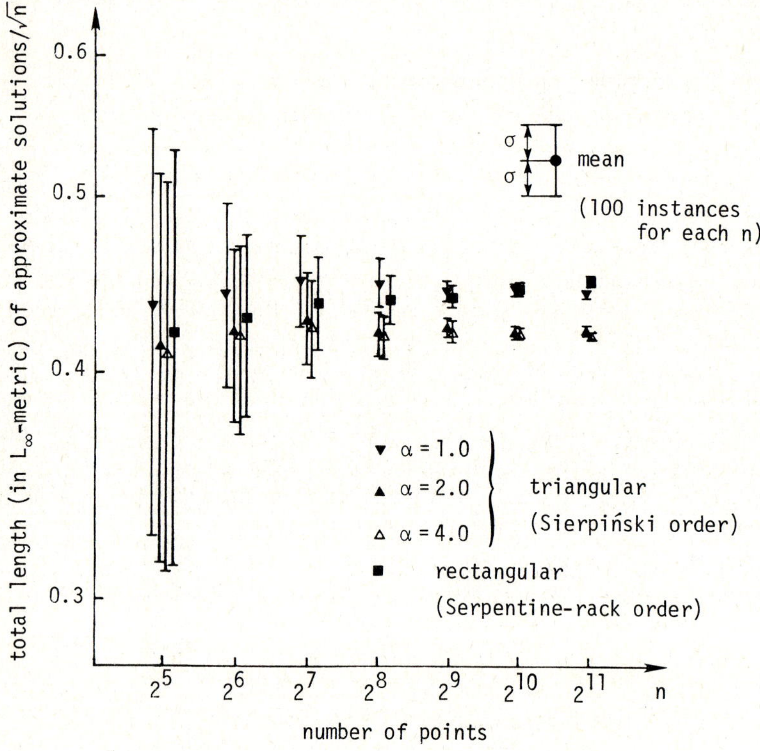

Fig.7.2. Comparison of the total lengths (in L_∞-metric) of approximate solutions to the minimum-weight perfect matching problem by various buckets

The application of triangular buckets with the Sierpiński order to the minimum-weight perfect matching problem will be self-evident. Fig.7.2 shows the distribution of the total lengths of approximate solutions to the minimum-weight perfect matchings for randomly uniformly scattered points by means of various buckets, where it is observed that, the finer the triangular buckets are, the better are the approximations obtained --- at the sacrifice of the increase in computation time, of course. Practically, the choice of α=2.0 ($N=\alpha^2 n$; n: number of points, N: number of buckets) will be the best in view of the trade-off between the quality of the solution and the necessary computation time (in the case of α=2.0, the

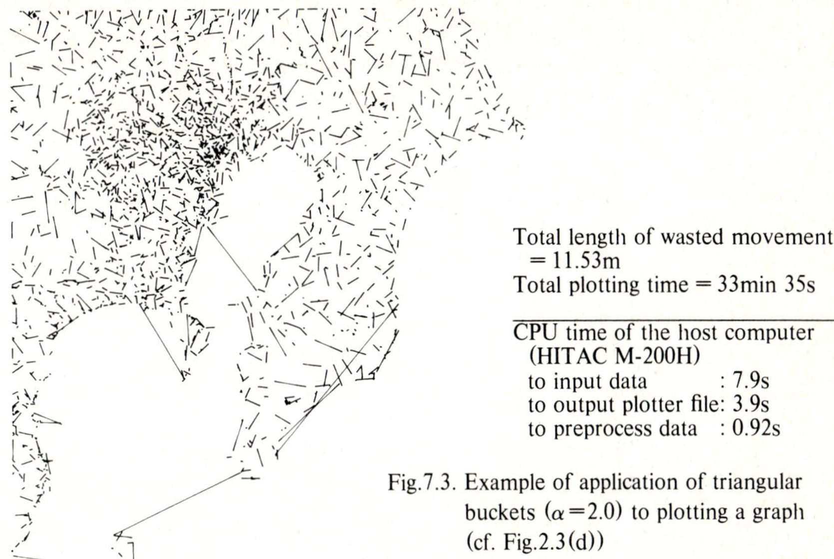

Fig.7.3. Example of application of triangular buckets ($\alpha=2.0$) to plotting a graph (cf. Fig.2.3(d))

computation time is about twice as large as that for the best rectangular buckets; cf. Section 2). It is worth noting that, at the limit of $\alpha=\infty$, this bucket solution coincides with that obtained by J. J. Bartholdi and L. K. Platzman's method which directly uses the Sierpiński space-filling curve [Bartholdi-Platzman 1983]. In Fig.7.3, one of the many experiments of applying the triangular buckets with the Sierpiński order to the plotting problem is shown, where the graph drawn is the same as that in Fig.2.3(a), and Fig.7.3 is to be compared with Fig.2.3(d). The experiments show that the triangular buckets afford the results of nearly the same quality as those obtained with the rectangular buckets, and that the necessary computation time is slightly greater for the former than for the latter.

It is as natural to build a binary tree on the triangular buckets as it is to build a quaternary tree on the rectangular buckets. In fact, the quaternary incremental algorithm using rectangular buckets in Section 3 can easily be modified to the "binary" incremental algorithm using triangular buckets for constructing the Voronoi diagram. However, there are several alternative methods of re-scrambling the bucket order to improve the efficiency of the algorithm. The performance, i.e., the speed, of the binary incremental algorithm for the Voronoi diagram is, as far as the computational experiments so far done show, largely the same as that of the quaternary incremental, and no significant difference has yet been found.

More details about the use of triangular buckets will be published elsewhere before long [Imai-Sanae-Iri 1984].

Acknowledgements

The authors owe much of the results included in this paper to their collaborators. Especially, the authors' former colleagues, Messrs. S. Matsui and T. Ohya (now in Central Research Institute of Electric Power Industry), did essential contribution to the development of basic ideas and algorithms for the matching and plotting, and those for the Voronoi diagrams. The cooperation of Messrs. A. Suzuki, I. Kokubo, H. Segawa and H. Sanae should also be greatly acknowledged. Many kinds of geographical data for computational experiments became available by the courtesy of the Geographical Survey Institute of the Ministry of Construction of the Government of Japan. Discussions with Professor D. Avis of McGill University, Professor T. Koshizuka of the University of Tsukuba and Professor A. Okabe of the University of Tokyo have been the constant source of motivation and stimulus for the authors. Last but not least, the authors are grateful to the editor of this book, Professor G. T. Toussaint of McGill University, for kindly encouraging them to write this paper and suggesting many improvements on the original manuscript.

References

Aho, A. V., Hopcroft, J. E., and Ullman, J. D. (1974): *The Design and Analysis of Computer Algorithms.* Addison-Wesley, Reading, Massachusetts.

Avis, D. (1983): A Survey of Heuristics for the Weighted Matching Problem. *Networks,* Vol.13, pp.475-493.

Avis, D., and Bhattacharya, B. K. (1983): Algorithms for Computing *d*-Dimensional Voronoi Diagrams and Their Duals. *Advances in Computing Research* (F. P. Preparata, ed.), Vol.1, JAI Press, London, pp.159-180.

Bartholdi, J. J., III, and Platzman, L. K. (1982): An $O(N \log N)$ Planar Travelling Salesman Heuristic Based on Spacefilling Curves. *Operations Research Letters,* Vol.1, pp.121-125.

Bartholdi, J. J., III, and Platzman, L. K. (1983): A Fast Heuristic Based on Spacefilling Curves for Minimum-Weight Matching in the Plane. *Information Processing Letters,* Vol.17, pp.177-180.

Bentley, J. L. (1980): Multidimensional Divide-and-Conquer. *Communications of the ACM,* Vol.23, pp.214-229.

Bentley, J. L., and Friedman, J. H. (1979): Data Structures for Range Searching. *Computing Surveys,* Vol.11, pp.397-409.

Bentley, J. L., and Maurer, H. A. (1980): Efficient Worst-Case Data Structures for Range Searching. *Acta Informatica,* Vol.13, pp.155-168.

Bentley, J. L., Weide, B. W., and Yao, A. C. (1980): Optimal Expected-Time Algorithms for Closest Point Problems. *ACM Transactions on Mathematical Software,* Vol.6, pp.563-580.

Denardo, E. V., and Fox, B. L. (1979): Shortest-Route Methods: 1. Reaching, Pruning, and Buckets. *Operations Research,* Vol.27, No.1, pp.161-186.

Devroye, L. (1981a): On the Average Complexity of Some Bucketing Algorithms. *Computers and Mathematics with Applications,* Vol.7, pp.407-412.

Devroye, L. (1981b): Recent Results on the Average Time Behavior of Some Algorithms in Computational Geometry. *Computer Science and Statistics* (W. F. Eddy, ed.), Proceedings of the 13th Symposium on Interface, Springer-Verlag, pp.76-82.

Dial, R., Glover, F., Karney, D., and Klingman, D. (1979): A Computational Analysis of Alternative Algorithms and Labeling Techniques for Finding Shortest Path Trees.

Networks, Vol.9, No.3, pp.215-248.

Dijkstra, E. W. (1959): A Note on Two Problems in Connexion with Graphs. *Numerische Mathematik*, Vol.1, pp.269-271.

Dobkin, D., and Lipton, R. J. (1976): Multidimensional Searching Problems. *SIAM Journal on Computing*, Vol.5, pp.181-186.

Edahiro, M., Iri, M., and Asano, T. (1984a): An Algorithm for Range-Search Problem and Its Practical Efficiency (in Japanese). *Proceedings of the 1984 Spring Conference of the Operations Research Society of Japan*, 1-C-7, pp.115-116.

Edahiro, M., Iri, M., and Asano, T. (1984b): An Algorithm for Range-Search Problem and Its Practical Efficiency (An extended English version of [Edahiro-Iri-Asano 1984a]). In preparation.

Edahiro, M., Kokubo, I., and Asano, T. (1983a): A New Point-Location Algorithm with Buckets and Its Practical Efficiency (in Japanese). *Proceedings of the 1983 Autumn Conference of the Operations Research Society of Japan*, B-7, pp.71-72.

Edahiro, M., Kokubo, I., and Asano, T. (1983b): A New Point-Location Algorithm and Its Practical Efficiency --- Comparison with Existing Algorithms. *Research Memorandum RMI* 83-04, Department of Mathematical Engineering and Instrumentation Physics, University of Tokyo.

Finkel, R. A., and Bentley, J. L. (1974): Quad-Trees: A Data Structure for Retrieval on Composite Keys. *Acta Informatica*, Vol.4, pp.1-9.

Fox, B. L. (1978): Data Structures and Computer Science Techniques in Operations Research. *Operations Research*, Vol.26, pp.686-717.

Green, P. J., and Sibson, R. (1978): Computing Dirichlet Tessellation in the Plane. *The Computer Journal*, Vol.21, pp.168-173.

Horspool, R. N. (1979): Constructing the Voronoi Diagram in the Plane. *Technical Report SOCS* 79.12, McGill Univeristy, Montreal, Canada.

Imai, H. (1982): OPLOT User's Manual. *Research Memorandum RMI* 82-05, Department of Mathematical Engineering and Instrumentation Physics, University of Tokyo.

Imai, H., and Iri, M. (1982): A New Shortest-Path Algorithm and Its Practical Efficiency (in Japanese). *Proceedings of the 1982 Spring Conference of the Operations Research Society of Japan*, 2C-2, pp.183-184.

Imai, H., and Iri, M. (1984): Practical Efficiencies of Existing Shortest-Path Algorithms and a New Bucket Algorithm. *Journal of the Operations Research Society of Japan*, Vol.27, No.1, pp.43-58.

Imai, H., Iri, M., and Murota, K. (1984): Voronoi Diagram in the Laguerre Geometry and Its Applications. To appear in *SIAM Journal on Computing*.

Imai, H., Sanae, H., and Iri, M. (1984): Use of Triangular Buckets in Computational Geometry. In preparation.

Iri, M., Murota, K., and Matsui, S. (1981a): Linear-Time Approximation Algorithms for Finding the Minimum-Weight Perfect Matching on a Plane. *Information Processing Letters*, Vol.12, pp.206-209.

Iri, M., Murota, K., and Matsui, S. (1981a): Linear-Time Approximation Algorithms for Finding the Minimum-Weight Perfect Matching on a Plane. *Information Processing Letters*, Vol.12, pp.206-209.

Iri, M., Murota, K., and Matsui, S. (1981b): Linear-Time Heuristics for the Minimum-Weight Perfect Matching on a Plane with an Application to the Plotter Algorithm. *Research Memorandum RMI* 81-07, Department of Mathematical Engineering and Instrumentation Physics, University of Tokyo.

Iri, M., Murota, K., and Matsui, S. (1982): An Approximate Solution for the Problem of Optimizing the Plotter Pen Movement. *Proceedings of the 10th IFIP Conference on System Modeling and Optimization*, 1981, New York, Lecture Notes in Control and Information Science 38, "System Modeling and Optimization" (R. F. Drenick and F. Kozin, eds.), Springer-Verlag, pp.572-580.

Iri, M., Murota, K., and Matsui, S. (1983): Heuristics for Planar Minimum-Weight Perfect Matching. *Networks*, Vol.13, pp.67-92.

Iri, M., Murota, K., and Ohya, T. (1984): A Fast Voronoi-Diagram Algorithm with Applications to Geographical Optimization Problem. *Proceedings of the 11th IFIP Conference on System Modelling and Optimization*, 1983, Copenhagen, Lecture Notes in Control and Information Science 59, "System Modelling and Optimization" (P. Thoft-Christensen, ed.), Springer-Verlag, pp.273-288.

Iri, M., and Taguchi, A. (1980): The Determination of the Pen-Movement of an XY-Plotter and Its Computational Complexity (in Japanese). *Proceedings of the 1980 Spring Conference of the Operations Research Society of Japan*, P-8, pp.204-205.

Iri, M., et al. (1983): Fundamental Algorithms for Geographic Data Processing (in Japanese). *Technical Report* TR-83-01, Operations Research Society of Japan.

Kirkpatrick, D. (1983): Optimal Search in Planar Subdivisions. *SIAM Journal on Computing*, Vol.12, pp.28-35.

Lawler, E. L. (1976): *Combinatorial Optimization: Networks and Matroids.* Holt, Rinehart and Winston, New York.

Lee, D. T. (1976): On Finding k-Nearest Neighbours in the Plane. *Technical Report*, Department of Computer Science, University of Illinois.

Lee, D. T. (1980): Two-Dimensional Voronoi Diagram in the L_p-Metric. *Journal of the Association for Computing Machinery*, Vol.27, pp.604-618.

Lee, D. T., and Drysdale, R. L., III (1981): Generalization of Voronoi Diagrams in the Plane. *SIAM Journal on Computing*, Vol.10, pp.73-87.

Lee, D. T., and Preparata, F. P. (1977): Location of a Point in a Planar Subdivision and Its Applications. *SIAM Journal on Computing*, Vol.6, No.3, pp.594-606.

Lee, D. T., and Wong, C. K. (1980a): Voronoi Diagrams in L_1 (L_∞) Metrics with 2-Dimensional Storage Applications. *SIAM Journal on Computing*, Vol.9, pp.200-211.

Lee, D. T., and Wong, C. K. (1980b): Quintary Trees: A File Structure for Multidimensional Data Base Systems. *ACM Transactions on Database Systems*, Vol.5, pp.339-353.

Lipton, R. J., and Tarjan, R. E. (1977): Applications of Planar Separator Theorem.

Proceedings of the 18th IEEE Annual Symposium on Foundations of Computer Science, San Juan, pp.162-170.

Lipton, R. J., and Tarjan, R. E. (1980): A Separator Theorem for Planar Graphs. *SIAM Journal on Applied Mathematics*, Vol.36, pp.177-189.

Lueker, G. (1978): A Data Structure for Orthogonal Range Queries. *Proceedings of the 19th IEEE Annual Symposium on Foundations of Computer Science*, Ann Arbor, pp.28-34.

Matsui, S., Murota, K., and Iri, M. (1981): A Utility Program for Determining the Plotter Pen Movement (in Japanese). *Proceedings of the 22nd Annual Conference of the Information Processing Society of Japan*, 1H-4, pp.973-974.

Ohya, T., Iri, M., and Murota, K. (1984a): A Fast Voronoi-Diagram Algorithm with Quaternary Tree Bucketing. *Information Processing Letters*, Vol.18, No.4 (1984), pp.227-231.

Ohya, T., Iri, M., and Murota, K. (1984b): Improvements of the Incremental Method for the Voronoi Diagram with Computational Comparison of Various Algorithms. To appear in *Journal of the Operations Research Society of Japan*.

Papadimitriou, C. H. (1977): The Probabilistic Analysis of Matching Heuristics. *Proceedings of the 15th Annual Allerton Conference on Communication, Control and Computing*, pp.368-378.

Pape, U. (1974): Implementation and Efficiency of Moore-Algorithms for the Shortest Route Problem. *Mathematical Programming*, Vol.7, pp.212-222.

Preparata, F. P. (1981): A New Approach to Planar Point Location. *SIAM Journal on Computing*, Vol.10 pp.473-483.

Reingold, E. M., and Tarjan, R. E. (1981): On a Greedy Heuristic for Complete Matching. *SIAM Journal on Computing*, Vol.10, pp.676-681.

Sanae, H. (1984): *Applications of Triangular Bucketing Techniques to Computational Geometry* (in Japanese). Graduation Thesis, Department of Mathematical Engineering and Instrumentation Physics, University of Tokyo.

Shamos, M. I. (1975): Geometric Complexity. *Proceedings of the 7th ACM Annual Symposium on Theory of Computing*, Albuquerque, New Mexico, pp.224-233.

Shamos, M. I., and Hoey, D. (1975): Closest-Point Problems. *Proceedings of the 16th IEEE Annual Symposium on Foundations of Computer Science*, Berkeley, pp.151-162.

Supowit, K. J., Plaisted, D. A., and Reingold, E. M. (1980): Heuristics for Weighted Perfect Matching. *Proceedings of the 12th ACM Annual Symposium on Theory of Computing*, Los Angels, pp.398-414.

Suzuki, A., Iri, M., and Murota, K. (1983): Geographical Optimization Problems and Their Solutions II (in Japanese). *Proceedings of the 1983 Autumn Conference of the Operations Research Society of Japan*, B-8, pp.73-74.

Toussaint, G. T. (1980): Pattern Recognition and Geometrical Complexity. *Proceedings of the 5th IEEE International Conference on Pattern Recognition*, Miami Beach, pp.1324-1347.

Toussaint, G. T. (1982): Computational Geometric Problems in Pattern Recognition. *Pattern Recognition Theory and Applications* (J. Kittler, K. S. Fu and L. F. Pau, eds.), D. Reidel Publishing Company, pp.73-91.

Willard, D. E. (1982): Polygon Retrieval. *SIAM Journal on Computing*, Vol.11, pp.149-165.

COMPUTATIONAL GEOMETRY, G.T. Toussaint (Editor)
© Elsevier Science Publishers B.V. (North Holland), 1985

Minimum Decompositions of Polygonal Objects

J. Mark KEIL

Department of Computational Science,
University of Saskatchewan,
Saskatoon,Canada, S7N 0W0

Jörg-R. SACK

School of Computer Science,
Carleton University,
Ottawa, Canada, K1S 5B6

1. Introduction

In artificial intelligence, image processing, and pattern recognition, objects are frequently described by polygons. These polygons and their properties are the center of attention of the rapidly growing field of computational geometry. Using computational geometry techniques one is often able to obtain elegant and efficient algorithms to solve certain geometrical problems. In some cases lower bounds have been derived giving us ideas about the intrinsic complexity of problems. Geometrical problems occur frequently in such areas as pattern recognition, image processing, computer graphics and VLSI. The large quantities of data produced by today's devices require algorithms handling these data to be designed efficiently.

Given a geometrical problem an efficient solution is often dependent on the nature of the objects we are dealing with. We illustrate this point by considering the problem of determining the convex hull of an arbitrary set of points versus that of a simple polygon connecting these points. A simple polygon is ***convex*** if all its internal angles are less than $180°$ as in figure 1.1. The ***convex hull*** of a set of points is defined as the minimum area convex polygon enclosing the points. The vertices of the convex hull are points of the set. The convex hull of a polygon is defined analogously. The convex hull of a set of n points can be determined in O(nlogn) time and this is optimal [4,17,49,53]. However, linear time algorithms for computing the convex hull of a simple polygon exist [7,12,18,25,34,45].

If the objects are given as polygons their specific polygonal properties may also lead to more efficient solutions. Before proceeding we define several such properties. Refer to figure 1.2. A ***polygon*** P is specified by its vertices $(p_1,...,p_n)$ listed in clockwise order. A ***polygonal chain*** $C_{i,j}$ is a sequence of consecutive vertices $p_i,...,p_j$ of P. A polygonal chain is ***monotone*** with respect to a line l if the projections of p_k, k=i,...,j on l are ordered in exactly the same way as the vertices in $C_{i,j}$. A polygon is ***monotone*** if there exists a line l such that the boundary of P can be partitioned into two

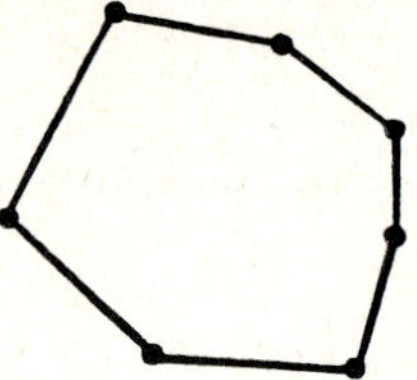

Figure 1.1 A Convex Polygon

polygonal chains $C_{i,j}$ and $C_{j,i}$ which are monotone with respect to l. A point x in P is *visible* from a point y in P, if the open line segment joining x and y lies entirely inside P. A polygon P is ***star-shaped*** if there exists at least one point x inside P from which the entire polygon is visible. A polygon is ***weakly visible*** from an edge pq of P if for every point x in P there exists a point y in pq that is visible from x. A polygon is ***edge-visible*** if there exists an edge from which the polygon is weakly visible. A ***spiral polygon*** is a simple polygon whose boundary chain contains precisely one concave subchain.

It appears that algorithms for such structured polygons are easier to write, thus easier to prove correct, and often exhibit lower space/time complexities. For example, the ***triangulation problem*** is to partition an n-vertex polygon into non-overlapping triangles. Linear-time algorithms for triangulating star-shaped, monotone, edge-visible and convex polygons exist [12,15,48,51,52]. An arbitrary simple polygon can be triangulated in O(nlog(n)) time [10,11,15]. However, this has not yet been shown to be optimal, i.e. no non-trivial lower bound on this problem exists. Other examples of problems for which fast algorithms exist on restricted polygons include point inclusion in convex or star-shaped polygons [49], medial axis of a convex polygon [43], and intersections involving convex or star-shaped polygons [9,35].

The usual strategy for solving many of these problems on general polygons is to decompose the polygon into simple component parts, solve the problem on each component using a specialized algorithm, and then combine the partial solutions [1,15,35,48]. This strategy of decomposing polygonal objects into simpler components is one motivation for the development of efficient ***decomposition techniques.***

These decomposition techniques have also received considerable attention in pattern recognition and image analysis [13,42]. Pattern recognition techniques extract information from an object in order to describe, identify or classify it. A usual strategy for recognizing a general polygonal object is to decompose it into simpler components, then identify the components and their interrelationships and use this information to determine the shape of the object [42]. The types of simpler components that have been considered include triangles[3], rectangles [14],trapezoids [2], convex polygons [13,47], spiral polygons [13], star-shaped polygons [5] and monotone polygons [50].

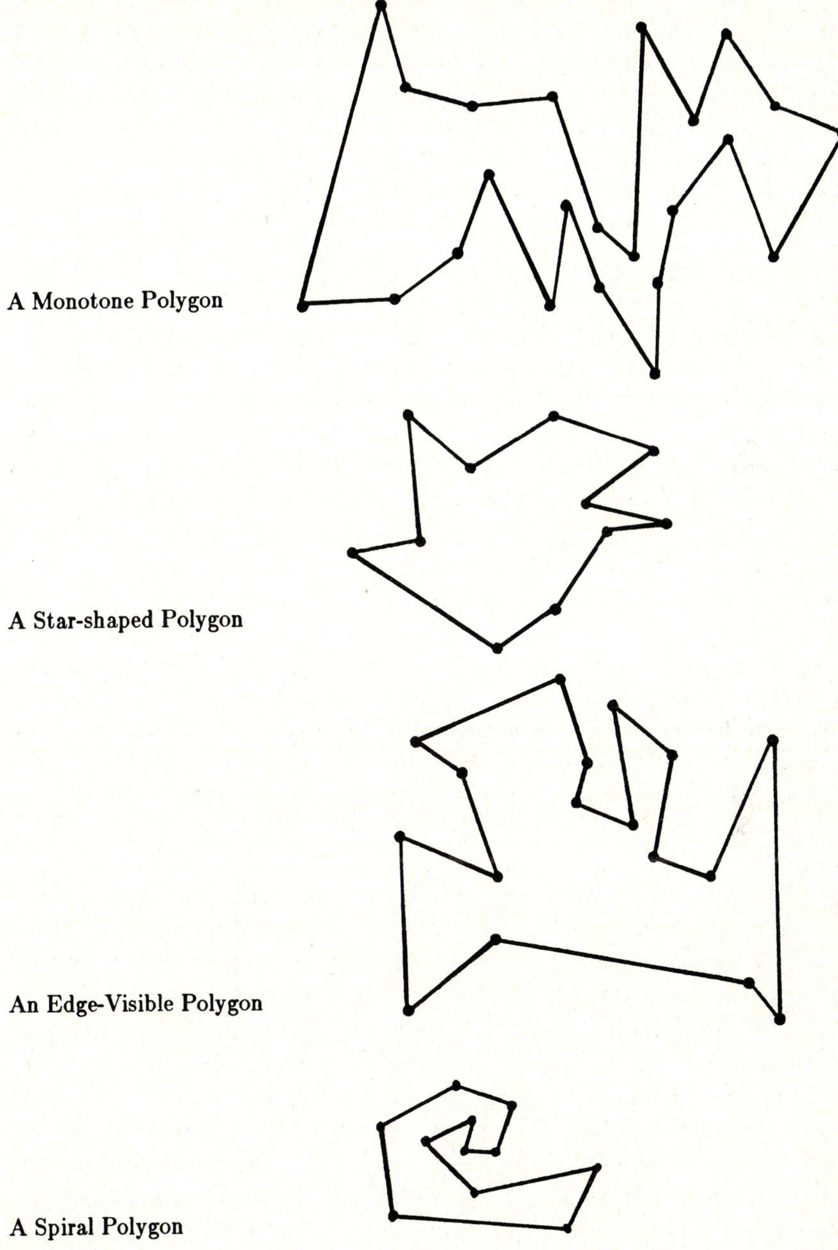
A Monotone Polygon
A Star-shaped Polygon
An Edge-Visible Polygon
A Spiral Polygon

Figure 1.2

We can distinguish between two kinds of decomposition techniques: procedure-directed and component-directed. In a *procedure-directed* decomposition technique, a procedure is specified that decomposes the given object into smaller components. The specific nature of the components may vary. In a *component-directed* approach the object is decomposed into well-defined smaller components. Component-directed techniques are the most common and we will consider only them in this paper.

A decomposition is called a *partitioning*, if the object is decomposed into non-overlapping pieces. Algorithms to partition arbitrary simple polygons into convex, star-shaped, monotone, edge-visible components have been developed. If overlapping pieces are allowed we call the decomposition a *covering*.

Decomposing a polygonal object into simpler component parts can be done with or without introducing additional vertices which are commonly called *Steiner points*. Whether or not Steiner points are allowed often makes a big difference in the solution of a problem. Figure 1.3 shows decompositions of simple polygons into the minimum number of convex polygons with and without Steiner points.

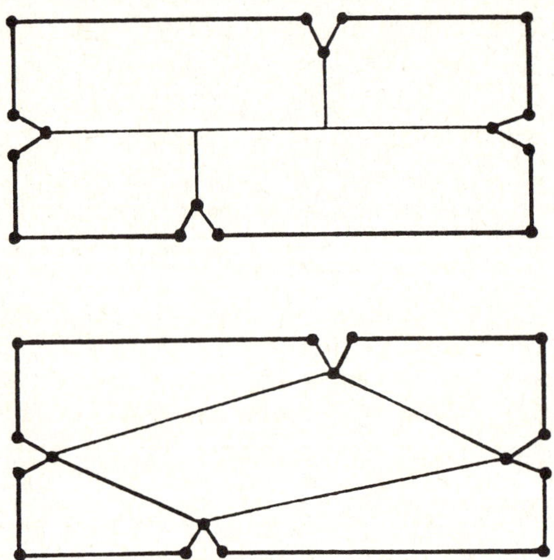

Figure 1.3 Convex Decompositions With and Without Steiner Points

If the interior angle at a vertex is reflex the vertex is called a *notch*. Experimental observation from graphics [36] and pattern recognition [41] shows that, in practice, the number of notches in a polygon is much smaller than the number of vertices. We let N denote the number of notches in a polygon and we describe and analyze decomposition algorithms with respect to both n and N.

In this paper we are primarily concerned with techniques which provide decompositions which are minimal in some sense. We may be interested in decomposing a polygonal object into the minimum number of some component type. In some applications a decomposition that minimizes the total length of the internal edges used to form the decomposition is useful. We measure length in the usual Euclidean metric until section 4 when for technical reasons we switch to the discretized Euclidean metric. A minimum edge length decomposition can be quite different from a minimum number decomposition for the same component type.

Polygonal objects can be further classified according to certain other characteristics. A ***rectilinear polygon*** is a simple polygon whose edges are parallel to two given orthogonal directions. Section 2 considers decompositions of rectilinear polygons. We may encounter polygons that contain holes. Holes are nonoverlapping "islands", simple polygons, inside the main polygon. The complexity of a decomposition problem usually increases if we allow the polygon to contain holes. We consider decompositions of polygons with holes in section 4.

2. Decomposition of Rectilinear Polygons

In many applications the objects encountered are rectilinear polygons. In image processing the boundaries of objects are often stored on a grid which usually implies that digitized images are rectilinear polygons. VLSI designs are also often stored on a grid and typically contain many rectilinear polygons. In this section we consider decompositions of rectilinear polygons, first when Steiner points are allowed and then when Steiner points are disallowed.

2.1. Decompositions allowing Steiner Points

Rectangles are the natural component type when decomposing rectilinear polygons allowing Steiner points. Ferrari, Sankar and Sklansky [14] have developed an algorithm for decomposing a rectilinear polygon into the minimum number of rectangles. Their algorithm is based on the following theorem.

Theorem

A rectilinear polygon can be partitioned into N - L + 1 rectangles, where N is the number of notches and L is the maximum number of nonintersecting chords that can be drawn either vertically or horizontally between two notches.

The theorem allows them to reduce the decomposition problem to that of finding the maximum number of independent vertices in the intersection graph of the vertical or horizontal chords between notches. The graph's matching properties are then used to solve the decomposition problem in $O(N^{5/2})$ time. Lipski [31] relates this decomposition problem to that of finding a Manhatten path and is able to develop an $O(n^{3/2}$ lognloglogn) time solution.

Lingas et al [30] mention process control (stock cutting), automatic layout systems for VLSI (channel definition) and architecture (partitioning into offices) as applications for the problem of decomposing a rectilinear polygon into rectangles while minimizing the total length of the edges used to form the decomposition. To solve this minimum "ink" problem they proceed by inducing a grid over the polygon by extending the edges of the polygon into lines. They then build up partial solutions and use dynamic programming to achieve an $O(n^4)$ time algorithm. If the rectilinear polygon is restricted to the shape of a histogram they are able to provide an $O(n^3)$ time algorithm.

2.2. Decompositions without Steiner Points

Without Steiner points quadrilaterals rather than rectangles become the natural component type for the decomposition of rectilinear polygons. Even though arbitrary simple polygons do not necessarily admit decomposition into convex quadrilaterals Kahn,Klawe and Kleitman [21], in what they refer to as *quadrilaterization*, proved that it is always possible to decompose a rectilinear polygon into convex quadrilaterals without using Steiner points. In the remainder of this section quadrilaterizations are understood to be convex. Sack [44] has recently developed an O(nlogn) time algorithm for quadrilaterizing an arbitrary simple rectilinear polygon. To find a quadrilaterization of an arbitrary simple polygon he begins by partitioning the rectilinear polygon into specific monotone polygons. See figure 2.1. Although arbitrary monotone polygons do not admit quadrilaterizations, these specific monotone polygons do. These monotone polygons, like arbitrary monotone rectilinear polygons, starshaped and edge-visible rectilinear polygons, can be quadrilaterized in linear time [46]. It is an open problem whether an arbitrary rectilinear polygon can be quadrilaterized in less than O(nlogn) time. Since any quadrilaterization of a rectilinear polygon results in the same number of quadrilaterals it is not meaningful to speak of a decomposition into the minimum number of quadrilaterals.

We instead turn to our other minimality criterion and present an $O(n^4)$ time algorithm for finding the minimum edge length quadrilaterization of a rectilinear polygon. Note that a brute force approach would require examining in worst case an exponential number of possible quadrilaterizations for a given rectilinear polygon. See figure 2.2.

The main tool used for solving minimum edge length polygonal decomposition problems is dynamic programming [22,23,24,30] and we demonstrate the use of it here. We let W(i,j) be the weight of the minimum weight quadrilaterization of the subpolygon P_{ij} of P cut off from p_i to p_j (i.e. $p_i,p_{i+1},\ldots,p_j$). The idea is to build up minimum weight quadrilaterizations of larger and larger subpolygons and solve the problem using dynamic programming (see Figure 2.3). Let Q(i,j) store the vertices of the quadrilaterization that contains the diagonal p_ip_j in the minimum weight quadrilaterization of P_{ij}. Define d_{ij} to be the Euclidean distance from p_i to p_j if p_i and p_j are visible from each other and $+\infty$ otherwise.

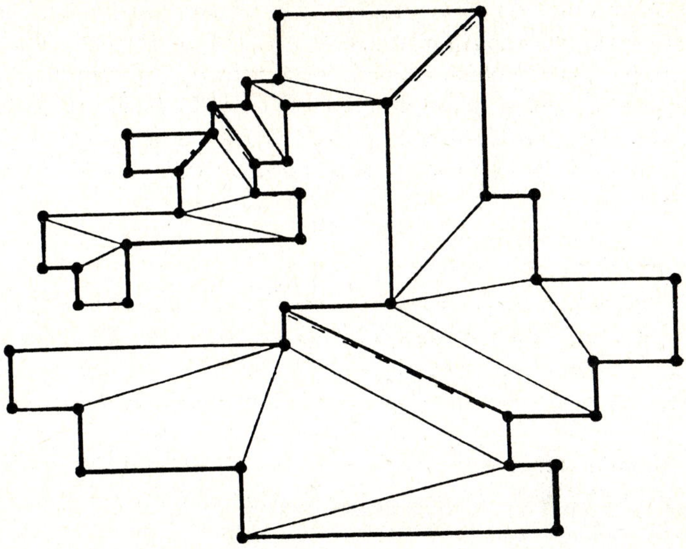

Figure 2.1 A Quadrilaterization. The dashed segments partition the polygon into monotone components.

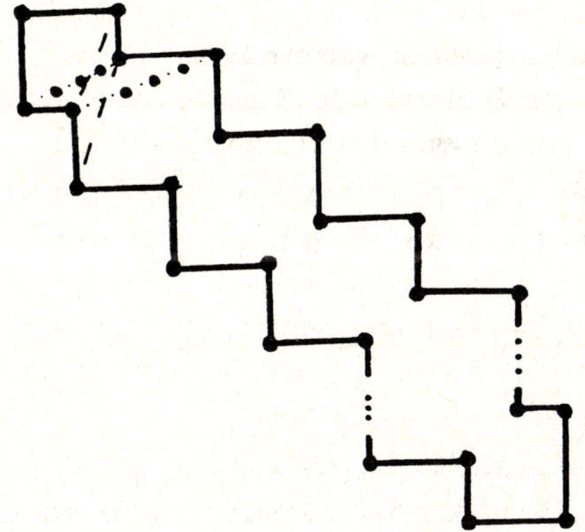

Figure 2.2 There can be an exponential number of quadrilateralizations of a rectilinear polygon.

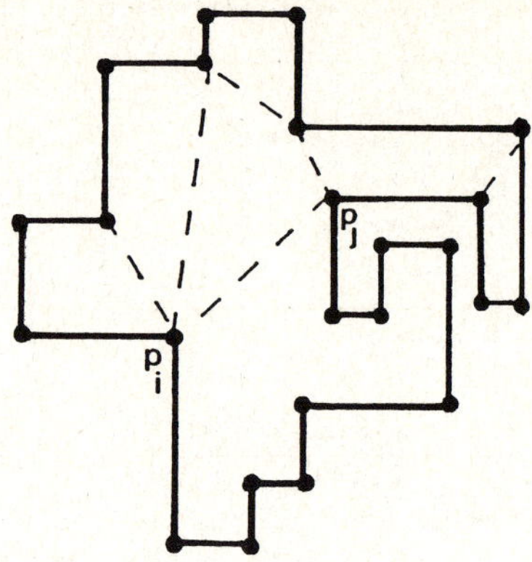

Figure 2.3 A Quadrilaterization of a subpolygon P_{ij}.

Algorithm Minimum Weight Quadrilaterization
Input: **A rectilinear polygon P**
Output: **The weight of the minimum weight quadrilaterizition and the quadrilaterization itself are stored in W(1,n) and Q respectively.**

```
begin
for k := 1 to n-1 do
        for i := 1 to n-k do
        begin
                j := i + k
                begin
                        if k = 1 then W(i,j) := 0
                        else
                        if k is even then W(i,j) := +oo
                        else
                        begin
                                W(i,j) := d_ij + min(W(i,l) + W(l,m) + W(m,j))
                                (* where the minimum is taken over all l and m
                                such that i<l<m<j, and p_i, p_l, p_m, p_j form a convex
                                quadrilateral *)
                                Q(i,j) := (l,m)
                                (* indices of vertices exhibiting the minimum are
                                stored in Q *)
                        end
                end
        end
end.
```

To print out all the diagonals of the minimum weight quadrilaterization call the procedure PRINT(1,n).

Procedure PRINT(i,j)
begin
if p_ip_j is not an edge of P *then* output segment p_ip_j
where Q(i,j) = (l,m) *do*
    call PRINT(i,l); *call* PRINT(l,m); *call* PRINT (m,j)
end.

Complexity Analysis

The values of d_{ij} can be computed in a preprocessing step in $O(n^2)$ time. There are $O(n^2)$ W(i,j) to be calculated. Each of these requires examining $O(n^2)$ pairs of candidates for p_l and p_m. The total run time of the algorithm is therefore $O(n^4)$.

3. Decomposition of Simple Polygons

In this section we consider the decomposition of arbitrary simple polygons, first when Steiner points are allowed and then when Steiner points are disallowed.

3.1. Decompositions allowing Steiner Points

The first result in this section concerns the partitioning of a polygonal region into a minimum number of trapezoids with two horizontal sides. Triangles with a horizontal side are considered to be trapezoids with two horizontal sides one of which is degenerate. This problem is closely related to VLSI artwork data processing systems of Electron-Beam Lithography for VLSI microfabrication. In such systems the layout is stored as a set of polygonal regions per layer which should be partitioned into fundamental figures since the aperture of a pattern generator is restricted. Trapezoids have been used as fundamental figures [2].

Asano and Asano [2] have developed an $O(n^3)$ time algorithm for the problem of partitioning a polygonal region into the minimum number of trapezoids with two horizontal sides. They solve the problem by introducing the concept of minimally effective chords. A chord s of a polygon P is minimally effective if either (i) s joins two edges of P collinear with s, or (ii) s is a horizontal chord joining two vertices of P and containing no other vertices of P. The following theorem provides the basis for their algorithm.

Theorem

The minimum number of trapezoids needed to partition a polygon P is n - h - d - 1, where h is the number of horizontal edges of P and d is the cardinality of a maximum independent set of the intersection graph of the minimally effective chords of P.

They then show that the class of intersection graphs of minimally effective chords of polygons without holes is the class of circle graphs, and use an efficient algorithm for finding a maximum independent set of a circle graph to solve the decomposition problem.

A more difficult problem is to decompose a polygon into the minimum number of convex components. Chazelle and Dobkin [8,9] exhibit an $O(n + N^3)$ time algorithm for this problem. They introduce the concept of an X-pattern from which minimal decompositions may be generated. An Xk-pattern is a particular interconnection of k notches which removes all reflex angles at the k notches and creates no new notches. The idea is to use as many X-patterns as possible. It turns out that determining whether some given notches can be combined to form an X-pattern seems too involved and they introduce a more constained type of interconnection which they call a Y-pattern. Using dynamic programming they are able to construct Y-patterns in polynomial time and with some further refinements achieve the $O(n + N^3)$ time bound.

Decomposition problems allowing Steiner points seem quite difficult. There are no futher results known concerning the decomposition of simple polygons into the minimum number of some component type. Nor are there any results known concerning the decomposition of simple polygons into components while minimizing the total internal edge length when Steiner points are allowed.

3.2. Decompositions without Steiner points

We first consider the decomposition of polygons into the minimum number of components and then the decomposition of polygons while minimizing the total internal edge length.

3.2.1. Minimum Number Decompositions

This section considers four types of fundamental components: convex, spiral, star-shaped and monotone polygons.

3.2.1.1. Convex Decompositions

Of all the polygon decomposition problems, the problem of decomposing a polygon into the minimum number of convex components has received the most attention. However, until recently, no polynomial time exact solution was known to the problem thus motivating the development of approximation algorithms. Feng and Pavlidis [13] describe an $O(N^3 n)$ time algorithm which does not generally yield a minimum decomposition. Schachter's [47] O(Nn) time decomposition algorithm, based on the Delaunay triangulation, also fails to guarantee a minimum decomposition. Chazelle [10] develops an O(nlogn) time divide and conquer algorithm that finds a decomposition that contains less than 4.333.. times the optimum number of components. Recently, Greene [19] has developed an O(nlogn) time algorithm that is guaranteed to find a decomposition that contains less than or equal to 4 times the optimum number of components. Note however that any convex decomposition that does not contain unnecessary edges will be less than of equal to 4 times the minimum decomposition. This is true since each edge can eliminate at most two notches and each notch requires at most two edges to eliminate it in any convex decomposition that does not contain unnecessary edges.

Recently, Greene [19] discovered an $O(N^2n^2)$ time exact algorithm for the convex decomposition problem. Independently, Keil [23] developed an algorithm for the problem that runs in $O(N^2 \text{nlogn})$ time. This result employs a general technique for reducing the size of the state space in a dynamic programming formulation in order to achieve the polynomial time bound. The idea is to build up minimum decompositions of larger and larger subpolygons P_{ij}. In general there can be an exponential number of minimum decompositions of a subpolygon. It turns out that the only edges of a minimum decomposition of a subpolygon P_{ij} that affect the decomposition's ability to form part of the global solution are the edges adjacent to the base p_ip_j. The minimum decompositions of a subpolygon can be placed in equivalence classes based on the angles the base p_ip_j forms with the two adjacent edges in the decomposition. By storing only one representative for each of these equivalence classes the complexity is reduced so that a dynamic programming approach can achieve the $O(N^2 \text{nlogn})$ time bound.

3.2.1.2. Spiral Decompositions

Recall that a spiral polygon is a simple polygon whose boundary chain contains precisely one concave subchain. That is, a spiral polygon has precisely one set of adjacent notches. Feng and Pavlidis [13] give an algorithm for decomposing a simple polygon into spiral polygons. Their algorithm does not generally yield a decomposition into the minimum number of spiral components. Keil [22] develops a dynamic programming algorithm to this problem that is similar to his algorithm for the convex decomposition problem. The algorithm may require $O(n^3 \text{logn})$ time to find the minimum spiral decomposition.

3.2.1.3. Star-shaped Decompositions

A star-shaped polygon is a simple polygon in which the entire polygon is visible from at least one (possibly interior) fixed point of the polygon. Avis and Toussaint [5] give an O(nlogn) algorithm that finds a decomposition into at most n/3 components. Their algorithm first triangulates the polygon P and then 3-colors it. A 3-coloring is an assignment of one of three colors to each vertex, such that so two vertices with the same color are adjacent. Consider the set S $= \{s_1,s_2,...,s_j\}$ of the vertices with a given color. P can be decomposed into j nonoverlapping starshaped polygons $P_1,P_2,...,P_j$ where $P_k = \{x|x$ is a vertex of P and x is adjacent to s_k in the triangulation of P$\} \cup \{s_k\}$, $1 \leq k \leq j$. This algorithm does not generally yield a minimum decomposition.

Keil [23] develops an $O(n^6N^2\log n)$ time algorithm for decomposing a simple polygon into the minimum number of star-shaped components. He employs the same general dynamic programming technique that he used for the convex decomposition problem. Again the idea is to build up solutions to larger and larger subpolygons P_{ij}. In general, there can be an exponential number of minimum star-shaped decompositions of a subpolygon. It turns out that there are situations where no minimum decomposition of a subpolygon can be developed into a global minimum decomposition. The solution is to introduce the concept of pseudo star-shaped polygons. A pseudo star-shaped subpolygon P_{ij} has the property that there exists a point x in P but not in P_{ij} so that every point of P_{ij} can be seen from x through the edge v_iv_j. The algorithm proceeds by dynamic programming while keeping only one star or pseudo star-shaped minimum decomposition of each of a number of equivalence classes of decompositions at each subpolygon.

3.2.1.4. Monotone Decompositions

Recall that a monotone polygon contains two extreme vertices in a prefered direction such that they are connected by two polygonal chains monotonic in the prefered direction. Lee and Preparata [26] give an O(nlogn) time algorithm for decomposing a simple polygon into monotone polygons that does not generally yield a minimum decomposition. Keil [22] develops an $O(Nn^4)$ time algorithm that uses the same dynamic programming approach he uses for the convex decomposition algorithm. The algorithm relies on the fact that there are only a polynomial number of prefered directions with repect to which a subpolygon can be monotone.

3.2.2. Minimum Edge Length Decompositions

A decomposition of a simple polygon into triangles while minimizing the total internal edge length of the decomposition is known as the minimum weight triangulation of the polygon. If instead of a triangulating a polygon we triangulate an arbitrary set of points in the plane, no polynomial time algorithm is known for the minimum weight triangulation problem [16]. The minimum weight triangulation of a simple polygon can be found in $O(n^3)$ time [24]. The algorithm, due to Klincsek, uses standard dynamic programming.

Keil [23] uses dynamic programming to develop an $O(N^2n^2\log n)$ time algorithm for the problem of decomposing a polygon into convex polygons while minimizing the total internal edge length. Figure 3.1 illustrates the difference between a decomposition into the minimum number of convex pieces and a decomposition into convex pieces that minimizes the total internal edge length. Solutions are built up for larger and larger subpolygons. It is not sufficient to keep only minimum solutions for each subpolygon but rather decompositions are kept that are minimum given that they include certain edges that affect their ability to merge into larger decompositions. Independently, Greene [20] has noticed that his algorithm for the convex minimum number problem [19] can be adapted to yield an $O(N^2n^2)$ time algorithm for the convex minimum edge length problem.

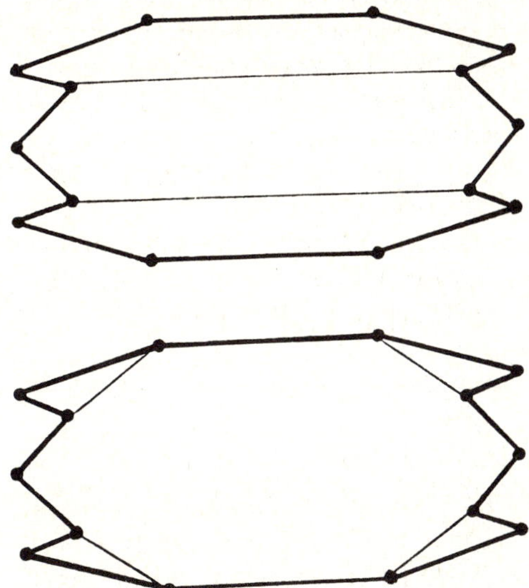

Figure 3.1 Minimum Number and Minimum "Ink" Convex Decompositions

Keil [22] is able is use his dynamic programming approach to decomposition problems to develop an $O(n^4 \log n)$ time algorithm for finding a minimum edge length decomposition into spiral polygons, an $O(N^2 n^5 \log n)$ time algorithm for finding a minimum edge length decomposition into star-shaped polygons and an $O(Nn^4)$ time algorithm for finding a minimum edge length decomposition into monotone polygons.

4. Decomposition of Polygons with Holes

When we allow a polygon to contain holes the problem of decomposing the polygon in some minimal way seems to become more difficult. Recently, several of these problems have been shown to be NP-hard [16] and thus unlikely to admit efficient algorithms.

Let F be a boolean formula in conjuctive normal form with at most 3 variables per clause where $V = \{v_1,\ldots,v_l\}$ are the variables and $C = \{c_1,\ldots,c_m\}$ are the clauses. 3SAT is the problem of determining whether a given formula F is satisfiable. A formula is said to be planar if the bipartite graph $G(F) = (V \cup C,E)$ with $E = \{(v_i,c_j) | v_i$ or $\bar{v}_i$ is a literal in $c_j\}$ is planar. P3SAT is the problem of determining whether a given planar formula is satisfiable. Both 3SAT [16] and P3SAT [27] are known to be NP-complete. Many problems in this section are shown to be NP-hard with transformations from either 3SAT or P3SAT.

In the introduction we distinguished between decompositions that partition a polygonal object into a given type of component and decompositions that cover a polygonal object with a given type of component. In this section the difference between partitioning and covering is important.

4.1. Rectilinear Polygons with Holes

The results in this section all concern the decomposition of rectilinear polygons with holes into rectangles allowing Steiner points. The first problem is that of covering a rectilinear polygon with holes with the minimum number of rectangles. Masek [33] has shown that this problem is NP-complete with a transformation from 3SAT.

If we consider a partition, rather than a covering, then the problem of decomposing a rectilinear polygon into the minimum number of rectangles becomes polynomial. Lipski et al [32] have developed an $O(n^3)$ time algorithm for the problem that makes use of bipartite graph matching techniques. Ohtsuki [37] and Ohtsuki et al [38] have improved this time and exhibit an $O(n^{5/2})$ algorithm.

If the holes are allowed to be points rather than rectilinear polygons, the nature of the problem changes again. Lingas [28] uses a transformation from P3SAT to show that the problem of partitioning a rectilinear polygon with rectilinear polygon or point holes into the minimum number of rectangles is NP-complete.

The minimum edge length minimality criterion has also been considered. Lingas et al [30] use transformations from P3SAT to show that the problems of decomposing a rectilinear polygon, with either rectilinear holes or rectilinear and point holes, into rectangles while minimizing total edge length are NP-complete. Lingas [29] has developed an $O(n^4)$ heuristic for this problem that yields a partition whose length is not greater that 41 times the length of an optimal partition.

4.2. Arbitrary Polygons with Holes

A number of problems involving decomposing arbitrary polygons with holes into a given component type have been studied and found to be NP-hard. We first consider decompositions allowing Steiner points and then decompositions without Steiner points.

4.2.1. Decompositions allowing Steiner Points

The first problem is that of partitioning a polygon with polygon holes into the minimum number of triangles. Lingas [28] uses a transformation from P3SAT to show that this problem is NP-hard. This problem arises in applications such as manipulation of LSI artwork data and image processing thus motivating the search for good approximation algorithms. Asano, Asano and Ohsuga [3] present an O(nlogn) time approximation algorithm that is guaranteed to find a decomposition that contains at most 4 times the number of triangles in an optimal decomposition.

Another problem that arises in the manipulation of VSLI artwork data processing is that of partitioning a polygon with holes into a minimum number of trapezoids with two horizontal sides. As before triangles with a horizontal side are considered to be trapezoids with two horizontal sides one of which is degenerate. Asano and Asano [2] study this problem and show it to be NP-complete using a transformation from the problem of finding a maximum independent set of a straight-lines-in-the-plane graph. They also develop an O(nlogn) approximation algorithm which uses only horizontal chords to partition a polygonal region. This algorithm finds a decomposition which contains not more than 3 times the number of trapezoids in a minimum decomposition.

Both the covering and partitioning versions of the problem of decomposing a polygon with holes into a minimum number of convex components are NP-hard. O'Rourke and Supowit [40] use a transformation from 3SAT to show the covering problem is NP-hard while Lingas [28] uses a transformation from P3SAT to show the partitioning problem is NP-hard. While neither problem is known to be in NP, O'Rourke [39] has proven that the covering problem is decidable.

The problems of covering a polygon with holes with the minimum number of spiral or star-shaped components are also NP-hard. O'Rourke and Supowit [40] use a transformation from 3SAT to prove this.

Use of the minimum edge length minimality crition does not make the convex partitioning problem easier. Lingas et al [30] use a transformation from P3SAT to show the problem is NP-hard.

4.2.2. Decompositions without Steiner Points

When Steiner points are disallowed both the covering and partitioning versions of the problems of decomposing a polygon with holes into the minimum number of convex, spiral and star-shaped components are NP-complete. O'Rourke and Supowit [40] show that the covering problems are NP-complete. The convex partitioning problem is shown to be NP-complete by Lingas [28]. Keil [22] modifies the proofs of O'Rourke and Supowit to show that the spiral and star-shaped partitioning problems are NP-complete. Using a similar proof he also shows that the monotone partitioning problem is NP-complete.

The only minimum edge length result concerns partitioning into convex polygons. Keil [22] uses a complex transformation from P3SAT to show this problem to be NP-complete.

5. Other Decompositions

So far all of the decomposition methods have used only set union to relate the decomposition to the original polygonal object. If we also allow set difference we open an entire class of problems which have been called sum/difference decompositions [50]. Are there polynomial time algorithms for representing a polygonal object by the sum and difference of the minimum number of some given type of component? Batchelor [6] describes a decomposition into a tree of convex polygons. The root is the convex hull of the polygon, the second level nodes represent the convex hulls of concavities in the polygon, the third the convex hulls of concavities of concavities, and so on.

There are a number of results concerning procedure-directed decomposition techniques. In most of these there is no obvious minimality criterion being used so we consider these techniques to be outside the scope of this paper.

6. REFERENCES

[1] N. Ahuya, R.T. Chien,R. Yen and N. Birdwell, *Interference detection and collision avoidance among three dimensional objects*, Proc. First Annual National Conference on Artificial Intelligence, Stanford, California, 1980, 44-48.

[2] T. Asano and T. Asano, *Minimum Partition oj Polygonal Regions into Trapezoids*, Proc. 24th Annual Symposium on Foundations of Computer Science, 1983, 233-241.

[3] T. Asano, T. Asano and Y. Ohsuga, *Partitioning polygonal regions into triangles*, Papers of Technical Groups of IECE, CAS83-98, 31-36.

[4] D. Avis, *On the Complexity of Finding the Convex Hull of a set of Points*, Discrete Applied Mathematics, 4, 1982, 81-86.

[5] D. Avis and G.T. Toussaint, *An efficient algorithm for decomposing a polygon into star-shaped polygons*, Pattern Recognition, 13, 6, (1981), 395-398.

[6] B.G. Batchelor, *Hierarchical shape description based upon convex hulls of concavities*, Journal of Cybernetics,10 (1980),205-210.

[7] B.K. Bhattacharya and H. El Gindy,*A new linear convex hull algorithm for simple polygons*, Technical Report, School of Computer Science, McGill University, March 1981.

[8] B. Chazelle and D. Dobkin, *Decomposing a polygon into its convex parts*, Proceedings of the 11th Annual ACM Symposium on Theory of Computing, 1979, 38-48.

[9] B. Chazelle, *Computational Geometry and Convexity*, Ph.D. Thesis, Yale University, 1980.

[10] B. Chazelle, *A theorem on polygon cutting with applications*, Proc. 23rd Annual Symposium on Foundations of Computer Science,Chicago, 1982,339-349.

[11] B. Chazelle and J. Incerpi, *Triangulating a polygon by Divide-and-conquer*, Proc. 21st Annual Allerton Conference on Communication, Control and Computing, 1983.

[12] H. El Gindy, D. Avis and G.T. Toussaint, *Appications of a two dimensional hidden-line algorithm to other geometric problems*, Computing, 31, 1983, 191-202.

[13] H. Feng and T. Pavlidis, *Decomposition of Polygons Into Simpler Components: Feature Generation for Syntactic Pattern Recognition*, IEEE Trans. on Computers C-24 (1975), 636-650.

[14] L. Ferrari, P.V. Sankar and J. Sklansky, *Minimal rectangular partitions of digitized blobs*, Proc. 5th International Conference on Pattern Recognition,

Miami Beach, 1981, 1040-1043.

[15] M.R. Garey, D.S. Johnson, F.P. Preparata and R.E. Tarjan, *Triangulating a simple polygon*, Information Processing Letters,7 (1978), 175-179.

[16] M.R. Garey and D.S. Johnson, *Computers and Intractability: A Guide to the Theory of NP-Completeness*, San Francisco, CA, Freeman, 1979.

[17] R. Graham, *An efficient algorithm for determining the convex hull of a planar set*, Information Processing Letters, 1 (1972), 132-133.

[18] R.L. Graham and F.F. Yao, *Finding the convex hull of a simple polygon*, Journal of Algorithms, in press.

[19] D. Greene, *The Decomposition of Polygons into Convex Parts*, Manuscript, Xerox PARC, 1982.

[20] D. Greene, Private Communication, May 1982.

[21] J. Kahn, M. Klawe, and D. Kleitman, *Traditional galleries require fewer watchmen*, SIAM J. Alg. Disc. Meth., Vol. 4, 2, June 1983, 194-206.

[22] J.M. Keil, *Decomposing Polygons into Simpler Components*, Ph.D. thesis, Department of Computer Science, University of Toronto, 1983.

[23] J.M. Keil, *Decomposing a Polygon into Simpler Components*, to appear in SIAM Journal on Computing.

[24] G.T. Klincsek, *Minimal Triangulations of Polygonal Domains*, Annals of Discrete Math., 9, 1980, pp. 121-123.

[25] D.T. Lee, *On finding the convex hull of a simple polygon*, Tech. Rep. No. 80-03-FC-01, Dept. Elec. Engr. and Comp. Sci., Northwestern University, 1980.

[26] D.T. Lee and F.P. Preparata, *Location of a point in a planar subdivision and its applications*, SIAM J. on Comput., 6 (1977), 594-606.

[27] D. Lichtenstein, *Planar Formulae and their Uses*, SIAM J. on Comput., 11 (1982), 329-343.

[28] A. Lingas, *The power of non-rectilinear holes*, Proc. 9th Colloquium on Automata, Languages and Programming, Aarhus, 1982.

[29] A. Lingas, *Heuristics for Minimum Edge Length Rectangular Partition*, manuscript, MIT,1982.

[30] A. Lingas, R. Pinter, R. Rivest and A. Shamir, *Minimum Edge Length Partitioning of Rectilinear Polygons*, Proc. 20th Annual Allerton Conference on Communication, Control and Computing, 1982, 53-63.

[31] W. Lipski, *Finding a Manhattan path and related problems*, Tech. Rep. ACT-17 Coordinated Science Lab., University of Illinois at Urbana-Champaign, August 1979.

[32] W. Lipski, E. Lodi, F. Luccio, C. Mugnai and L. Pagli, *On two dimensional data organization II*, Fundamenta Informaticae, 2 (1979), 245-260.

[33] W.J. Masek, *Some NP-complete set covering problems*, unpublished manuscript, August 1979.

[34] D. McCallum and D. Avis, *A linear algorithm for finding the convex hull of a simple polygon*, Infor. Proc. Lett., 9 (1979),201-206.

[35] D.Y. Montuno and A. Fournier, *Detecting Intersections Among Star Polygons*, Tech. Pept. CSRG-146, University of Toronto, Toronto, Canada, Sept. 1982.

[36] W. Newman and R. Sproull, *Principles of Interactive Computer Graphics*, Second Edition, McGraw Hill, New York, 1979.

[37] T. Ohtsuki, *Minimum dissection of rectilinear regions*, Proc. of 1982 IEEE International Symposium on Circuits and Systems, Rome, 1982, 1210-1213.

[38] T. Ohtsuki, M. Sato,M. Tachibana and S. Torii, *Minimum partitioning of rectilinear regions*, Trans. of Information Processing Society of Japan.

[39] J. O'Rourke, *The complexity of computing minimum convex covers for polygons*, Proc. 20th Allerton Conf., Allerton IL, Oct. 1982.

[40] J. O'Rourke and K.J. Supowit, *Some NP-Hard Polygon Decomposition Problems*, IEEE Trans. on Information Theory, Vol. IT-29, 2, (1983), 181-190.

[41] T. Pavlidis, *Analysis of set patterns*, Pattern Recognition, 1 (1968), 165-178.

[42] T. Pavlidis, *Structural pattern recognition*, Springer Series in Electrophysics Vol. 1, Springer Verlag Berlin, Heidelberg, New York, 1977.

[43] F.P. Preparata, *Medial Axis of a Convex Polygon*, in Steps into Computational Geometry, F.P.Preparata, ed., Coordinated Science Lab., Tech. Rept. R-760, 1977, 6-9.

[44] J.R. Sack, *An O(nlog(n)) algorithm for decomposing simple rectilinear polygons into convex quadrilaterals*, Proc. 20th Annual Allerton Conf. on Communication, Control and Computing, 1982, 64-75.

[45] J.-R. Sack, *A simple hidden-line algorithm for simple polygons*, School of Computer Science, Carleton University, unpublished manuscript, 1983.

[46] J.-R. Sack and G.T. Toussaint, *A linear time algorithm for decomposing rectilinear star-shaped polygons into convex quadrilaterals*, Proc. 19th Annual Allerton Conf. on Communication, Control and Computing, Urbana, 1981, 21-30.

[47] B. Schachter, *Decomposition of polygons into convex sets*, IEEE Transactions on Computers, C-27, 1978, 1078-1082.

[48] A.A. Schoone and J. van Leeuwan, *Triangulating a star-shaped polygon*, Tech. Rept. RUV-CS-80-3, University of Utrecht, April 1980.

[49] M.I. Shamos, *Computational Geometry*, Ph.D. Thesis, Yale University, 1978.

[50] G.T. Toussaint, *Pattern recognition and geometrical complexity*, Proc. Fifth International Conf. on Pattern Recognition, Miami Beach, 1980, 1324-1347.

[51] G.T. Toussaint, *A New Linear Algorithm for Triangulating Monotone Polygons*, Technical Report SOCS-83.9, School of Computer Science, McGill University, June 1983.

[52] G.T. Toussaint and D. Avis, *On a Convex Hull Algorithm for Polygons and its Application to Triangulation Problems*, Pattern Recognition, 15, 1, 1982, 23-29.

[53] A. Yao, *A lower bound to finding convex hulls*, internal report, Computer Science Department, Stanford University, 1979.

COMPUTATIONAL GEOMETRY, G.T. Toussaint (Editor)
© Elsevier Science Publishers B.V. (North Holland), 1985

A FRAMEWORK FOR COMPUTATIONAL MORPHOLOGY

David G. Kirkpatrick*
IBM Research Laboratory
San Jose, California 95193

John D. Radke
Department of Geography
Wilfred Laurier University
Waterloo, Canada N2L3C5

This paper outlines a new methodology for describing the "internal structure" (or "skeleton") of planar point sets. The methodology, which is based on parameterized measures of neighbourliness, gives rise to a spectrum of possible internal shapes. Applications to the analysis of both point set and network patterns are described.

1. INTRODUCTION

The relatively young field of computational geometry has had a significant and steadily increasing impact on the development and unification of techniques in pattern recognition. It has also made it possible to discuss with greater uniformity and precision the inherent complexity of many pattern recognition tasks. Toussaint [T3] surveys all but the most recent literature in computational geometry related to pattern recognition. He emphasizes the fundamental geometrical structures underlying a broad range of pattern recognition applications along with the complexity issues that arise in the construction and manipulations of these structures.

Toussaint is careful to point out that certain aspects of pattern recognition problems are not adequately captured by geometrical properties alone. In fact, he

---

*On leave from the Computer Science Department, University of British Columbia, Vancouver, B.C., V6T1W5, Canada

pays little attention to statistical attributes of patterns, even those that attempt to aggregate fundamental geometrical properties. We follow Toussaint in focusing our attention on the purely geometrical attributes of patterns. We assume that in practice these will be analyzed in various ways in an attempt to characterize the pattern at hand. It is understood that even the exact solution of the geometrical problems that arise in our discussions constitute only partial, approximate, or heuristic solutions of the associated pattern recognition problems.

Toussaint [T3] coins the term *computational morphology* to describe the situation where "a computational geometrical structure is intended to extract the *form* of the input." This paper can be viewed as a presentation of new tools for the development of computational morphology. Our results and techniques were developed originally in the context of geographical pattern recognition [R]. However, as we hope to make clear, they have somewhat broader application.

We seek to characterize the shape or form of input sets. It seems appropriate to start by asking what is meant by shape. However, since this question appears to lack an explicit answer, it is more fruitful to ask what we intend or expect to hypothesize or conclude from knowledge about the form of an input set. On one hand, we might like to know if two or more input sets (some of which could be artificially generated) have the same or similar forms. Alternatively, we might inquire whether the form of a single input set is regular or homogeneous, and if not, isolate the irregularities or homogeneous components. These correspond to what we might call external and internal aspects of form. At a deeper level (and aided by these more superficial notions), we might wish to speculate on the process or processes that give rise to a particular input set (*cf.* [BG]).

Our attention will be restricted to the form exhibited by fixed point sets and, indirectly, networks defined on these point sets. To set our objectives a bit more precisely, we will define the external and internal shapes of a given point set as follows. We say that the *external shape* of a point set is the form exhibited by identifying the "essential" extreme points of the set and, among these, joining "essential" neighbours. Secondly, the *internal shape* of a point set is exhibited by identifying the "essential" points of the set and, among these, joining "essential" neighbours. (Toussaint [T3] uses the terms "shape hull" and "skeleton" to describe what are fundamentally the same notions.) Of course, the value and validity of our definitions hinge on our choice of interpretation for the terms "essential." Nevertheless, they do serve to strip away the aesthetic aspects of form, leaving us with purely combinatorial structures. The problem remains to find approximate geometrical properties which will form the basis of precise definitions of the terms "essential."

We do not wish to suggest that internal and external aspects of shape are clearly distinguishable. Imposing such a division may even be counterproductive. For example, the problem of *clustering*, that is partitioning point sets into clusters based on some notion of similarity or proximity, relates to both the external and internal shape of the underlying point set. Nevertheless, for the sake of clarity we will assume that it is meaningful to talk about either the internal or external aspects of shape in isolation.

Underlying a great deal of the work surveyed by Toussaint [T3] are various notions of neighbourhood. We can think of the diameter, minimum bonding box and convex hull as characterizing the global neighbourhood of a point set. Similarly, the set of nearest neighbour pairs, minimum spanning tree, Gabriel graph, relative neighbourhood graph, and Delaunay triangulation all describe local notions of neighbourliness. Exploiting the well known relationships between these global (respectively, local) descriptors of neighbourliness, we can view these structures as providing a kind of spectrum of progressively more detailed descriptions of the external (respectively, internal) shape of the underlying point set.

A central objective of this paper is to develop measures of neighbourliness that permit us to construct an essentially continuous spectrum of descriptors of internal shape. Our work was inspired by a conceptually similar technique, known as α-shapes [EKS], for describing the external shape of a point set. We describe two specific measures of neighbourliness. Beyond that, we set out some of the properties, including computational properties, of our measures and describe applications to the analysis of both point set and network patterns. We have recently become aware of the work of Urquhart [U] who takes a similar approach in the development of clustering algorithms. See also the work of Jarvis [J] and Fairfield [F] for related approaches to the description of external shape.

In the next section, we describe in more detail some of the geometrical structures and related work in computational morphology which support or have inspired our work. Section 3 sets out our notion of β-skeletons, some applications of which are described in Section 4. Our paper concludes with a discussion of possible refinements and extensions of our techniques.

2. THE SHAPE OF POINT SETS

As explained in the introduction, we are concerned with the development of tools to aid in the description and analysis of shape of both point sets and embedded networks. Our emphasis will be on what we have defined as internal shape, yet our approach will in a number of ways parallel a recently developed technique for the description of external shape. Thus, for the sake of completeness, we will begin our description of related work with a review of some of the geometrical structures that arise in the characterization of the external shape of point sets.

For the remainder of this section, we will assume for the sake of simplicity that S denotes a fixed set of n points in the plane. Most of the techniques and algorithmic results discussed here have analogs in higher dimensions. A complete treatment of these would take us beyond the intended scope of this paper. However, the reader is encouraged to visualize these extensions.

2.1 External Shape

Consideration of the form of any input set reveals an obvious relationship between the intrinsic complexity of the data and the complexity of its descriptions. The more random a point pattern is, the more complex is its accurate description. Our objective, of course, is to find incomplete descriptions that capture the essence of the data set. For this reason, it is natural that the available descriptors of form vary in their intrinsic complexity, and that no single such descriptor appears to be best for all possible applications.

The simplest geometrical constructs for describing the external shape of a point set are the *minimum bounding box* and *minimum bounding circle.* These both admit $0(n \log n)$ algorithms in the plane [FS,SH,TB]. A natural generalization of the minimum bounding circle is the *minimum bounding ellipse.* All of these have found quite widespread use as descriptors of shape (*cf.* [T3]).

The next natural step, from the viewpoint of descriptive complexity, is to consider the convex polygon of minimum area that contains the entire point set. This is well known as the *convex hull* of the point set. Toussaint devotes a significant portion of his survey [T3] to the description of the convex hull and its applications in pattern recognition (see also [T1]). Convex hulls in two and more dimensions are extremely well-studied objects with numerous properties that can be exploited in pattern recognition applications.

As descriptors of shape convex hulls have two notable (and complimentary) weaknesses. On one hand, the convex hull may provide a too detailed description of the shape of point set, *i.e.*, not all of the geometrical extreme points are "essential." On the other hand, it is constrained by definition to be convex, *i.e.*, it will fail to identify points which, though not geometrically extreme, are extreme in the pattern. As Toussaint [T3] points out, the external shape of a point set (what he terms the *shape hull*) may be disconnected and may have a disconnected compliment.

A recent paper of Medek [Me] addresses the first of these weaknesses by providing a new definition of hull or boundary of a point set that is sensitive to the density of the points. Madek's boundary is not necessarily convex. However, its vertices form a superset of the convex hull vertices. Furthermore, it has the unpleasant property that points can cease to qualify as boundary points on the addition of points to the interior of the set.

An alternative approach to escape the weaknesses of convex hulls is presented by Edelsbrunner *et al.* [EKS]. They propose a natural generalization of convex hulls that they call α-hulls. The α-hull of a point set is based on the notion of generalized discs in the plane. For arbitrary real valued α, a *generalized disc of radius* $1/\alpha$ is defined as follows:

(i) if $\alpha>0$, it is a (standard) disc of radius $1/\alpha$;

(ii) if $\alpha<0$, it is the compliment of a disc of radius $1/\alpha$; and

(iii) if $\alpha=0$, it is a half plane.

The *α-hull* of a point set S is defined to be the intersection of all closed generalized discs of radius $1/\alpha$ that contain all of the point S. The convex hull of S is precisely the 0-hull of S. The family of α-hulls includes the smallest enclosing circle (when

α=1/radius (S)), the set S itself (for all α sufficiently small) and an essentially continuous set of enclosing regions in between these extremes.

Edelsbrunner *et al.* also define a combinatorial variant of the α-hull, called the *α-shape* of a point set, which can be viewed without serious misunderstanding as the boundary of the α-hull with curved edges replaced by straight edges. Unlike the family of α-hulls, the family of distinct α-shapes has only finitely many members. These provide a spectrum of progressively more detailed descriptions of the external shape of a given point set. To quote Edelsbrunner *et al.* [EKS], ""relatively large" α tend to produce a rather crude shape of the points ... whereas smaller α reveal more and more details." Thus, the problem described earlier of finding an appropriate descriptor of shape is reduced to the problem of choosing an appropriate value of the parameter α. Figure 1 (copied from [EKS]) illustrates the α-shape of a point set for two different values of α.

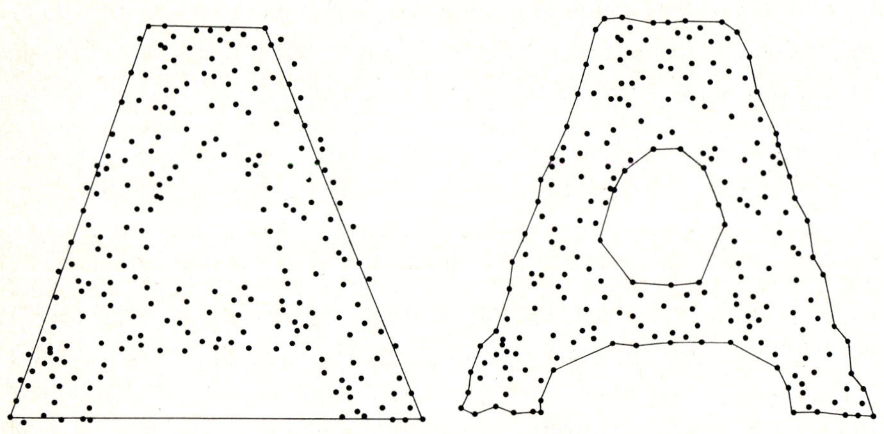

Figure 1 (from [EKS])

Two different α-shapes of a common point set.

2.2 Internal Shape

In the various schemes that we considered for describing external shape, the identification of "essential" extreme points is intimately linked with the process of determining their interconnections. This partitioning of the underlying point set does not seem to be implicit in existing schemes for the description of internal structure and we will not attempt to impose any geometrical criterion to determine which data points are "essential" In practice, it is often possible to assert that certain points in a data set are more "essential" than others by employing nongeometrical knowledge of the data.

As is the case with external shape, there exists a discrete family of nested, and progressively more detailed, structures that have been used to describe the internal structure of point sets. The most elementary structure is the *closest pair* of the point set. This gives a rough measure of the density of the point set but obviously fails to capture most other characteristics of internal structure. Probably the most sparse structure that could be viewed as a realistic descriptor of internal structure is the set of *all nearest neighbour pairs.* We call the graph formed by linking each vertex to its nearest neighbour the *nearest neighbour graph* (NNG). The NNG of a point set, like the single closest pair, can be found in $0(n \log n)$ time, and this is known to be asymptotically optimal in the worst case [SH].

If we insist that the internal structure of a point set be described by a connected graph, then it is natural to turn to the (Euclidean) *minimum spanning tree* (MST) (the spanning tree that minimizes the total edge length) as a minimal descriptor. Once again, an $0(n \log n)$ algorithm is available [SH,CT]. Toussaint [T3] points out the importance and widespread use (*cf.* [Z]) of the MST as a descriptor of the "skeleton" of a point pattern, but goes on to suggest that a supergraph of the MST, known as the relative neighbourhood graph, "appears to be superior." The *relative neighbourhood graph* (RNG), as defined by Toussaint [T3] (a minor variant of the original definition of Lankfard [L]), is given by the set of edges describing relative neighbours within the point set. Two points x and y are said to be *relative neighbours* in S if

$$d(x,y) \leq \min\{\max\{d(x,z),d(y,z)\} \mid z \in S\} .$$

Graphically, x and y are relative neighbours if their "lune of influence," depicted in Figure 2a, is *empty*, *i.e.*, contains no point of S in its interior. This lune is formed by the intersection of the two circles of radius d(x,y) centered at x and y, respectively. An 0(n log n) algorithm for the construction of the RNG has been discovered recently by Supowit [Su]. See [T2,TM] for a discussion of algorithmic issues and applications involving the RNG.

Conceptually similar to the RNG is another planar descriptor of internal shape known as the Gabriel graph [MS]. The *Gabriel graph* (GG) on a point set is formed by the set of all Gabriel neighbours. Two points x and y are *Gabriel neighbours* in S if their "circle of influence," depicted in Figure 2b, is empty. This is the unique circle of radius d(x,y)/2 passing through both x and y. It is trivial to see that the RNG is subgraph of the GG. The latter has a straightforward 0(n log n) time construction. These together with other properties and applications are described in [MS].

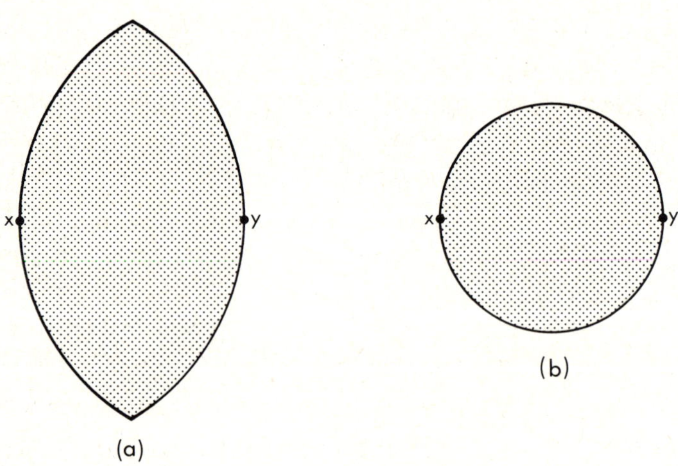

Figure 2
The "lune of influence" (a) and "circle of influence" (b) of a point pair.

A unifying property of all of the preceding structures, that has had a direct impact on their efficient construction, is the fact that they are all subgraphs of the Delaunay triangulation of the underlying point set. The *Delaunay triangulation* (DT) has a number of equally interesting defining properties. Ignoring degeneracies, two of these properties are

(i) it is the combinatorial dual of the well-studied Voronoi diagram of the point set [SH]; and

(ii) it is the unique triangulation satisfying the following equiangular property [S1]: every convex quadrilateral in DT is subdivided into triangles so as to maximize the minimum of the resulting angles.

For our purposes, it is of interest to define the Delaunay triangulation as the set of edges joining Voronoi neighbours. Two points x and y are *Voronoi neighbours* if there exists a point z in the plane for which $d(x,z)=d(y,z)=\min\{w,z) \mid w \in S\}$. Equivalently, this can be expressed as, there exists an empty circle passing through x and y.

The Delaunay triangulation of a point set (and the associated Voronoi diagram) have a large number of interesting properties and applications [SH,LS,GB,GKLN,Law,S2,TBP]. Indeed these structures have had one of the greatest unifying effects of all those studied in computational geometry (the subject matter at hand being a notable example). Obviously, the DT is a maximal planar description of internal structure. It is reasonable to assert that if we are interested in descriptors that are planar then we can restrict our attention to subgraphs of the DT. (This is equivalent to saying that a necessary condition for neighbourliness of a pair of points is the existence of an empty circle passing through both.)

Figure 3 illustrates the NNG, MST, RNG, GG, and DT for a fixed point set. We have seen that, in general, $NNG \subseteq MST \subseteq RNG \subseteq GG \subseteq DT$, providing a discrete hierarchy of internal shape descriptors. The defining properties of the RNG, GG, and DT suggest a unification (along the lines of that provided earlier by α-shapes) based on the notion of empty neighbourhoods. Unfortunately, the MST does not admit such a local characterization (the linking of points in a MST may depend on the existence and position of other points that are, in no sense, neighbours of the points in question). This results from the simultaneous requirement of connectivity and lack of cycles. Of course, some point patterns, *e.g.*, regular grids, simply do not have useful tree-like descriptions. So, despite the fact that we are eliminating from the start one of the

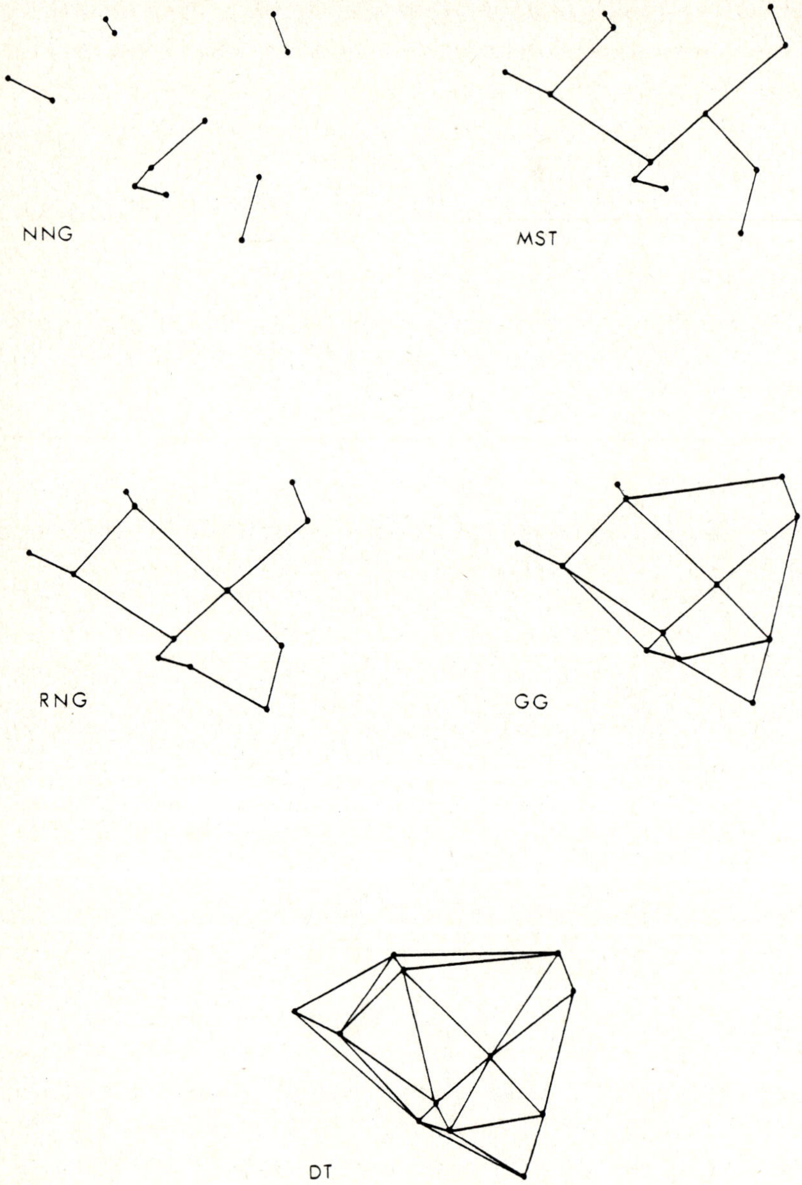

NNG

MST

RNG

GG

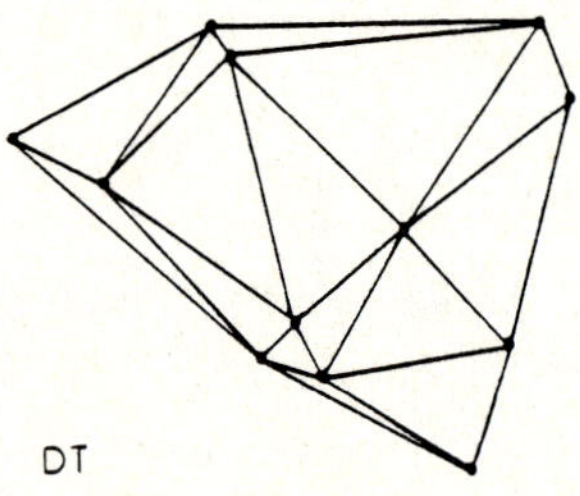

Figure 3

NNG, MSG, RNG, GG and DT of a fixed point set.

most widely used descriptors of internal shape, we seek a unification and generalization of the above hierarchy based solely on the simple idea of progressively larger excluded neighbourhoods.

In the next section, we present a more detailed description of our hierarchy of internal shapes, which we call β-skeletons, based on a parameterized notion of neighbourhood. Before doing so, we note that we have tentatively set the NNG and DT as rough minimal and maximal descriptors of internal shape. However, in practice both of these limits can be breached. For example, we will apply our techniques to the analysis of empirical networks, and we will make no assumptions about either the connectivity or planarity of such structures.

3. β-SKELETONS

As we described in the preceding section, we are motivated to seek a hierarchy of descriptors of internal shape based on a measure of neighbourliness related to the structure of empty neighbourhoods. Such a hierarchy serves two purposes. It allows us to visualize a spectrum of internal shapes of various edge densities. The entirety of this spectrum, including, in particular, the transitions between adjacent structures, provides an added dimension for the representation of structure. The second advantage is that it serves as a kind of benchmark with the aid of which empirical networks can be analyzed and, to some extent, compared. These applications will be explored and illustrated in more detail in Section 4.

3.1 Abstract Description

For a given pair of x,y, we consider a continuous family of neighbourhoods indexed by a single positive real valued parameter β. We denote by $N(x,y,\beta)$ the neighbourhood with index β. It is assumed that

$$N(x,y,\beta_1) \subseteq N(x,y,\beta_2)$$

for all $0 \leq \beta_1 \leq \beta_2$. $N(x,y,0)$ is taken to be some appropriately small neighbourhood of the point pair x,y, whose emptiness defines a necessary condition for the pair to be considered neighbourly. Similarly, as β approaches ∞, $N(x,y,\beta)$ describes some appropriately large neighbourhood of x and y whose emptiness defines a sufficient condition for the pair to be considered neighbourly (equivalently, whose nonemptiness defines a necessary condition for the pair to be considered non-neighbourly). (See Urquhart [U] for a similar definition based on what he calls regions of influence.)

Assuming nothing else about these indexed neighbourhoods, we can say that two points x and y are *β-neighbours* in the set S if $N(x,y,\beta)$ contains no point of S, other than x or y, in its interior. We define the *β-skeleton* of a point set S to be the set of edges joining β-neighbours in S. We say that the *β-value* of the pair (x,y) is the largest β such that x and y are β-neighbours. This permits a more precise notion of neighbourliness, namely, x and y are *(β_1,β_2)-neighbours* if $\beta_1 \leq \beta\text{-value}(x,y) \leq \beta_2$. The *$\beta$-skeleton spectrum* of S is the complete graph on the vertex set S each edge of which is labelled with the β-value of the corresponding point pair. As its name suggests, the β-skeleton spectrum of S encodes the entire family of β-skeletons for the point set S.

3.2 Neighbourliness and Distance

Probably the first notion of indexed neighbourhoods that comes to mind is one based on distance. For example, one might define points x and y to be β-neighbours if $d(x,y) \leq \beta$. This seems to capture the intuitive idea that points that are close to one another are neighbourly. There is no doubt that this intuition is close to the truth in many applications, notably clustering. However, it should be observed that relying on absolute distance alone as a measurement of neighbourliness can have unacceptable consequences. First, note that neighbourliness, if it is to be broadly applied, ought to be a scale independent notion. (One trusts that the internal shape of a point pattern is independent of scale.) Even though scale independence could be guaranteed by appropriate normalization, there remain problems created by variation in density (or internal scale). For example, if points are clustered, then a typical β-skeleton based on such a definition would be highly connected and nonplanar in some regions and perhaps disconnected as a whole. While this obviously reflects something of the structure of such a point set, it is insensitive to other important aspects of internal structure. Its lack of uniformity seems to be at odds with our understanding the term skeleton. Using distance alone ignores the relative positions of nearby points. For example, in Figure 4 distance alone would suggest that point b is more significant than point c as a neighbour of point x.

Figure 4

Which point, b or c, is more significant as a neighbour of x?

We choose to de-emphasize the role of distance in defining neighbourliness. The choice is based on the following two arguments (in addition to the discussion above).

(i) The significance of absolute distance is application dependent. Therefore, it should be combined with other measures of neighbourliness as dictated by the application at hand.

(ii) As we noted in Section 2, the geometrical structures that have been employed successfully in the characterization of internal shape are all scale independent. In keeping with those structures, we seek to take full advantage of the geometrical properties of the data, recognizing the importance of such intuitive notions as "interference" of points.

3.3 Lune-Based Neighbourhoods

Our first serious candidate for a family of indexed neighbourhoods arises from an attempt to subsume the NNG-MST-RNG-GG-DT hierarchy discussed earlier. For $\beta \geq 1$, we define $N(x,y,\beta)$ to be the intersection of the two circles of radius $\beta d(x,y)/2$ centered at the points $(1-\beta/2)x+(\beta/2)y$ and $(\beta/2)x+(1-\beta/2)y$, respectively.

When $\beta=1$, $N(x,y,\beta)$ corresponds exactly to the Gabriel neighbourhood of x and y, and when $\beta=2$, we get the "relative neighbourhood" of the RNG. As β approaches

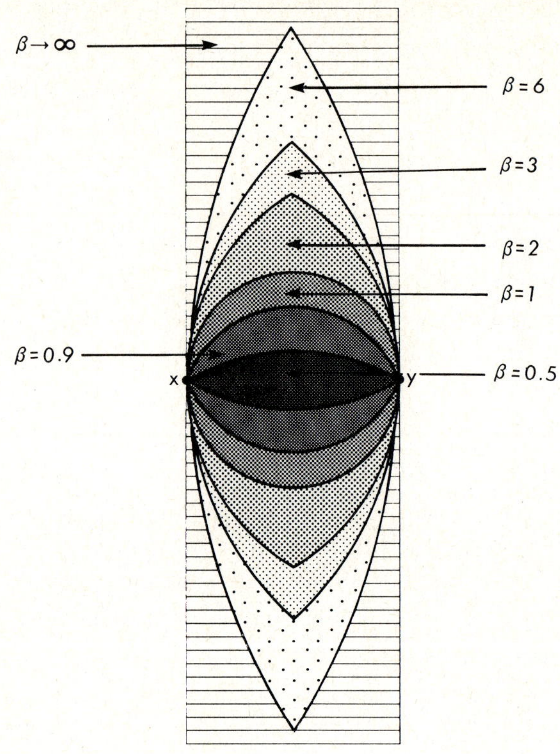

Figure 5

Lune-based neighbourhoods $N(x,y,\beta)$ for various $\beta>0$.

∞ the neighbourhood of x and y approximates the infinite strip formed by translating the line segment (x,y) normal to itself. If this corridor is empty, then it is clear that, except in degenerate cases, the edge (x,y) must belong to the MST of the point set.

We can extend this family of neighbourhoods in the other direction by defining $N(x,y,\beta)$, for $\beta \in [0,1]$, to be the intersection of the two circles of radius $d(x,y)/(2\beta)$ passing through both x and y. When $\beta=1$, this is consistent with the definition above. As β approaches 0, $N(x,y,\beta)$ approximates the line segment joining x and y. Thus, except in degenerate situations (three or more points colinear), all point pairs are β-neighbours under this scheme for β sufficiently small. Figure 5 illustrates the neighbourhoods $N(x,y,\beta)$ for various $\beta \geq 0$.

3.4 Circle-Based Neighbourhoods

There are, of course, numerous alternatives for the definition of neighbourhoods. (This was the motivation for introducing the idea in its abstract form.) An alternative that has a number of useful geometrical and computational properties is based on empty circles. In this case, when $\beta \geq 1$, we define $N(x,y,\beta)$ to be the union of the two circles of radius $\beta d(x,y)/2$ that pass through both x and y. As with the lune-based neighbourhoods, we define $N(x,y,\beta)$, when $\beta \in [0,1]$, to be the intersection of the circles of radius $d(x,y)/(2\beta)$ that pass through x and y. This family of neighbourhoods is illustrated in Figure 6.

Note that $N(x,y,1)$ again coincides with the Gabriel neighbourhood. Since $N(x,y,\beta)$ approximates the entire plane as β approaches ∞, we can conclude that, except for degenerate point sets, the 0-skeleton is complete, and the β-skeleton is devoid of edges for β sufficiently large.

3.5 Computational Issues

It should be clear that even in the most abstract setting, the β-value of an arbitrary point pair can be determined in $O(n)$ time (assuming that the test for inclusion in a fixed β-neighbourhood takes constant time). Thus, the complete β-skeleton spectrum can be constructed in $O(n^3)$ time at worst. In practice, the situation may be much better than this. Since, in both of our definitons, the β-skeleton is a subgraph of the GG (which, in turn, is a subgraph of the DT) when $\beta \geq 1$, we conclude that for all

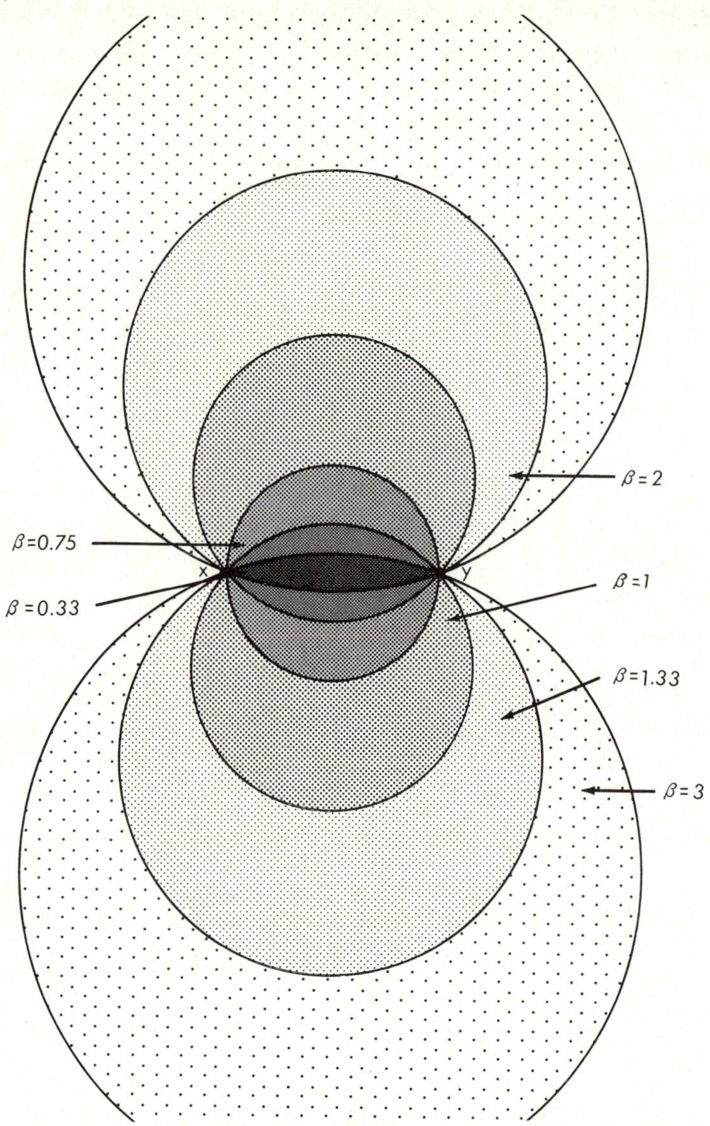

Figure 6

Circle-based neighbourhoods $N(x,y,\beta)$, for various $\beta>0$.

such β, at most 0(n) edges are possible candidates for being β-neighbours. Thus, the complexity of constructing β-skeletons for $\beta \geq 1$ is at worst $0(n^2)$.

A further reduction to 0(n log n) is possible for all $\beta \geq 1$ under our circle-based scheme. Here, we exploit the fact that not only are such β-neighbours edges of the DT, but, furthermore, if x and y are not β-neighbours, then their β-neighbourhood must be occupied by at least one of the Voronoi neighbours of either x or y. A more careful treatment of these issues, including methods for dealing with dynamically changing point sets, will be presented elsewhere.

4. APPLICATIONS AND ILLUSTRATIONS

In this section, we discuss and illustrate some of the applications of our techniques to the analysis of both point set and network data. All of our illustrations use the lune-based definition of β-skeletons.

4.1 Point Patterns

We propose to use β-skeletons, like the geometrical structures that preceded (and motivated) them, as descriptors of internal shape. The principle advantage of the family of β-skeletons over such fixed structures as the MST, RNG, or GG is its versatility. Depending on the choice of β, the β-skeleton of a point set will vary from a very sparse (typically empty) graph to a very dense (typically complete) graph. This variety, together with the knowledge of the incremental transitions that fill out this spectrum, provides an added dimension in the characterization of shape that cannot be captured by any single (unlabelled) structure. To illustrate this point, consider the sequence of β-skeletons given in Figure 7 (using the same underlying point set as Figure 3). Of course, this is merely a sampling from the complete family of β-skeletons (encoded in the single β-skeleton spectrum) but it nevertheless conveys some sense of its continuity.

If we turn our attention to point sets with a very regular internal structure, we can observe another interesting feature of β-skeletons. Suppose we have a set of points arranged in a regular hexagonal pattern (Figure 8a). For obvious reasons, the set of distinct β-skeletons is much smaller than one would expect for random point sets of comparable size, and the transitions between β-skeletons are more striking. Figure 8 illustrates the edges with three specific β-values as well as the first three

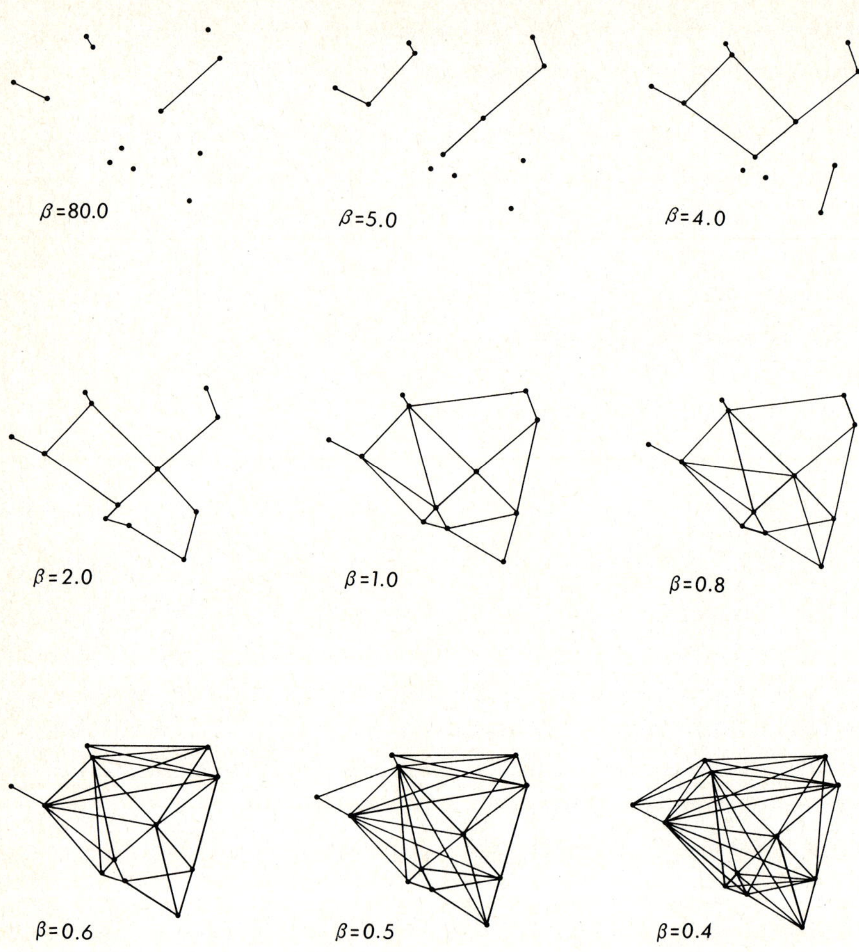

Figure 7

A sequence of β-skeletons on a fixed point set.

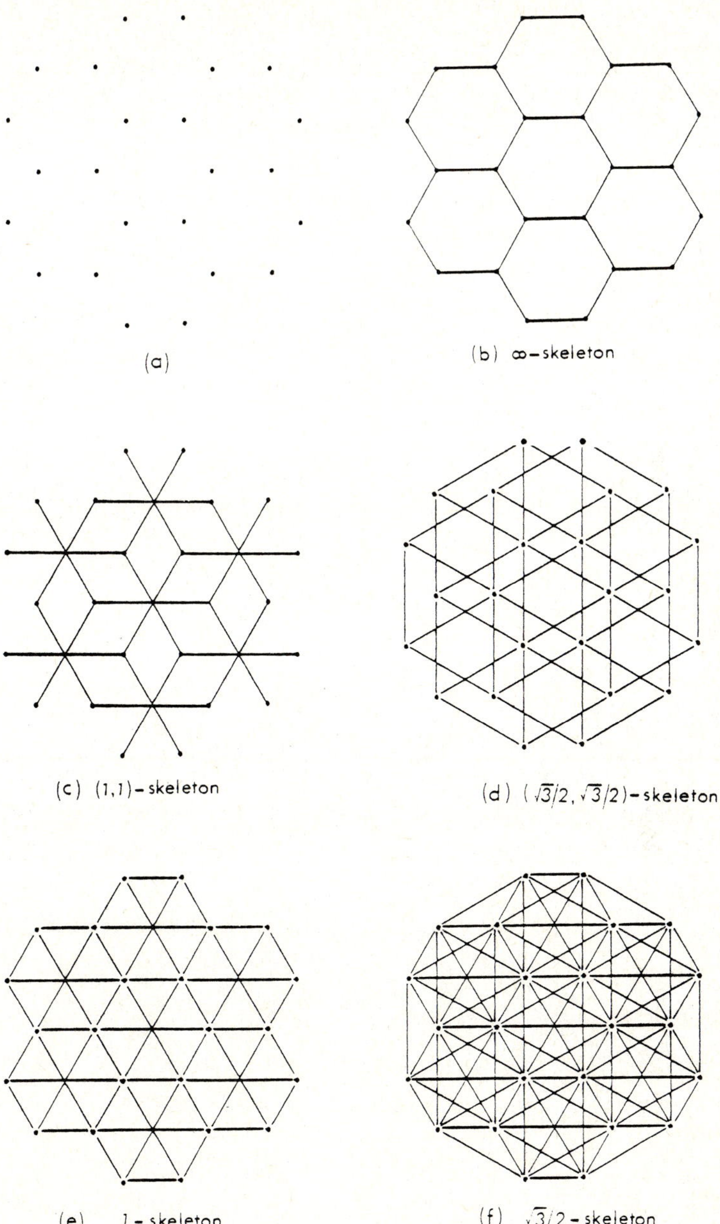

Figure 8

Skeletons of a hexagonal grid.

β-skeletons in the β-skelton hierarchy. Note that the order of appearance of edges is not strictly determined by edge length. This illustration suggests a number of ways in which β-skeletons might be used, perhaps in conjunction with distance measurements, as a measure of internal regularity of arbitrary point sets.

To help clarify both the similarities and differences between the notions of β-skeletons and α-shapes, we illustrate in Figure 9 a β-skeleton of the same point set that was used in Figure 1. What appears in Figure 9 is a tesselation of the interior of the shape formed by the underlying point set. It suggests that β-skeletons might be useful in the description of external shape as well.

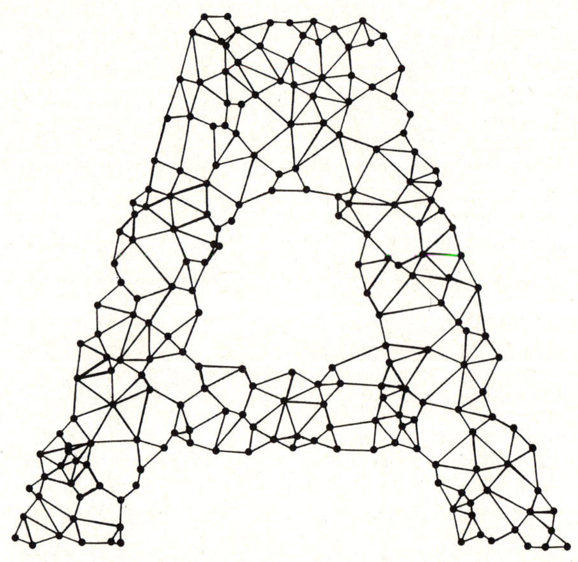

Figure 9

β-skeleton of a point pattern.

Two further applications of β-skeletons deserve a brief mention. The first concerns methods for "editing" (or "thinning") training data in nonparametric decision problems. Toussaint [T3] describes schemes that he calls Voronoi thinning and Gabriel thinning, respectively, in which objects X all of whose neighbours (Voronoi and Gabriel, respectively) are of the same class as X are deleted from the set of training data. Toussaint *et al.* [TBP] compare these schemes experimentally. It is natural to ask whether an analogous thinning scheme based on the notion of β-neighbourhoods might provide an even more versatile tool for this application.

Finally, we note the success with which the notion of Voronoi neighbours (and Delaunay triangulations) have been applied in surface fitting applications [Ba,Law,Mc]. We conjecture that β-values might be an even more potent tool for this application providing, as they do, a measure of the relative influence of various points.

4.2 Edge Patterns

Probably the most striking application of the notion of β-skeletons – indeed, the application for which the concept was first developed [R] – is to the analysis of the interpoint linkages that appear in empirical networks. Here, the β-skeleton spectrum serves as a continuous benchmark against which individual links of real networks can be compared. The basic idea is very straightforward. We start with the assumption (we could call it the null hypothesis) that all of the data points are equally significant and the neighbourliness is the sole (or dominant) factor determining interpoint connections. If this hypothesis is correct, then we would expect point pairs with the highest β-values to be connected and those with the lowest β-values to be free of connections. More generally, we would expect that the existence of edges would coincide roughly with some fixed β threshold. Of course, we expect that in real networks, the null hypothesis will be violated. By identifying links (either present or absent) that conflict the most with this hypothesis, it is possible to focus attention on those parts of a given network where forces other than neighbourliness must be at work. While, of course, this provides no direct explanation of what those other forces might be, it can be a step in the direction of a deeper understanding of the generation process of a given network.

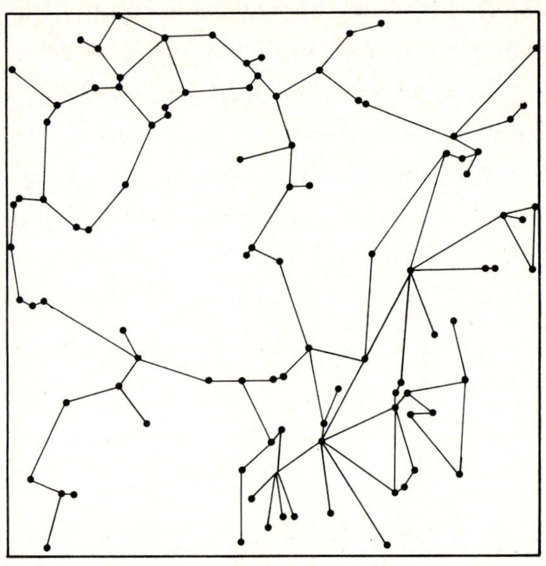

Figure 10

A hypothetical planar network.

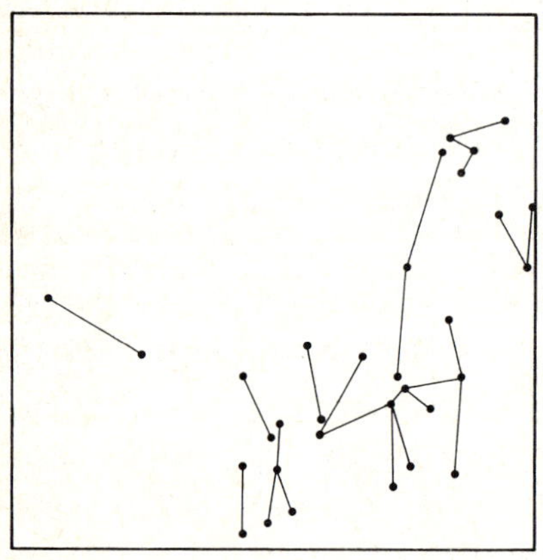

Figure 11

Links with lowest 20% of existing β-values.

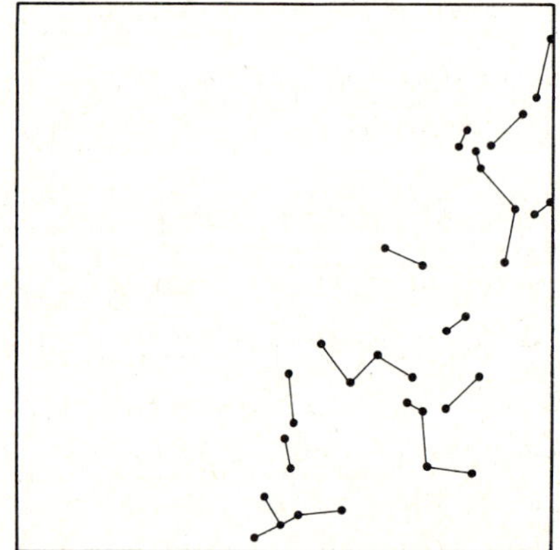

Figure 12

Missing links with β-value ≥5.

4.2.1 Artificially Generated Networks

To illustrate our methodology, consider the network depicted in Figure 10. The links in this network were generated by two quite distinct processes. In the upper left half of the figure points which were considered to be neighbourly were connected. In the lower right half a constrained stochastic process was used to form links. Examination of the β-values of the links in the resulting network reveals two quite distinct patterns. Figure 11 illustrates the links in the network possessing the lowest 20% of the existing β-values. We might call these the least expected links. Figure 12 illustrates the point pairs with high β-values that are not linked in the network. Clearly, virtually all of these exceptional links (or nonlinks) are concentrated in the lower right half of the figure.

4.2.2 Empirical Networks

The success in revealing the nonhomogeneity of the artificially constructed network above is not particularly surprising. The real test of our methodology is its ability to detect significant patterns in empirical networks. In the following, we consider the application of the same approach to the analysis of two planar networks (both highway networks) and one nonplanar (airline) network.

Figure 13 depicts the road network on Saltspring Island, British Columbia, where points represent major intersections and interconnections are drawn as straight lines. Figure 14 represents the edges with the lowest 20% of the existing β-values. Figure 15 illustrates the point pairs with high β-values that are not linked.

As we have suggested, the existence and pattern of these two types of exceptional edges can direct the questions of researchers concerned with the development of empirical networks of this type. Note that the edges of Figure 14 have a significant northwest to southeast bias in their direction. The fact that the mountainous terrain of Saltspring Island flows in a northwest to southeast direction seems to have had an obvious effect on the development of this transportation network. In Figure 15, the links run roughly orthogonal to those in Figure 14, reinforcing our conclusions. A significant number of the "missing links" of Figure 15 are "explained" by the intervening ocean.

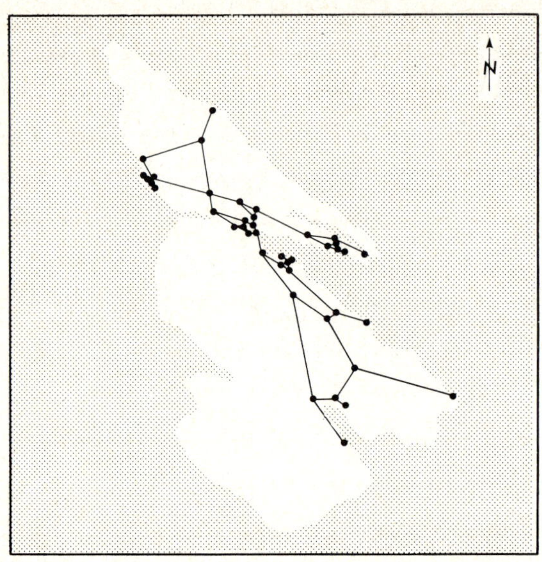

Figure 13

Road network on Saltspring Island, British Columbia.

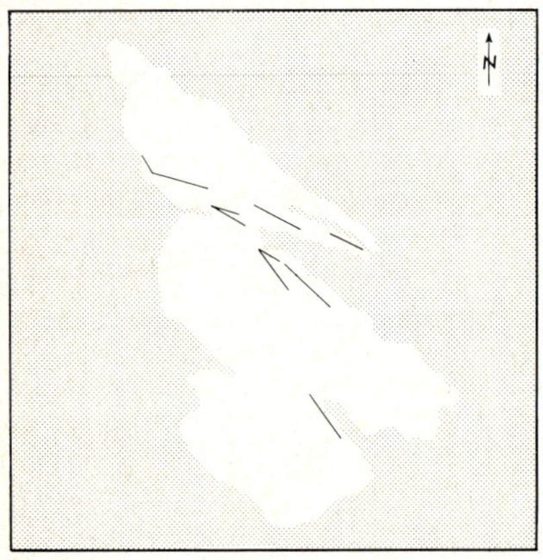

Figure 14

Links with lowest 20% of existing β-values.

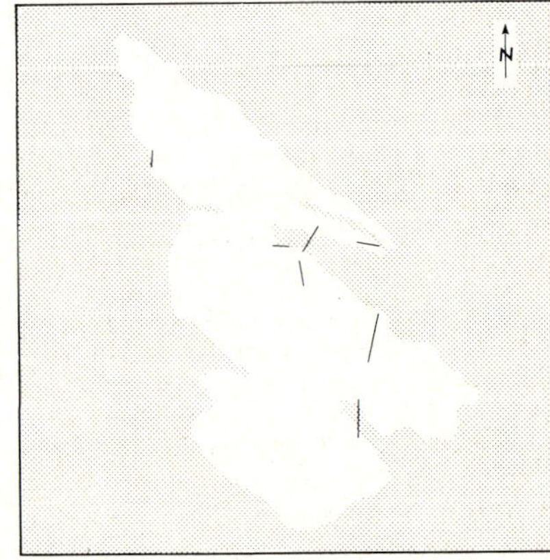

Figure 15

Missing links with β-value ≥5.

Our second planar example, illustrated in Figure 16, is the major road network in the Province of Saskatchewan. Unlike the Saltspring Island network, the Saskatchewan road network is built on a fairly homogeneous surface. Therefore, one would expect that nontopological forces, such as socioeconomic constraints, would have a greater influence on its pattern of development.

Figure 17 shows the pattern of links with the lowest 20% of β=values. Missing links with relatively high β-values are shown in Figure 18. As before, these two types of exceptional links have roughly orthogonal directional biases. The east to west bias of the links in Figure 17 is consistent with the direction of the major railroad lines in Saskatchewan. Recognition of such a connection prompts questions relating to the role played by the railroad in the development of the Saskatchewan road network.

The final empirical network that we consider here is nonplanar. It illustrates the major Canadian airline network, including both national and regional carriers (Figure 19). In contrast to our planar examples, the airline network is highly connected. Many of the links have quite low β-values. Figure 20 illustrates the links with the lowest 20% of the existing β-values. Not surprisingly, the majority of these span long distances in an east-west direction. Of course, these coincide with routes joining major cities in eastern Canada with their counterparts in the west. The missing links with high β-values, depicted in Figure 21, tend to pair up comparatively smaller (and less distant) centers.

In this example, the violation of our null hypothesis seems to have a number of apparent explanations. First, it is clear that all points (cities) are not equal with regard to the demands they impose on a transportation network. The existence of links is presumably more a function of demand (traffic) than neighbourliness. Indeed, the need for air connections seems to increase with distance (rather than decrease as is the case in other applications). It we assume that air traffic between cities is related to their size, then a better model for a network of this type (a more sophisticated null hypothesis) would involve a scheme for attaching relative weights to the points. This general idea is explored in some more detail in the next section.

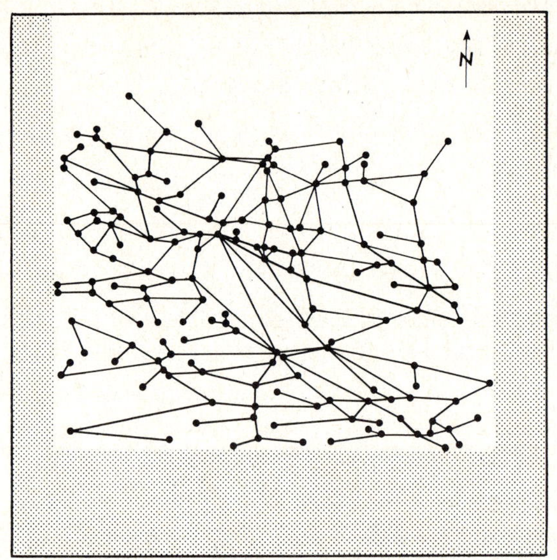

Figure 16

Major road network in Province of Saskatchewan.

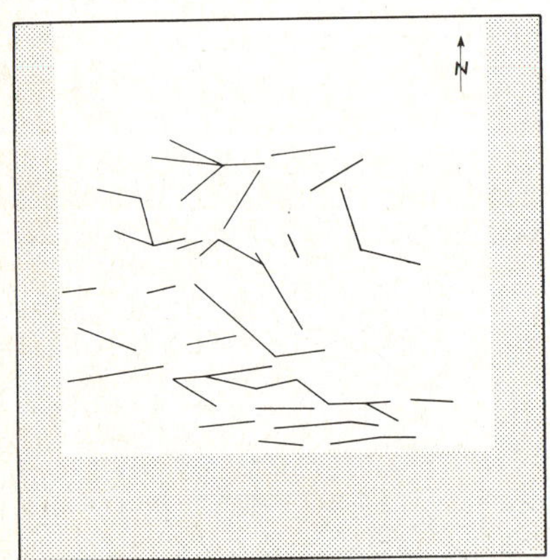

Figure 17

Links with lowest 20% of existing β-values.

Figure 18

Missing links with β-value ≥7.5.

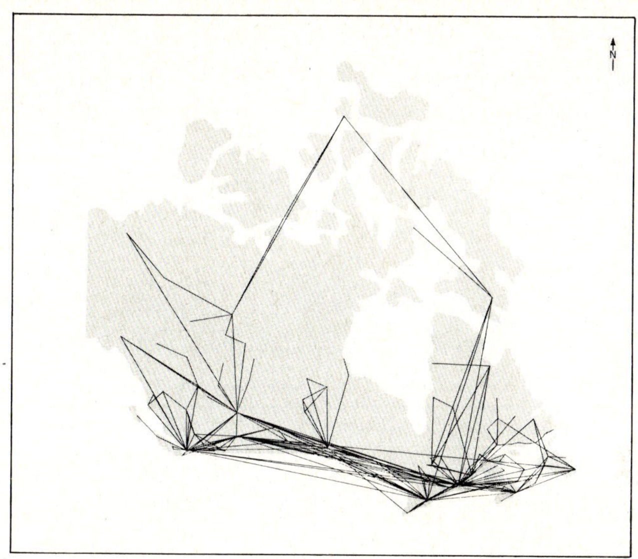

Figure 19

Major Canadian airline network.

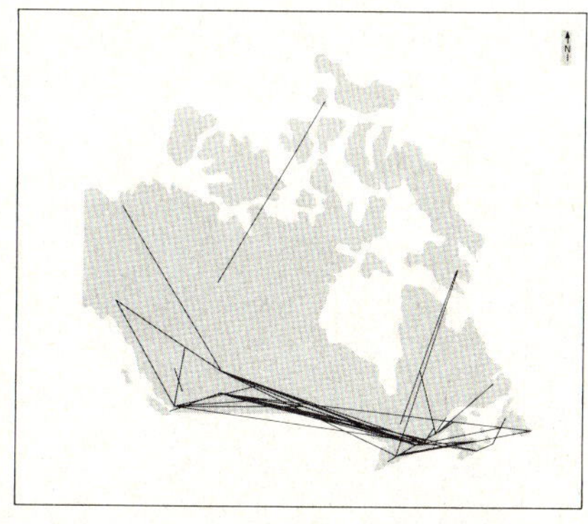

Figure 20

Links with lowest 20% of existing -values.

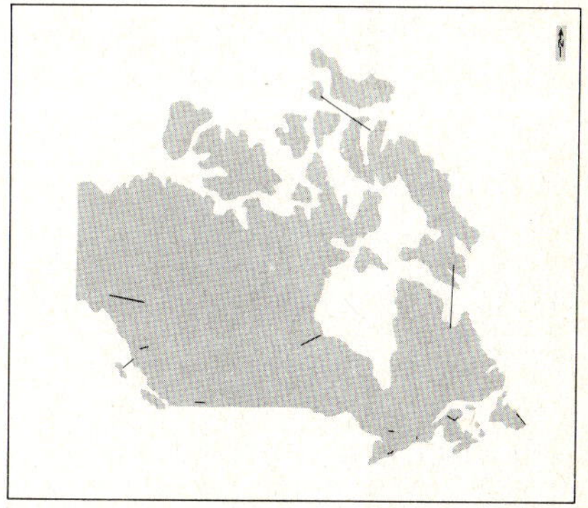

Figure 21

Missing links with β-value ≥20.

5. EXTENSIONS

We have set out the basic idea of β-skeletons for point sets in two dimensions. A number of their properties and applications have been explored, though we have not attempted an exhaustive treatment. In this concluding section, we sketch some of the ways in which the notion of β-skeletons can be extended.

5.1 Higher Dimensional Point Patterns

The most obvious, and perhaps the most pressing, extension of our ideas is to patterns in dimensions greater than two. Except for visualization and complexity considerations, we have not really exploited the assumption that point sets are two dimensional. We note that all of our geometrical constructs have natural analogs in dimensions greater than two (circles being replaced by higher dimensional spheres). However, the complexity issues that arise are far from trivial, and we leave their exploration to a later paper.

5.2 Objects Other Than Points

The notion of a Voronoi diagram, and in particular the concept of Voronoi neighbourhood, is quite easily generalized to objects other than points [LD,K]. It is natural to ask whether such generalizations apply to our concept of β-neighbourhood. Such an extension would allow us to talk, for example, about the neighbourliness of regions, avoiding one level of abstraction (replacing a region by its centroid) inherent in a number of current applications.

5.3 Weighted Point Sets

Our examination of airline networks points to the need for a measure of neighbourhood that does not treat all points uniformly. One possible approach is to adopt a gravity model (*cf.* [HCF]) in which points of greater weight (mass) have a proportionally greater attraction. This could be achieved by making the size of the β-neighbourhood of a point pair (x,y) be a decreasing function of the weights of x and y. For example, we could define a weighted Gabriel graph (β=1) using the following: points x and y are said to be *weighted Gabriel neighbours* if, for all other points z in S, z lies outside of $N(x,y,w(z)/(w(x)\cdot w(y)))$ (where w(p) denotes the weight of the point p). It is possible that some of the structure of weighted Voronoi diagrams [AE,Bo] could be used in this setting.

5.4 Temporal Structure

In practice, point sets and networks are often not static objects (*cf.* [JM]). We can imagine points and links being added (and possibly deleted) as a configuration "matures." What impact does this dimension of structure have on the notion of β-values? First, it is easy to see that the β-value of a point pair becomes a time dependent function, defined whenever the points coexist. If points are never deleted, then this function is monotone decreasing. In an empirical network, if might be significant to study the β-values of links at the time they are created or destroyed. This could reveal time dependent patterns that are undetectable in any static description of the network.

REFERENCES

[AE] Aurenhammer, F. and H. Edelsbrunner, "An optimal algorithm for constructing the weighted Voronoi diagram in the plane," *Report 109*, Institute for Information Processing, Technical University of Graz, 1982.

[Ba] Barnhill, R. E., "Representation and approximation of surfaces," in *Mathematical Software III*, J. R. Rice, ed., Academic Press Inc., 1977, pp. 69-120.

[Bo] Boots, B.N., "Weighting Theissen polygons," *Economic Geography* **56** (3), July 1980, pp. 248-259.

[BG] Boots, B. N. and A. Getis, "Probability model approach to map pattern analysis," *Progress in Human Geography* **1** (2), 1977, pp. 264-287.

[CT] Cheriton, D. and R. E. Tarjan, "Finding minimum spanning trees," *SIAM Journal of Computing* **5**, Dec. 1976, pp. 724-742.

[EKS] Edelsbrunner, E., D. G. Kirkpatrick and R. Seidel, "On the shape of a set of points in the plane," *IEEE Trans. Information Theory* **29** (4), July 1983, pp. 551-559.

[F] Fairfield, J., "Contoured shape generation: forms that people see in dot patterns," *Proc. IEEE Conf. on Systems, Man, and Cybernetics*, 1979, pp. 60-64.

[FS] Freeman, H. and R. Shapira, "Determining the minimum-area encasing rectangle for an arbitrary closed curve," *Comm. ACM* **18**, July 1975, pp. 409-413.

[GB] Getis, A. and B. N. Boots, *Models of Spatial Processes*, Cambridge University Press, Cambridge, 1978.

[GKLN] Gowda, I. G., D. G. Kirkpatrick, D. T. Lee and A. Naamad, "Dynamic Voronoi diagrams," *IEEE Trans. Information Theory* **29** (5), Sept. 1983, pp. 724-731.

[HCF] Haggett, P., A. D. Cliff and A. Frey, *Locational Analysis in Human Geography*, Arnold, London, 1977.

[JM] Johnson, W. A. and R. F. Mehl, "Reaction kinetics in processes of nucleation and growth," *Trans. American Inst. of Mining, Metallurgical and Petroleum Engineers* **135**, 1939, pp. 410-458.

[K] Kirkpatrick, D. G., "Efficient computation of continuous skeletons," *Proc. Twentieth Annual IEEE Symposium on Foundations of Computer Science*, Oct. 1979, pp. 18-27.

[Lan] Lankford, P. M., "Regionalization: theory and alternative algorithms," *Geographical Analysis* **1** (2), April 1969, pp. 196-212.

[Law] Lawson, C. L., "Software for C^1 surface interpolation," in *Mathematical Software III*, J. R. Rice, ed., Academic Press Inc., 1977, pp. 161-194.

[LD] Lee, D. T. and R. L. Drysdale, "Generalization of Voronoi diagrams in the plane," *SIAM J. Comput.* **10** (1), Feb. 1981, pp. 73-87.

[LS] Lee, D. T. and B. J. Schachter, "Two algorithms for constructing a Delaunay triangulation," *International Journal of Computer and Information Sciences* **9** (3), 1980, pp. 219-242.

[Mc] McLain, D. H., "Two dimensional interpolation from random data," *Computer J.* **19** (2), 1976, pp. 178-181.

[Me] Medek, V., "On the boundary of a finite set of points in the plane," *Computer Graphics and Image Processing* **15**, 1981, pp. 93-99.

[MS] Matula, D. W. and R. R. Sokal, "Properties of Gabriel graphs relevant to geographic variation research and the clustering of points in the plane," *Geographical Analysis* **12**, pp. 205-222.

[R] Radke, J., "Pattern recognition in circuit networks," Ph.D thesis, Department of Geography, University of British Columbia, November 1982.

[Sh] Shamos, M. I., "Geometric complexity," *Proc. Seventh ACM Symposium on the Theory of Computing*, May 1975, pp. 224-233.

[SH] Shamos, M. I. and D. Hoey, "Closest-point problems," *16th Annual IEE Symposium on Foundation of Computer Science*, Oct. 1975, pp. 151-162.

[S1] Sibson, R., "Locally equiangular triangulations," *The Computer Journal* **21**, 1978, pp. 243-245.

[S2] Sibson, R., "The Dirichelet tesselation as an aid in data analysis," *Scandinavian Journal of Statistics* 7, 1980, pp. 14-20.

[Su] Supowit, K. J., "The relative neighbourhood graph, with an application to minimum spanning trees," *J. ACM* **30** (3), July 1983, pp. 428-448.

[T1] Toussaint, G. T., "The convex hull as a tool in pattern recognition," *Proc. AFOSR Workshop in Communication Theory and Applications*, Provincetown, MA, Sept. 1978, pp. 43-46.

[T2] Toussaint, G. T., "The relative neighbourhood graph of a finite planar set," *Pattern Recognition* **12** (4), 1980, pp. 261-268.

[T3] Toussaint, G. T., "Pattern recognition and geometrical complexity," *Proc. Fifth International Conference on Pattern Recognition*, Dec. 1980, pp. 1324-1347.

[TB] Toussaint, G. T. and B. K. Bhattacharya, "On geometric algorithms that use the furthest-point Voronoi diagram," Technical Report No. SOCS-81.3, School of Computer Science, McGill University, Jan. 1981.

[TBP] Toussaint, G. T., B. K. Bhattacharya and R. S. Paulsen, "The application of Voronoi diagrams to nonparametric decision rules," *Proc. of Computer Science and Statistics: 16th Symposium on the Interface*, Atlanta, Georgia, March 1984.

[TM] Toussaint, G. T. and R. Menard, "Fast algorithms for computing the planar relative neighbourhood graph," *Proc. Fifth Symposium on Operations Research*, Köln, August 1980, pp. 425-428

[U] Urquhart, R., "Graph theoretical clustering based on limited neighbourhood sets," *Pattern Recognition*, **15** (3), 1982, pp. 173-187.

[Z] Zahn, C. T., "Graph-theoretical methods for detecting and describing gestalt clusters," *IEEE Trans. Computer* **C-20**, Jan. 1971, pp. 68-86.

COMPUTATIONAL GEOMETRY, G.T. Toussaint (Editor)
© Elsevier Science Publishers B.V. (North Holland), 1985

DISPLAY OF VISIBLE EDGES OF A SET OF CONVEX POLYGONS**

D. T. Lee* and I. M. Chen
Dept. of Electrical Engineering
and Computer Science
Northwestern University
Evanston, Illinois 60201, U.S.A.

We present an efficient algorithm for computing visible boundary edges of a set of m nonoverlapping convex polygons. The algorithm runs in time $O(m \log m + \Sigma_{i=1}^{m} \log n_i + F)$, where n_i is the number of vertices of polygon i and F the complexity of the visible scene. The method can be applied to solve three dimensional hidden-surface problem based on scan-line approach, namely, each scan-line plane intersects a set of objects in three-dimensional space and yields a set of nonoverlapping convex polygons.

Keywords: Hidden-surface elimination, raster-scan, convex polygon, analysis of algorithms.

1. INTRODUCTION

In producing computer generated pictures of three dimensional (3-D) objects on a 2-D display screen we must perform, among other things, the task of removing hidden surfaces of the objects in order to enhance comprehensibility of the scene. This task in computer graphics is known

---

*Supported in part by the National Science Foundation under Grants No. MCS-7916847 and MCS-8202359.

**An earlier version was presented in the International Computer Symposium 1982, Taichung, Taiwan, Dec. 1982, pp. 727-733.

as hidden-line or hidden-surface elimination problem. In the past two decades tremendous amount of effort has been devoted by computer scientists to address the problem. It was first attacked by displaying "Wire-frame" structures with hidden-lines eliminated or dashed on random-scan vector drawing devices. Recently with the advent of raster display devices which make computer-shaded pictures possible, researchers begin to consider the hidden-surface problem by displaying the objects' visible surfaces with various "tones", (while obscured surfaces removed) in order to produce more realistic pictures.

To generate an image of a 3-D scene we need a model to represent the objects in the scene. A convenient primitive to use as the basis for modeling an object is the polyhedron. By increasing the number of polygonal faces, a polyhedron can be constructed to satisfactorily approximate any solid object. Furthermore, this planar approximation offers a means of reducing hidden-surface computation to within reasonable bounds [1]. Because of its simplicity most of the recently devised fast hidden-surface algorithms have been based on the use of this approximation. These algorithms have been summarized and classified by Sutherland et al.[2] into three major categories, i.e., object space algorithms, image space algorithms and list priority algorithms. They propose sorting in the survey as a fundamental means of differentiating among the algorithms, whereas coherence is a primary property of scenes that all efficient algorithms use. Object space algorithms take a part of the scene such as an edge or a face and determine if it is visible. The portions of edges or faces that are invisible are then removed from the scene and the resulting picture is displayed on the screen. This type of algorithms (see, for example, [3-7]) do not take into account the resolution of the display device but solve the hidden-line problem to the precision of the CPU. Image space algorithms, on the other hand, operate over the pixels of the screen and determine the color of each one. They calculate only to the accuracy of the display device. (See, for example, [8-11]). The list priority algorithms, however, fall in the middle because they operate partly in each space, i.e., they do depth calculation in object space but finish in image space, calculating the picture only to the resolution of the display device (See, for example, [12-16]). For detailed description of these algorithms, the reader is referred to [2].

In this paper we shall disucss a scan-line oriented hidden-surface elimination algorithm which, according to the above classification, may be considered to fall into the last category, i.e., list priority. The algorithm presented is most suitable for situations when the same scene consisting of various objects needs to be displayed repeatedly for different view directions. In Section 2 we give some background on scan-line algorithms and the terminology to be used later on. In Section 3 we present the algorithm and conclude in Section 4 with some discussions about how the algorithm can be run in conjunction with a shader that performs calculations of shades or tones of visible surfaces.

2. BACKGROUND AND TERMINOLOGY

The input of a hidden-surface elimination algorithm usually consists of: i) a description of the "environment" in object coordinate system, ii) an observer's eye point and view direction and iii) a description of the screen on which the resultant picture is to be displayed. Preprocessing of the input is required. First, the data in object coordinate system is transformed into eye coordinate system with view direction being the positive Z-axis. Then a perspective or orthogonal transformation is performed which further transforms the scene into screen coordiante system, followed by a clipping operation to tailor the scene to screen size. All these opertions preserve the lines, faces and depth information. Hidden-surface algorithms are then applied to objects in screen coordinate system [1].

The raster scan devices usually scan the scene from top to bottom and from left to right in a fixed pattern. The scan-line algorithms are designed to generate the computed image line by line. Each scan-line is considered as a lateral "cut" through the scene where the scan-line plane cuts faces of the scene into segments. The hidden-surface problem then becomes a two-dimensional hidden-line problem which calls for the elimination of invisible segments created by each scan-line.

Each scan-line is then divided into smaller sections, called sample spans (Figure 1). Within each sample span, depth information is used to determine which segment (or face) in visible. Let m denote the number of polygons resulted from the intersection of all faces with the scan-line plane and n denotes the total number of edges. Since the number of sample spans is of order n and for each span there may be as many as m segments

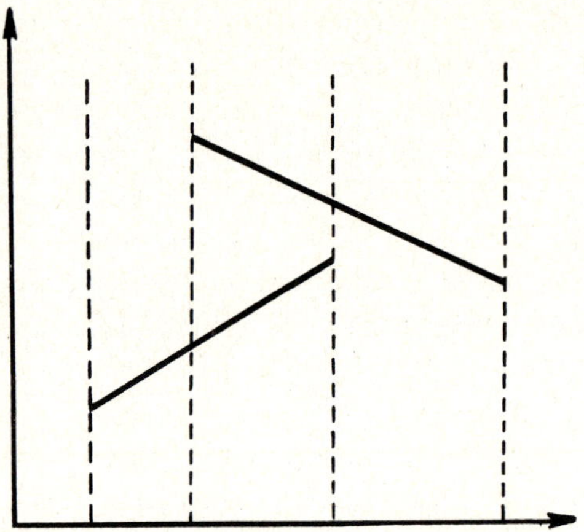

Figure 1
Sample Spans of Each Scan-line

for which the depth computation is needed in order to find the visible segments, the time required to solve the hidden-line problem is of order m n. As we shall show below, the time bound can be improved if the polygons satisfy certain properties.

Under most circumstances, the polygons are generally nonoverlapping and simple. We shall further assume that the polygons are convex. If n_i denotes the number of edges of polygon P_i, the algorithm to be presented below runs in time $O(m \log m + \Sigma_{i=1}^{m} \log n_i + F)$, where F is the number of visible edges or, more concisely, the output size. A similar problem with different objectives is studied by Edelsbrunner et al. [17]. From here on we shall address the problem of finding visible portion of the boundary edges of a set of nonoverlapping convex polygons from a given view point.

A few notation and terminology are in order. Each polygon P is represented by a list of its vertices $v_0, v_1, \ldots, v_{n-1}$ in counter-clockwise (CCW) cyclic order. A point q of an edge is said to be visible from p if the line segment $\overline{p,q}$ does not intersect any other edges of the polygons. For a view point q outside polygon P, a vertex v of P is called <u>supporting</u> if all the vertices of P lie on one side of the line determined by q and v, is called <u>concave</u> if the line segment $\overline{q\,v}$ intersects the interior of P and is called <u>reflex</u> otherwise (Figure 2). The vertex u is a <u>left</u> (or <u>right</u>) supporting vertex if all the vertices of P lie on the <u>right</u> (or <u>left</u> respectively) side of the directed line $\overrightarrow{q\,u}$. If there are more than one left or right supporting vertices we select the one that is closer to the view point. The left and right supporting vertices divide the edges of P into two chains, front and back. The front (back) chain of P with respect to q consists of those edges that are potentially visible (definitely invisible respectively) from q. An edge in the front chain is potentially visible because it may be obstructed by other polygons.

3. THE ALGORITHM

In this section we describe the algorithm in two parts. The first part deals with each polygon individually by locating its two supporting vertices, thereby obtaining potentially visible portion of the polygon. The second part then determines for each polygon whether any of its potentially visible edges is obstructed by other polygons from the depth information. In order to facilitate locating supporting vertices of each polygon P we shall store the vertices of P in a height-balanced doubly threaded binary search tree T(P) [18]. Each node in T(P) is associated with a vertex v_i and a directed boundary edge $E(v_i)$ emanating from v_i to v_{i+1}. Without loss of generality we assume that v_0 is the leftmost vertex in the tree T(P).

3.1 SUB-ALGORITHM FOR A SINGLE POLYGON

Let us consider the first part which is composed of two passes. The first pass uses a recursive procedure RXTRM on T(P) to find the right extreme point RX of P with respect to a directed line RVDLN, where RVDLN is a reference view direction determined by the view point q and an arbitrarily chosen vertex, say v_0. The right extreme point RX is defined to be the vertex on the <u>right</u> of RVDLN such that the line L passing through RX and parallel to RVDLN has the property that all the vertices of P lie on one side of L. (Again, if there is more than one such vertex, select the one closer to q.) The second pass determines the left extreme point

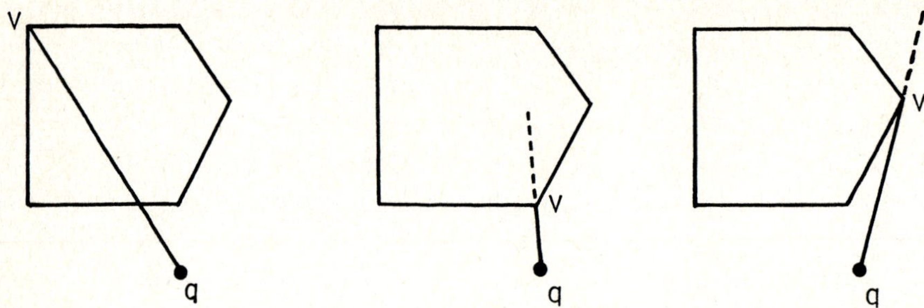

Figure 2

Illustration of Concave, Reflex and Supporting Vertices

LX of P which is defined similarly with RVDLN in reverse direction. In Figure 3a vertices 2 and 5 are right and left extreme points respectively.

Given a base directed line whose polar angle is assumed to be zero, and two directed lines L1 and L2, we say that the direction of L1 is <u>greater</u> than the directon of L2, denoted DIR(L1) > DIR(L2), if the polar angle of L1 is greater than that of L2, where the polar angle is measured CCW from the base directed line to the line. Thus, the base directed line is the "smallest". Without loss of generality we choose as the base the directed line of $E(v_0)$, the directed edge associated with the vertex $v_0 = \min(T(P))$. Thus, for the nodes in T(P) we have $DIR(E(v_{i+1})) > DIR(E(v_i))$, $i = 0,1,\ldots,n-2$.

We are now ready to give the procedure RXTRM to find RX. By calling RXTRM (root(T(P)) we go through the tree to search for the node v_k such that $DIR(E(v_k)) \geqslant DIR(RVDLN) > DIR(E(v_{k-1}))$ for some k. Referring to Figure 3b, we will start with node 2 and visit node 0 and then node 1 and find that node 2 is the right extreme point.

```
Procedure  RXTRM(v)
  if DIR(RVDLN) > DIR(E(v))
  then if RSON(v)= Λ then return (v+1)mod n
                     else RXTRM(RSON(v))
  else if LSON(v)= Λ then return v
                     else RXTRM(LSON(v))
```

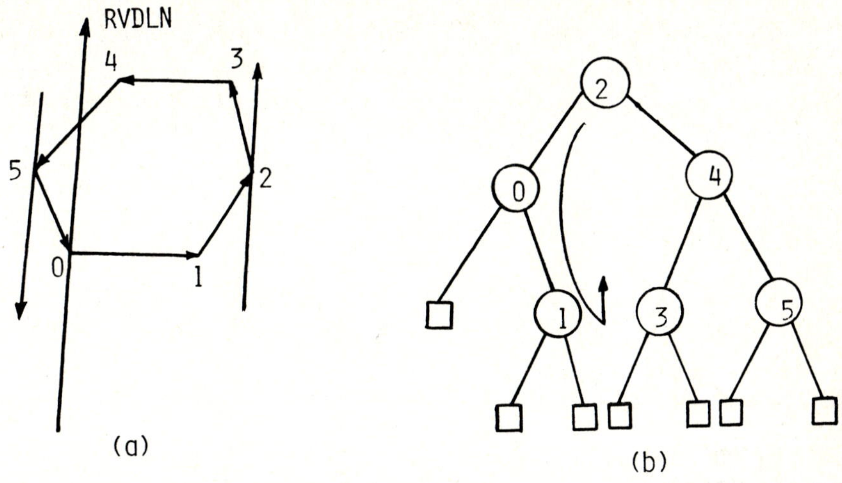

Figure 3
Search for the Extreme Points

Immediately, we can see that this procedure takes time of order logn, where n is the number of vertices in T(P). The left extreme point is found in a similar way by reversing the direction of RVDLN and replacing > in DIR(RVDLN) > DIR(E(v)) with ⩾ . After RX and LX have been determined, we begin the second pass to locate the right and left supporting vertices. Note that the supporting vertices must lie within the portion of vertices traversed CCW from LX and RX. For brevity, we simply call this portion of vertices as the vertices between LX and RX.

The procedure RSUPPORT given below starts from RX via call RSUPPORT(RX) and goes through nodes in T(P) to find the right supporting vertex RS in time O(log n). RSUPPORT is a modification of an algorithm due to Preparata [19] and works as follows. If RX is not right supporting, it must be concave. Let u be a reflex vertex between LX and RX. Then the right supporting vertex must lie between u and RX. With this observation in mind we shall start from RX, a concave vertex, to search for the right supporting vertex by first locating a reflex vertex u that lies between LX and RX. Considering the current node, which is RX initially, as a "leaf" we traverse via PARENT pointers up the tree to find this leaf's "predecessor" v in T(P). Note that in RSUPPORT the parent of the root of T(P) is defined to be the right-most node of T(P), i.e., v_{n-1} = max (T(P)). We shall call v a <u>pseudo predecessor</u> of the current node because it may not be the vertex immediately before the current node in the list of vertices of P. In Figure 4, node v1 is the pseudo prodecessor of RX. If v is right supporing, we are done. If v

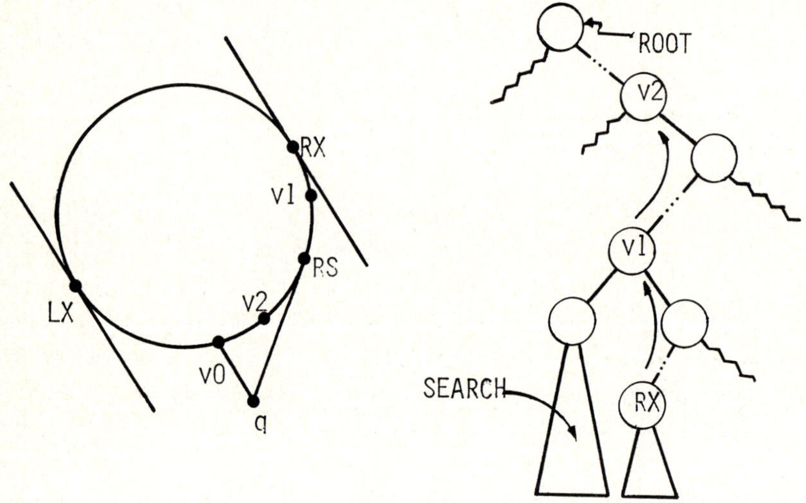

Figure 4

An Example Illustrating Procedure RSUPPORT.

is concave and between LX and RX, we repeat the "up" traversal process, with v being the last concave node. Otherwise, (i.e., either v is reflex or is concave but not between LX and RX), the right supporting vertex must lie between the current node and the last concave node (LASCCV), i.e., in the sub-tree rooted at the left son of the last concave node. We can then find the right supporting vertex simply by a tree search down the sub-tree in a manner similar to that in finding RX in T(P). As shown in Figure 4, since RX is not right supporting, we go up the tree to find its pseudo predecessor v1. But v1 is concave, so we go up the tree and find v2. Now that v2 is reflex, we conclude that the right supporting vertex must lie between v2 and v1 and search the vertex in the sub-tree rooted at LSON(v1).

```
Procedure RSUPPORT(v)
        begin
              if v is right supporting
              then return v
              else if v is root then CYCLE(PRED(v), v)
                                 else CYCLE(v, v)
        end

Procedure CYCLE(u, LASCCV)
        begin
              v = u
              vp = PARENT(v)
              while v is LSON(vp) do begin v = vp
                                           vp = PARENT(v)
                                     end
(Comment: vp is now the pseudo-predecessor of u)
              if vp is right supporting
              then return vp
              else if vp is concave and between LX and RX
                   then CYCLE(vp, vp)
                   else return SEARCH(LSON(LASCCV))
                        (Comment:  Locate right supporting vertex
                         in subtree rooted at LSON(LASCCV).)
        end
```

The procedure LSUPPORT finds the left supporting vertex in a similar way with left and right interchanged. Analysis of both LSUPPORT and RSUPPORT indicated that O(log n) time is sufficient to locate left and right supporting vertices. The proof of correctness is omitted.

We remark here that if perspective view is desirable, we can perform perspective transformation first on the input and then solve the hidden-line problem by moving the view point to negative infinity. In this case, all rays from the view point to the scene are parallel to the given view direction and right and left supporting vertices become right and left extreme points respectively. The second pass can thus be skipped.

3.2 SUB-ALGORITHM FOR MULTIPLE NONOVERLAPPING POLYGONS

Now we have identified left and right supporting vertices LS and RS of each polygon. The next task is to find the maximal visible range with respect to the view point for each of the m polygons. We shall make use of the depth information to determine those truely visible edges of the polygons. Since the vertices of P_i between the left supporting vertex $LS(P_i)$ and the right supporting vertex $RS(P_i)$ are "ordered" around the view point in decreasing polar angle, we can determine the visible edges by an m-way merge of m <u>sorted</u> lists in time $O(m \log m + N)$, where N is the total number of edges in the front chains of these m polygons. If N is large, this straightforward method can be too inefficient. In fact, the second term can be replaced with $\Sigma_{i=1}^{m} \log_2 n_i + F$, where n_i is the number of vertices of polygon P_i and F the total number of visible edges. (n_i can be further reduced to n_i' the number of vertices between $LS(P_i)$ and $RS(P_i)$, with a more sophisticated algorithm that does splitting and splicing operations on height balanced binary trees). To see this let us refer to Figure 5 in which only front chains of the polygons are shown and for the moment imagine that each chain is replaced with a line segment determined by $LS(P_i)$ and $RS(P_i)$, for $i = 1, 2, \ldots, m$. We shall first determine the visible segment among these m line segments. Once the visible portion of the line segment $\overline{LS(P_i), RS(P_i)}$ is found we can apply binary search again on the front chain of edges of P_i to obtain the visible edges. As we shall show below, the segments that are visible can be determined in $O(m \log m)$ time and the visible edges of polygon P_i determined in $O(\log n_i)$ time. Therefore, the time needed to display the visible edges is $O(m \log m + \Sigma_{i=1}^{m} \log n_i + F)$.

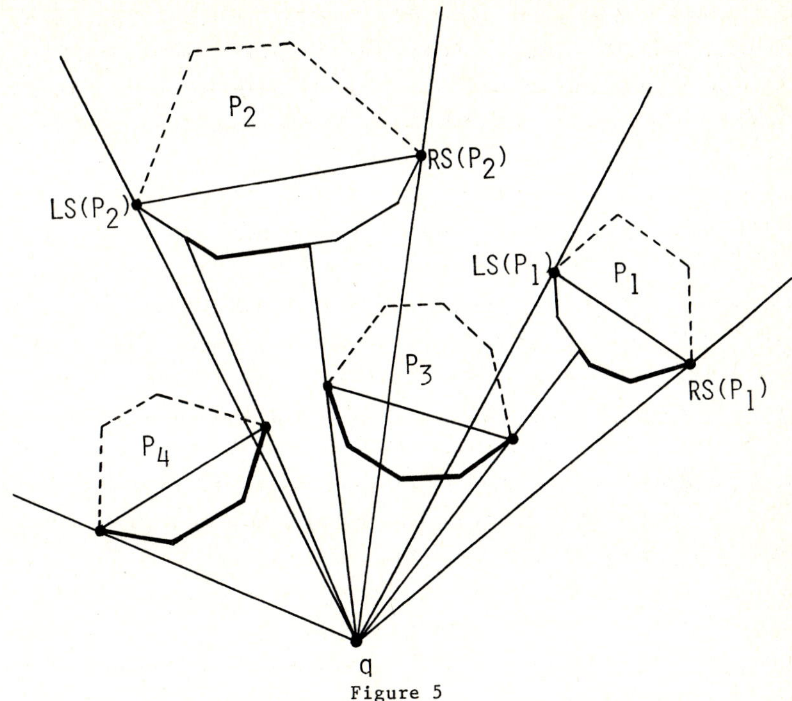

Figure 5

Illustration of Visible Edges of Multiple Polygons

Let us now describe the algorithm. We first choose arbitrarily a right supporting vertex and use as our base line the line determined by the supporting vertex and view point. Next we sort all supporting vertices (endpoints of the m line segments) according to the polar angle about the view point in CCW order. If two supporting vertices have the same angle we compare their depth to determine which one appears before the other in the sorted list. The line segments (and hence the polygons) that are cut by the base line are treated each as two individual segments. With this sorted list we shall make a linear scan by inspecting each vertex on the list one by one. If it is a right supporting vertex we insert the line segment and hence the corresponding polygon into a heap [20]; if it is a left supporting vertex, we delete its associated information from the heap. The heap is maintained according to depth so that the top element of the heap is always visible. Whenever a change is made at the top of the heap, visible polygon is updated. The right bound of a new visible segment and left bound of the last visible segment are obtained either from current supporting vertex or by performing a procedure CUT with current supporting line being a cutting line. The edges within the last visible range are displayed. When the scan is completed, all visible edges will have been displayed. In Figure 5, visible segments are shown in darkened lines.

The procedure CUT finds an edge in the front chain associated with currently the top element of the heap that intersects the cutting line and computes its intersection. CUT is obtained by modifying procedure RSUPPORT such that the edges in the front chain are classified into intersected, on the right side and on the left side with respect to the cutting line, instead of vertices being supporting, concave and reflex respectively. Deletions of elements from the heap are done with the aid of an array that keeps track of every polygon in the heap. Thus, it takes time O(log m) to maintain the heap for each update. The sub-algorithm takes $O(m \log m + \Sigma_{i=1}^{m} \log n_i + F)$ time.

In case of perspective mode operation, since the view point is now at infinity, the sorting step is performed on the set of parallel supporting lines from right to left.

4. CONCLUDING REMARKS

We have presented an efficient algorithm for computing visible boundary edges of a set of m nonoverlapping convex polygons from a view point. The set of nonoverlapping convex polygons is assumed to be the intersection of set of 3-D objects and a scan-line plane. In concept, the algorithm is similar to sample span processing but operates on polygons rather than on edges. The two supporting lines of each polygon define a sample span. Each sample span operation takes O(log m) time for maintaining a heap which makes use of object depth coherence (depth relationship among convex polygons remains the same).

The algorithm can prove to be useful when the intersections of the scene and scan-line planes are given as part of the input. Situations which warrant such an input may be when the scene is to be rotated about Y-axis or when the same scene is to be displayed repeatedly [21]. Since the algorithm calculates visible edges scan-line by scan-line, it can also be used in conjunction with a shading algorithm to produce a more realistic picture. The shader is invoked when we are ready to display visible edges which are part of polygonal faces of the object.

As pointed out by Sutherland et al. [2], depth coherence and scan-line coherence can be further used for subsequent scan-line to save some computation effort. For instance, if current scan-line intersects polyhedron P at the same faces as the previous scan-line does, then the supporting vertices will remain the same, except that their extent on X may be different, hence the supporting vertex computation can be saved.

The algorithm presented has been implemented in Pascal. Figures 6 and 7 are two output samples.

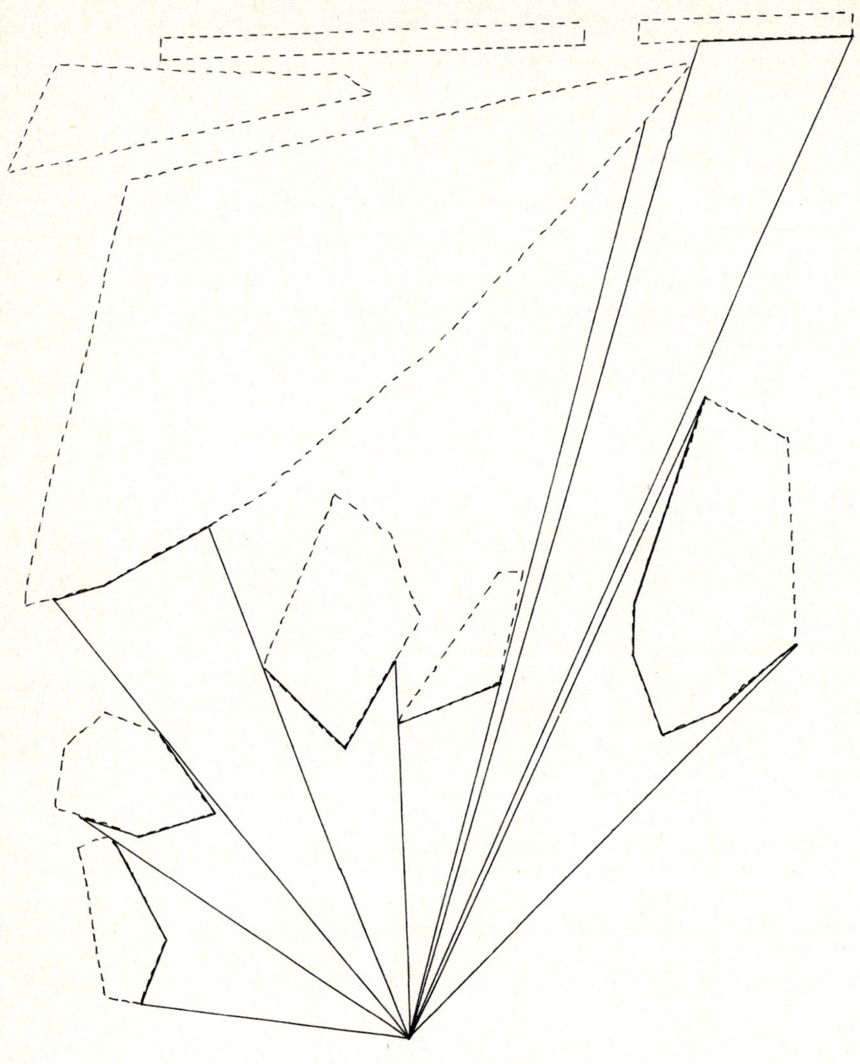

Figure 6
Computer Output Sample 1

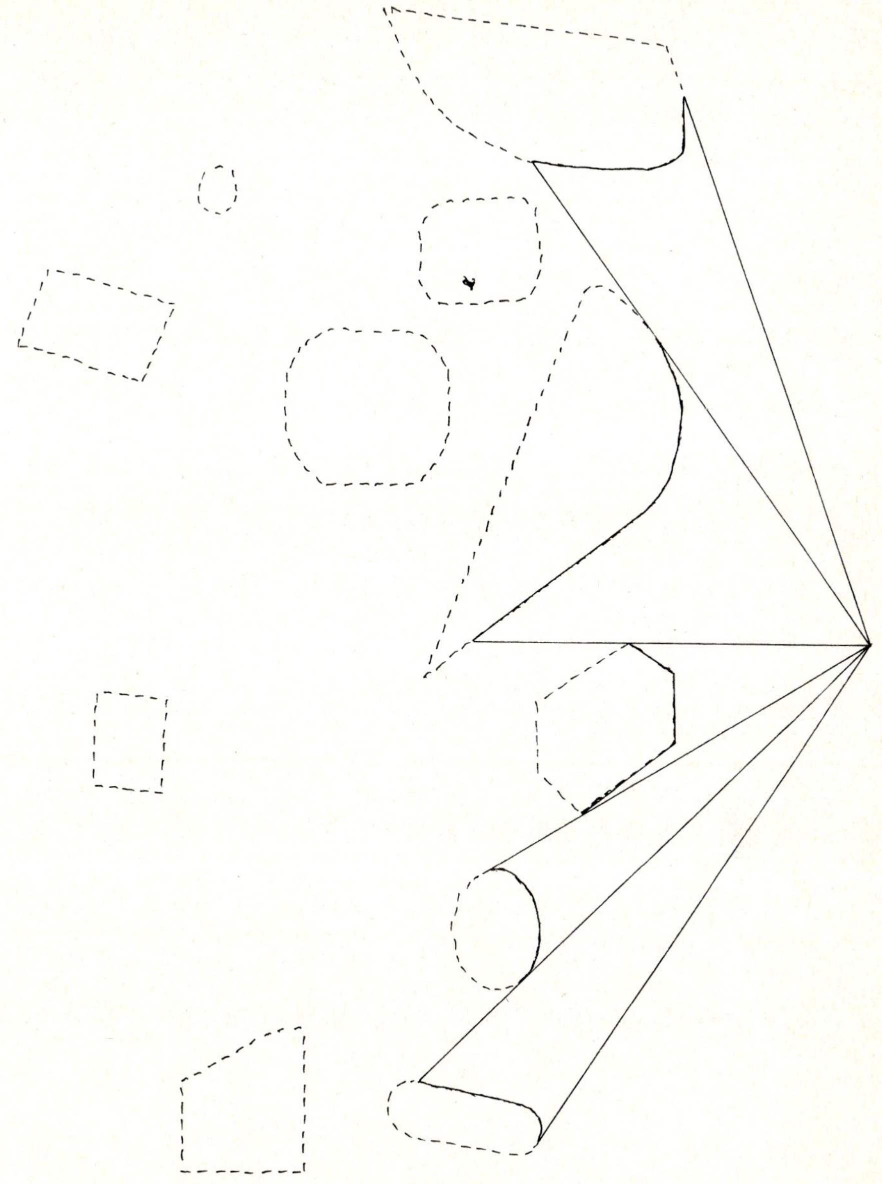

Figure 7
Computer Output Sample 2

References

1. Newman, W. M. and R. F. Sproull, Principles of Interactive Computer Graphics, 2nd edition, McGraw Hill, 1979.

2. Sutherland, J. E., R. F. Sproull, and R. A. Schumacker, "A Characterization of Ten Hidden-Surface Algorithms", Comput.Survey, March 1974, 1-55.

3. Freeman, H. and P. Loutrel, "An Algorithm for the Solution of the Two-Dimensional Hidden-line Problem", IEEE Trans. on Comput., Dec. 1967, 784-790.

4. Appel, A., "The Notion of Quantitative Invisibility and the Machine Rendering of Solids," Proc. ACM Nat. Conf., Thompson Books, 1967.

5. Galimberti, R. and U. Montanari, "An Algorithm for Hidden-line Elimination," Comm. ACM, April 1969, 206-211.

6. Loutrel, P. P., "A Solution to the Hidden-line Problem for Computer-drawn Polyhedra", IEEE Trans. Comput., March 1970, 205-213.

7. Franklin, W. R., "Combinatorics of Hidden Surface Algorithms", Ph.D. dissertation, Harvard Univ., 1978.

8. Warnock, J. E., "A Hidden-Surface Algorithm for Computer Generated Half-tone Pictures", Tech. Rep. TR-4-15, Univ. of Utah, 1969.

9. Watkins, G. S., "A Real-Time Visible Surface Algorithm", Tech. Rep. UTEC- CSc-70-101, Univ. of Utah, June 1970.

10. Archuleta, M., "Hidden Surface Line Drawing Algorithm", Tech. Rep. UTEC- CSc-72-111, Univ. of Utah, June 1972.

11. Hamlin, G. and C. W. Gear, "Raster-Scan Hidden Surface Algorithm Techniques", Proc. SIGGRAPH '77, July 1977.

12. Schumacker, R. A., B. Brand, M. Gilliland and W. Sharp, "Study for Applying Computer-generated Images to Visual Simulation", U. S. Air Force Human Resources Lab. Tech. Rep. AFHRL-TR-69-14, Sept. 1969.

13. Newell, M. E., R. G Newell and T. L. Sancha, "A New Approach to the Shaded Picture Problem", Proc. ACM Nat. Conf., 1972, p. 443.

14. Newell, M. E., "Solution to Hidden Surface Problem", Proc. ACM 1972 Annual Conf., 1972, 443-450.

15. Fuchs, H., Z. Kedem and B. F. Naylor, "On Visible Surface Generation by A Priori Tree Structure", Proc. SIGGRAPH '80, July 1980, 124-133.

16. Yao, F. F., "On the Priority Approach to Hidden-Surface Algorithms", Proc. IEEE 21st Annual Symp. on Foundations of Computer Science, Oct., 1980, 301-307.

17. Edelsbrunner, H., M.H. Overmars, and D. Wood, "Graphics in Flatland: A Case Study," in : Preparata, F. P. (ed.), Advances in Computing Research vol. 1, JAI Press, 1983, 35-59.

18. Knuth, D. E., The Art of Computer Programming, Vol. 3, Addison Wesley, 1973.

19. Preparata, F. P., "An Optimal Real-Time Algorithm for Planar Convex Hulls", Comm. ACM, July 1979, 402-405.

20. Aho, A., J. E. Hopcroft, and J. D. Ullman, The Design and Analysis of Computer Algorithms, Addison-Wesley, 1974.

21. Pavlidis, T., Algorithms for Graphics and Image Processing, Computer Science Press, 1982.

COMPUTATIONAL GEOMETRY, G.T. Toussaint (Editor)
© Elsevier Science Publishers B.V. (North-Holland), 1985

AN IMPLEMENTATION STUDY OF TWO ALGORITHMS FOR THE MINIMUM SPANNING CIRCLE PROBLEM

Robert C. Melville

Department of Electrical Engineering and Computer Science
The Johns Hopkins University
Baltimore, Maryland

1. INTRODUCTION AND PROBLEM STATEMENT

We consider the following Computational Geometry problem:

> Given a set of n points $\{p_1, p_2, \ldots, p_n\}$ in the plane, find the center and radius of the smallest circle such that no points is exterior to the circle.

Such a circle is called the "Minimum Radius Spanning Circle," because it "spans" the point set.

This problem is stated in the Operations Research literature [1] as the "Euclidean Distance Facility Location Problem" and has the following interpretation: A community of users desire the services of some facility. For example, radio receivers listening to a single transmitter. Where should the facility be located so as to minimize the maximum distance from the facility to any user. If we locate the users as points in the plane, then the center of the minimum spanning circle for this set is the optimal location for the facility. This interpretation of the problem motivates a restatement in algebraic terms:

$$\min_{(x,y)} \max_{j} \text{dist}(p_j, (x,y))$$

where dist is the Euclidean metric and (x,y) is the center of the minimum spanning circle.

Our paper develops two algorithms for this problem, and then describes the implementation of these algorithms as _working programs_ in the popular PASCAL language. This final implementation stage is often not mentioned in the published description of Computational Geometry, or other, algorithms. The reason is simple — in many cases, the algorithm has never been implemented! We point out two undesirable consequences of this lassez faire attitude towards programming:

1. Published algorithms, especially for Geometry problems, often contain errors ("bugs") particularly concerning limiting or degenerate cases. From a theoretical standpoint, these errors are more a nuisance than fundamental flaws. However, from an implementation standpoint, they are extremely bothersome. If one designs an algorithm with its ultimate implementation in mind, he is motivated to steer away from such troublesome problems and the final algorithm is often neater from a theoretical as well as from an implementation standpoint.
2. Algorithms are compared for speed using asymptotic (big "O") bounds, usually based on a worst-case analysis. Such comparisons can be extremely unrealistic and misleading if one wants to actually solve a large problem. First of all, the big "O" notation sweeps all constants under the rug, constants which might be quite significant for the values of "n" which will arise in a particular application. Also, a worst-case analysis may be inappropriate for the distribution of inputs in a particular application.

The first program we develop is an implementation of an algorithm described by Shamos [2]. A similar approach is described by Rademacher and Toeplitz [9]. This algorithm has a worst-case bound of $0(n^2)$ and works for any point set. The second program we develop is an implementation of a new algorithm for the minimum spanning circle problem described in [3]. The second algorithm runs in linear time, but requires that the convex hull of the set of data points be known in advance. If the hull is not known, $0(n \log n)$ time may be required to compute it.

The last part of our paper describes some simple experiments with these two algorithms.

2. THE SHRINKING ALGORITHM

2.1 Geometric preliminaries

The first algorithm which we present for the spanning circle problem is described in Shamos' "Notes on Computational

Geometry." Some of the key ideas of this algorithm appear in a very early paper by Chrystal [3]. Chrystal's paper is surprisingly modern in style and must be one of the earliest examples of a combinatorial analysis of the running time of an algorithm.

Let C be a circle such that each of $p_1, p_2, \ldots, p_n$ is properly inside C. Clearly, C is not the minimum spanning circle for the set. Imagine that C touches exactly one of the data points and that all the others are still inside — again C is not the minimum spanning circle (we assume here that $n \geq 2$). What if C touches two points? In this case, C is the minimum spanning circle if and only if these two points are the endpoints of a diameter of C. We call points which are on the perimeter of a spanning circle "contact points" and points which are properly inside "spanned points." The observation above can be summarized as:

> If some two contact points of a spanning circle C, define a diameter of C, then C is a minimum spanning circle.

However, this is not the whole story. The minimum spanning circle for a point set may also be determined by three points. For example, consider an isosoles triangle. All three vertices are contact points for the minimum spanning circle whose center is on the angle bisector between the two equal sides. In general:

> A spanning circle, C, is minimum if and only if the contact points for C do not fit properly in some semi-circle of C.

The proof is by contradiction. We noted above that if the set of contact points is empty or a singleton, then C may be "shrunk" to a smaller spanning circle. Suppose then, that there are at least two contact points and that the contact points fit properly in a semi-circle of C. Construct a line, L, through the center of C such that all the contact points lie to one side of L. Now shift L towards the contact points until it touches two of them. This is illustrated in Figure 1a and Figure 1b, where $\hat{L}$ denotes the shifted line through p_i and p_j. Construct the perpendicular bisector of segment p_i — p_j . Note that the center of C is on this bisector. Now shift the center of C along the bisector towards $\hat{L}$. The circle shrinks but p_i and p_j remain contact points. The shrinking process may continue until p_i and p_j become a diameter as illustrated in Figure 1c or until a new contact point, p_k , is encountered as illustrated in Figure 1d. In either case, a smaller spanning circle, $\hat{C}$ has been constructed.

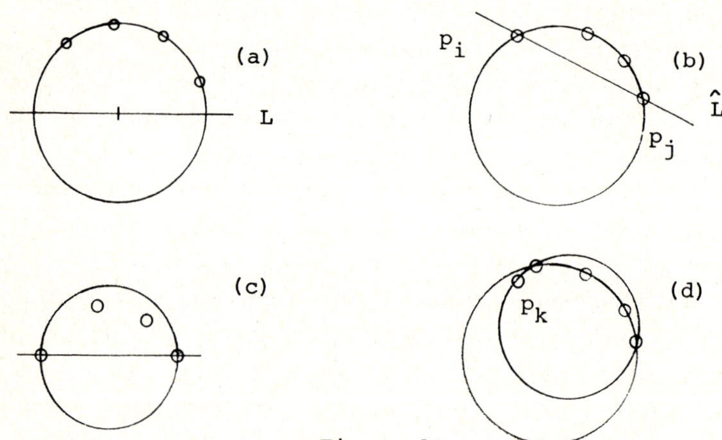

Figure 1
Illustrations for proof of shrinking lemma.

The constructive proof above is the basis for an algorithm. Get an initial spanning circle with at least two contact points. If the set of contact point does not fit in a semi-circle, then stop — the circle is minimum; Otherwise, construct $\hat{L}$ as above and shrink the circle to a new, smaller, spanning circle.

2.2 Development of the algorithm

There are three major components in an implementation of the above algorithm.

1. How to find an initial spanning circle;
2. How to test if the set of contact points fits in a semicircle of the current spanning circle, and, if so, how to compute the separating line $\hat{L}$;
3. How to shrink to a smaller spanning circle.

Another concern is an argument that the algorithm eventually terminates, and in what running-time bound.

This program does not require any sophisticated data structures. We assume that the set of points $\{p_1,p_2,...,p_n\}$ is given as the first n entries in an array:

```
type
        Point = record x,y:real end;
        Point Array =
        array [0...nmax] of Point
```

where nmax is a known upper bound of the size of n. The 0th array location will be used as a temporary variable.

A line such as $\hat{L}$ above will be represented as a normal vector and a distance from the origin. We impose the restriction that the normal vector have length 1. Moreover, if ν is the normal vector for line L and α is the distance from the origin, then $\alpha\nu$ should be the perpendicular projection of the origin onto L. Geometrically, this means that ν points "away" from the origin. The only ambiguous case is when L goes through the origin. In this case either ν or $-\nu$ will work. Notice that in the following declaration we make no distinction between a point and a vector.

```
        Line =
        record
             dir : Point;
             dst : real
        end
```

where the dir field is the normal vector and the dst field is the distance from the origin.

There are three possibilities for a non-empty set of contact points on a circle, C. The contact points may be "properly separable" (Figure 2a) because they fit properly in a semi-circle of C; they may be "improperly separable" (Figure 2b) because two of them define a diameter of C; or, finally, "inseparable" (Figure 2c). We introduce a user-defined scalar type to distinguish these three cases:

```
              SepType = ( proSep , impSep , inSep )
```

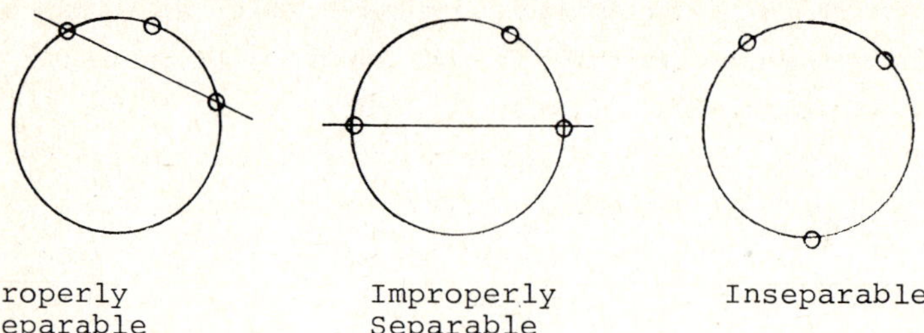

Figure 2
The three cases of separability.

The separability test is, itself, an interesting problem however we defer this to a subroutine in order to keep the paper to a reasonable size:

```
procedure TestSep(
    pts : Point Array;
    a , b : integer;
    center : Point; radius : real;
    var sep : SepType; var H : Line )
```

The procedure assumes that $1 \le a \le b \le n$ and that pts[a],...,pts[b] are the contact points on the circle of radius radius centered at center. In other words, distance(pts[k],center)=radius for $a \le k \le b$. The procedure determines whether the point set is properly separable, improperly separable or inseparable, and sets sep appropriately. In case the point set is not inseparable, H returns a separating line as described above.

We discuss briefly two strategies for the separability test. Suppose that the points _do_ fit in a semi-circle and that E is a point on the perimeter of the circle but not in the semi-circle containing the contact points. Using E as pole, stereographically project the contact points onto a line as in Figure 3. Now, in linear time, find the end-points of the set of projected points and inverse-project them back onto the circle. A line through these two inverse images will be a separating line. However, what if the points are not separable and how did we get such a convenient point E in the first place? The trick is to compute _three_ poles E_1, E_2, E_3 which are the vertices of an equilateral triangle. If there is a semi-circle containing all the data points, then at least one of these poles will not be in that semi-circle. Compute the projection described above from each pole. If all three poles fail to give a separating line, then the point set is inseparable. Since the stereographic and inverse-stereographic image of a single point can be computed in time $0(1)$, this technique runs in time linear in the number of contact points. This information is returned in the output variable sep. In the first two cases, a separating line is computed and returned in the output variable H .

The second technique is simpler. First, apply a translation which brings center to the origin. For each (translated) point pts[k] compute a rotation to bring pts[k] up to (0,radius) . Then test if all the other data points, after translation and rotation, lie to one side of the y-axis. If the point set is not inseparable, then there is some choice of k such that when pts[k] is rotated up to (0,radius), all the other points will fit in one of the half-planes determined by the y-axis. In the worst case, this technique could require $0(m^2)$ time where m is the number of contact points.

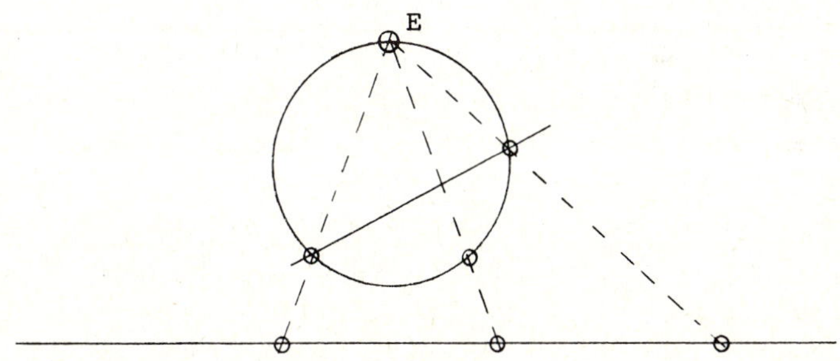

Figure 3
Separability test by stereographic projection.

Assuming the separability test as a subroutine, we can outline the minimum spanning circle procedure:

<u>procedure</u> MSC(
pts : Point Array; n : integer,
<u>var</u> center : Point; <u>var</u> radius : real);
(*Compute the center center and the radius radius of the minimum spanning circle of the set of points pts[1],pts[2],...,

```
pts[n].*)
var
     a , b : integer,
     sep : SepType;
     H : Line;
begin
     (*--Get initial spanning circle --*)
     (*-- Test separability of initial set of contact points--*)
          TestSep( pts , a , b , center , radius , sep , H );
     while sep = proSep do begin
          (*--Shrink current circle to a smaller spanning
          circle --*)
     end
end
```

In the remainder of this section we will fill in the commented sections of code in the above outline.

It is not difficult to find an initial spanning circle. Let the origin be the initial center and let the initial radius be the maximum distance from the origin to any data point. The circle so defined is clearly a spanning circle which goes through at least one point. However, by accident, the initial circle may go through more than one data point. The code segment we give below keeps track of all known contact points for the current radius. If the next point inspected is at a distance = radius from the origin then it is added to the current set of contact points. If the next point inspected is at a distance > radius then radius is updated and the set of contact points is reset to include just this more recently inspected point. We can state this more precisely as an invariant relation on the program variables:

$1=a\le b\le k\le n$

$\text{distance(pts[i],center)=radius}:a\le i<b$

$\text{distance(pts[i],center)<radius}:b\le i<k$

The code segment terminates with {pts[a],...,pts[b-1]} the contact set for the initial spanning circle. We use a procedure swap(i,j) which interchanges pts[i] and pts[j] . Also a function dist is used to compute the Euclidean distance between two points.

```
(*Get initial spanning circle and its contact points*)
center .x:=0.0; center.y:=0.0;
a:=1;b:=2;k:=2;
radius:=dist(center,pts[1]);
while k<=n do begin
     NewDist:=dist( center , pts[k]);
     if NewDist=radius
     then begin
          swap( b,k);b:=b + 1
     end else
          if NewDist>radius then begin
               swap( 1,k);b:=2;
               radius:=NewDist
          end;
     k:=k + 1
end
```

If, by accident, the initial set of contact points is not properly separable, then the while loop will execute zero times and the program will terminate with center and radius already computed.

However, the most likely situation is that the initial spanning circle has only one contact point so TestSep will return sep = proSep. In this case, H is tangent to the initial circle at the one contact point. The general iteration of the while loop is that H intersects the current spanning circle in two points which do not yet define a diameter. Let C denote the current spanning circle and D denote the circle such that $H \cap C$ is a diameter of D . If D is a spanning circle, then it is the minimum spanning circle, since two contact points of D define a diameter. Here is a slightly more subtle observation:

> Even if D is not a spanning circle, any data point properly inside of D may be "discarded" in the sense that it cannot be a contact point of the minimum spanning circle of the whole point set.

This is an important observation because it lets us eliminate at least one point from consideration at each iteration and, therefore, place an O(n) bound on the number of iterations. This lemma is illustrated in Figure 4 below. All the contact points, except the two which define H , are inside of D and may be discarded.

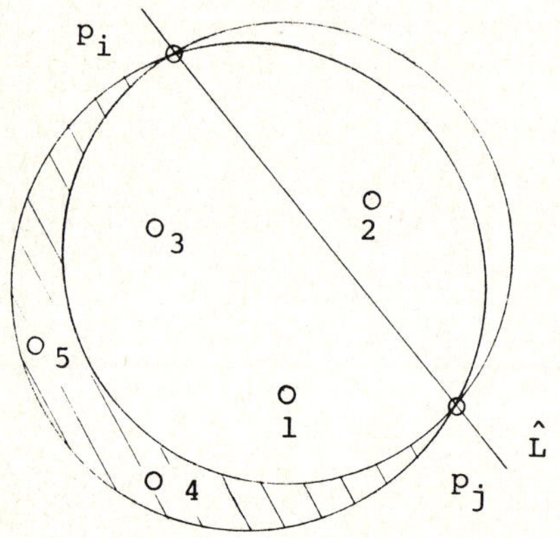

Figure 4
Illustration of shrinking step and discardable points.

Which points may not be discarded? The two contact points which define H and any data point inside C but outside D . In other words, all data points in the lune region of Figure 4.

We define cenH to be the perpendicular projection of center onto H . Let radH be the distance from cenH to $H \cap C$. In other words, cenH is the center of the circle D mentioned above. Note that radH may be computed as:

$$\text{radH} = \sqrt{\text{radius}^2 - \text{dist(center,cenH)}^2}$$

Once the center and radius of D have been computed, we can classify the data points as inside, on, or outside of this circle. As explained above, any data point which is inside of D may be eliminated from consideration at the current and all future iterations since it cannot be a contact point of the minimum spanning circle. Our program moves these discarded points to an index ≥ 1 and $< a$. Contact points for D are moved to an index $\geq a$ and $< b$. Finally, points with an index

$\geq$ b and $\geq$ n are those points in the lune region of Figure 4. These points are potential contact points at subsequent iterations.

```
(*Find center and radius of circle determined by
     intersection of H with current spanning circle.*)
cenH:=center,
Project( H , cenH );
radHsq:= sqr( radius ) - sqr( dist( center , cenH ));
radH := sqrt( radHsq );
(*Classify all points as inside, on, or outside of
     this circle *)
b := a;
k := a;
while k < = n do begin
     if dist(cenH,pts[k])=radH
     then begin
          swap( b , k );b:= b + 1
     end
     else if dist( cenH , pts[ k ])<radH
               then begin
                    (*Move pts[ k ] to 'discard' region *)
                    swap( b , k ); swap( a , b );
                    a:= a + 1;b := b + 1
               end;
     k := k + 1
end
```

The next operation is to shrink the current spanning circle. There are two cases. If the lune region of Figure 4 has no data points in it, then D is the minimum spanning circle determined by its two contact points as a diameter. The code segment above would terminate with b=n+1 in this case, since all the data points are either inside or on D . Otherwise, for each point pts[k] such that $b \leq k \leq n$, there is a unique circle C_k which goes through pts[k] , through $D \cap C$ and which has its center on the line through center and cenH.

This is most easily understood as a continuous motion in which center moves along a line perpendicular to H towards cenH . As center moves towards cenH, radius decreases terminating with center = cenH and radius = radH . This shrinking circle would pass through each data point in the lune. Again, we defer the computation of the shrinking circle to a subroutine:

```
procedure Shrink(
     H : Line;
     cenH : Point; radH : real;
     sh : Point;
     var NewCenter : Point; var NewRadius)
```

Here sh is a point in the lune which determines a new circle with center NewCenter and radius New Radius. This computation is simplified by using vector methods. Note that cenH=center+ αH.dir where H.dir si the normal vector for line H. NewCenter= center+βH.dir for some β, $|\beta|<|alpha|$. The correct value of β is determined by the condition that the distance from NewCenter to sh equal the distance from NewCenter to $H \cap C$.

The program calls Shrink for each point in the lune and records which point in the lune gives the new circle of maximum radius. This circle will be a spanning circle smaller than the current spanning circle. A loop is used to collect all the data points which are contact points of this new spanning circle (by accident, there may be more than one contact point). Finally, TestSep is called to evaluate the separability of the new set of contact points. In the first call to Shrink, only the radius is important so pts[0] is passed as a placeholder.

```
     if b = n+1
     then begin
          (*Circle is determined by a diameter*)
          sep := impSep;
          radius := radH; center := cenH
     end else begin
          MaxRad := radH;
          for K := b to n do begin
               Shrink( H ,cenH , radHsq , pts[ k ] , pts[ 0 ] , NewDist
               if NewDist>MaxRad then begin
                    MaxRad := NewDist;
                    swap( b , k )
```

```
          end
     end;
     Shrink( H ,cenH , radHsq , pts[ b ] , center , radius );
     (*--Determine new set of contact points--*)
     b := a;
     for k := a to n do
          if dist( pts[ k ] , center ) = radius then begin
               swap( b , k ); b := b + 1
          end;
     TestSep( pts , a , b-1 , center , radius , sep , H )
end
```

2.3 Running time of the shrinking algorithm

In this section we show that the worst-case running time of the shrinking algorithm is $0(n^2)$ for any set of n points. This is much better than the $0(n^3)$ bound for the algorithms of Bass and Schubert [10], but not as good as the $0(n \log n)$ algorithm due to Toussaint and Bhattacharya [11] or the $0(n)$ algorithm due to Megiddo [12]. The excellent asymptotic bounds of these last two algorithms must be balanced against the rather complex programming required.

If the sterographic projection technique is used for the separability test, then the $0(n^2)$ bound on the running time is almost immediate. The key observation is that there can be atmost n shrinking steps. If the shrinking circle never has more than two contact points,then the algorithm will terminate after two shrinking steps with these two contact points as a diameter. On the other hand, whenever the circle contacts three or more points which are separable, then all but two of these contact points are discarded, as shown in Figure 4. In other words, all but the last two shrinking steps reduce the number of points which must be considered.

But how expensive is each shrinking step? The first separability test mentioned runs in time linear in the number of points tested. Also, the scan through all potential "next" contact points (in the lune region of Figure 4) requires time proportional to the number of points scanned. Again, $0(n)$. Therefore, the entire algorithm is $n*0(n) = 0(n)$.

The situation is a bit different with the second separability test described above, since this test runs in time quadratic in the number of points tested. However, the observation about discarding points comes to the rescue. Let s_i be the number of points tested for separability at the ith shrinking step of the algorithm (using the second separability test). Note that $\max(0, s_i - 2)$ points are discarded at the ith step. Intuitively, if some s_i is large, then that separability test is expensive, however many points have been discarded so there cannot be too many more iterations. On the other hand, if there are many iterations, then each s_i must be small. The worst-case time spent in the separability test is the maximum value of the sum

$$\sum c s_i^2$$

where c is a constant. However

$$\sum_i s_i < 3n$$

This condition limits the maximum value of the first sum to $O(n^2)$, so the separability testing is not the dominant cost, even when using the second method.

3. The Rolling Algorithm

Our second approach to the problem is much different. We present an approximation algorithm which computes a convex region which is sure to contain the center of the minimum spanning circle. The area of this region may be made as small as desired allowing the location of the center to be approximated more and more accurately. The accuracy with which the center is located is a parameter of the algorithm and should be chosen with the application in mind. For example, if the minimum spanning circle is to be drawn on a graphic output device, then there is no reason to locate the center of the circle to an accuracy greater than one-half of a pixel.

The second algorithm also differs from the first in that it works only for a convex polygon, rather than an arbitrary set of points in the plane. Observe that any contact point on the minimum spanning circle must be a vertex of the convex hull so the first algorithm above would still work if presented with only the vertices of the convex hull of the point set. The second algorithm assumes that the convex hull has already been computed. With this assumption, it is able to find the center and radius of the minimum spanning circle within any desired fixed accuracy in time proportional to the number of hull vertices presented to it. This is much better than the $0(n^2)$ bound for the first algorithm. The problem of computing the convex hull has been well studied and several efficient algorithms are known [5],[6].

3.1 Overview of the Algorithm

The input to the algorithm is a sequence of n points which should form the vertices of a convex polygon in clockwise order. Again, no fancy data structure is necessary. Let the points be given in order as n entries in an array.

Let c^* denote the exact center of the minimum spanning circle. Likewise, r^* denotes the exact radius. How can we find a good approximation to c^* ? Consider the following definition of c^*

$$c^* = \bigcap_{\nu \varepsilon V} \mathrm{Circle}(\nu, r^*)$$

where $\mathrm{Circle}(\nu,r^*)$ is the circle of radius r^* centered at ν and V is the set of vertices of the input polygon. This observation motivates the following definition of D_r for any $r \geq 0$.

$$D_r = \bigcap_{\nu \varepsilon V} \mathrm{Circle}(\nu, r)$$

Alternatively, D_r may also be defined as

$$D_r = \{p \varepsilon R \times R \mid \max_{\nu \varepsilon V} \mathrm{dist}(p,\nu) \leq r\}$$

Note that D_r is convex since it is the intersection of convex figures. Moreover, if $r = r^*$ then $D_r = \{c^*\}$ and if $r<r^*$ then $D_r = \emptyset$. Figure 5 shows the construction of D_r for some $r>r^*$. Some explanation of the term "Rolling Algorithm" is in order. Imagine that the polygon vertices are poles and that a metal ring of radius r encircles these poles. If $r>r^*$, then the ring is free to "roll" around the poles. As the ring makes one trip around, the center of the ring will trace out exactly the perimeter of D_r.

Our algorithm approximates the answer by constructing a nexted sequence of convex regions, D_{r_i}, all of which contain c^*. Moreover, by making a good guess at c^* at each step, we can insure that $\text{area}(D_{r_{i+1}}) \leq 5/9\ \text{area}(D_{r_i})$ so that these areas are decreasing at an exponential rate.

Several lemmas are used to argue that the algorithm always converges to the right answer. All of these, except (3), follow easily from the definitions given above.

Assertion (3) is discussed in reference [7].

(1) For any $r \geq r^*$, $c^* \varepsilon D_r$;

(2) $r_1 > r_2 \geq r^*$ implies that $D_{r_2} \subseteq D_{r_1}$;

(3) Any line through the centroid of a convex figure divides the figure into two convex regions whose worst-case ratio of areas is 4:5;

(4) Let c_i be a point inside D_{r_i}, where $r_i > r^*$. Let r_{i+1} be the maximum distance from c_i to a vertex of the polygon. Then c_i is a point on the boundary of $D_{r_{i+1}}$.

From an algorithmic standpoint, the most interesting part of our construction is to show how to compute D_r in time $0(n)$ where n is the number of vertices of the polygon.

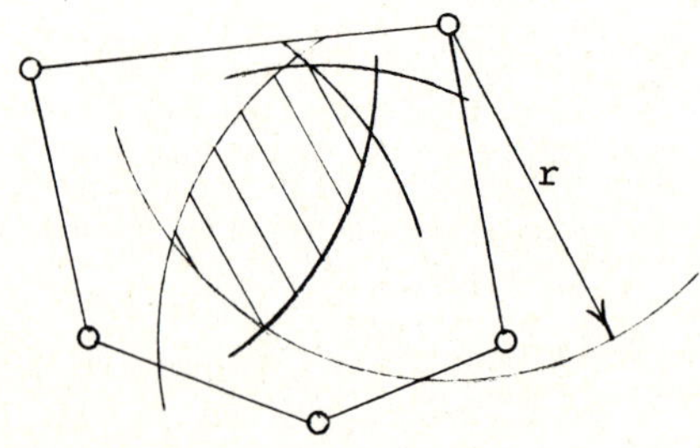

Figure 5
Construction of D_r.

3.2 Fast Computation of D_r .

Let r be an upper bound on the radius of the MSC, i.e., $r \geq r^*$. A vertex of the polygon, x , is a contact point at radius r means that there is a radius-r spanning circle through x . The following fact is a simple consequence of the definition of a contact point:

If $r_1 > r_2 \geq r^*$ then the set of radius-r_2 contact point is a subset of the set of radius-r_1 contact points.

As a limiting case, the original vertices are all contact points of a straight line, that is $r = \infty$. We assume that the vertices are available in clockwise order around the perimeter of the polygon. Let succ(x) denote the clockwise neighbor vertex of vertex x . Given $r \geq r^*$, we want to identify quickly the radius-r contact vertices. Our algorithm will produce these again in clockwise order as a sub-sequence of the input sequence. Therefore we may use the output of one iteration as input to the next iteration.

In the development of the first algorithm, we gave a sufficient condition for a point to be "discarded", that is, removed from the set of points under consideration because it could not be a contact point of the minimum spanning circle. We make a similar observation for the second algorithm. Let $circ_r(x,y)$ denote the radius-r circle through points x and y . Note that there may be two such circles — the context will clarify which one is intended. Suppose that x,y,z are consecutive vertices of a convex polygon and that y is inside $circ_r(x,z)$ where $circ_r(x,z)$ is the circle whose center is not on the same side of the line through x and z as y is. Then y is not a radius-r contact point. Intuitively, the "curvature" of the circle is greater than the "curvature" of the section of the polygon perimeter from points x to z and point y must fall inside. Note that the circle through x and z need not be a spanning circle of the entire polygon.

The first part of our algorithm is to find a pair of consecutive contact points at radius r . Let C be a spanning circle of radius r through point A_1 of the polygon. For A_2 a vertex different from A_1 , imagine "swinging" the circle through an angle θ around A_1 as a pivot until the circle first contacts A_2. Since A_2 is inside C, we are sure that the swinging circle contacts A_2 and that $0 \leq \theta < 180$. Consider choices of A_2 in clockwise order from A_1 – succ(A_1), succ(succ(A_1)),...,etc. and chose that A_2 which minimizes θ. The procedure could be repeated, using A_2 as pivot, to generate the next contact point. However, since there may be $0(n)$ contact points, we might spend $0(n^2)$ time using this procedure to find all the contact points.

We describe instead a more subtle linear-time algorithm which makes one clockwise trip around the perimeter and identifies all radius-r contact points as a sub-sequence. Our technique is a bit like algorithms which find the convex hull of a simple polygon in linear time. This problem has turned out to be quite tricky [8], so we try to give a careful correctness proof for out construction.

Our algorithm constructs a sequence $(s_1, s_2, \ldots, s_k)$ of vertices which may grow and shrink like a stack. Upon termination,

$$(s_1, s_2, \ldots, s_k) = (A_1, A_2, \ldots, A_M, A_1)$$

where M is the total number of radius-r contact points.

Suppose that contact points $A_1, A_2, \ldots, A_N$ have been identified and that there is still a contact point A_{N+1} after A_N. The stack will contain $A_1, A_2, \ldots, A_N$ as a prefix followed by some of the vertices between A_N and A_{N+1}. The stack may grow and shrink as we work from A_N to A_{N+1}, however once A_{N+1} has been placed on the stack it will not be removed.

The crucial information about the stack is summarized in invariant ST below:

ST:

1 STACK=$(s_1, s_2, \ldots, s_k)$

2 $k \geq 2$

3 s_{I-2} is inside $circ_r(s_{I-1}, s_I)$ for $3 \leq I \leq k$

4 There exists N , $2 \leq N \leq k$, such that $s_1, s_2, \ldots, S_N$ are contact points. The Invariant ST is established by the assignment:

$$(s_1, s_2) := (A_1, A_2)$$

where A_1 and A_2 were described above. The algorithm advances clockwise as long as succ(s_k) is inside $circ_r(s_{k-1}, s_k)$. However, suppose that succ(s_k) is outside $circ_r(s_{k-1}, s_k)$. We know that point s_k is not a contact point and may be skipped. In the algorithm, this means that s_k is deleted from the current sequence of vertices so that succ(s_k) becomes the new successor of s_{k-1}. Now the program retreats counter-clockwise, popping the stack, until succ(s_k) is again inside of $circ_r(s_{k-1}, s_k)$. The retreat must terminate because at worst, the program will back up to $circ_r(A_{N-1}, A_N)$ which is a spanning circle. Once succ(s_k) comes back inside, the program again starts advancing clockwise. Of course, succ(s_k) is not necessarily a contact point and the algorithm may later back up over this current succ(s_k). We trace the execution of the algorithm on the section of a polygon perimeter illustrated in Figure 6 (at end of paper).

A "Pidgin" PASCAL description of the rolling algorithm is given below:

```
STACK:= (A1,A2);
(*Invariant ST established*)
DP:= (center of initial circle);
while s[k] ≠ s[1] do begin
    while succ(s[k]) inside circ(r,s[k-1],s[k]) and (s[k]≠ s[1])
    do begin
        push center of circ(r,s[k-1],s[k]) onto DP;
        push succ(s[k]) onto STACK
    end
    if s[k]≠ s[1] then begin
        z: = succ(s[k]);
        while z outside circ(r,s[k-1], s[k]) do begin
            succ(s[k-1]): = z;
            pop STACK;
            pop DP
        end
    end
    (*Invariant ST holds now*)
end
```

3.3 Running Time Analysis

We argue that the above algorithm requires time $O(n)$ to select the radius-r contact points from an n vertex polygon.

When advancing clockwise, the algorithm is always stacking a point it has not considered before. Since there are at most n vertices, the total cost for advancing is $O(n)$ total.

Now consider the retreating action in which the algorithm "backs up" past points it has already placed on the stack. The test to determine if the algorithm should back up costs $O(1)$, since it requires only checking that one point is inside or outside of a known circle. Whenever the algorithm backs up, it eliminates a point as a possible contact point. Let us "charge" the $O(1)$ of the back up step to the said eliminated point. Again, since at most $O(n)$ points may be eliminated, the total cost of retreating is bounded by $O(n)$.

3.4 The complete algorithm

The complete algorithm consists of several iterations of the ro-ling procedure described, generating a decreasing sequence of radii:

$$r_0 > r_1 > r_2 > r_3 \ \ldots$$

and a corresponding sequence of nested convex regions:

$$D_{r_0} \supseteq D_{r_1} \supseteq D_{r_2} \quad \ldots$$

The key property of the D_{r_i} is the following. Let c_{i+1} be any point in the interior of D_{r_i}. Let r_{i+1} be the maximum distance from c_{i+1} to a vertex of the polygon. Then $r_i > r_{i+1} \geq r^*$. In other words, the interior of D_{r_i} is the locus of all points which are centers of spanning circles with radii less than r_i. Chosing c_{i+1} carefully will insure that the algorithm converges quickly.

The Area Centroid of a convex figure has the following physical interpretation. If the figure were to be cut out of sheet metal, it would balance on a pin point located under the area centroid. Moreover, the area centroid of a convex figure has the following interesting property (see [7]). Let L be any line through the area centroid of a convex figure. This line partitions the figure into two convex pieces, call them L_A and L_B. Then:

$$\max_L \frac{\text{area } L_A}{\text{area } L_B} \le \frac{5}{4}$$

This fact fits neatly into our algorithm. Let c_{i+1} be the area centroid of D_{r_i}. In linear time, determine the maximum distance from c_{i+1} to a vertex. Generate $D_{r_{i+1}}$ by "rolling" the circle of radius r_{i+1}, initially centered at c_{i+1}. Each iteration, as explained above, takes linear time. Note also that c_{i+1} is a point on the perimeter of $D_{r_{i+1}}$, because the perimeter of $D_{r_{i+1}}$ is the set of points which achieve a distance of r_{i+1} to some polygon vertex. Because $D_{r_{i+1}}$ is convex, there exists a line of support, L_{i+1}, of $D_{r_{i+1}}$ through c_{i+1}. Clearly, $D_{r_{i+1}}$ is entirely contained in one of the two convex pieces of D_{r_i} defined by L_{i+1}, therefore the area of $D_{r_{i+1}}$ cannot be more than 5/9 of the area of D_{r_i}. This tells us that the areas of the D_r's are decreasing in geometric progression. In most computer implementations of floating point arithmetic, real numbers are approximated by a rational fraction limited to a certain fixed precision. After enough iterations, the area of D_{r_i} becomes less than the smallest discernible difference between two real number representations, and we might as well stop the iteration. The number of such iterations is a function of the (fixed) precision of the real number representation. In this sence, we say that our algorithm runs in linear time over-all, because it stops after a fixed number of iterations, each of which costs $O(n)$.

The initial center, c_0, may be taken to be the area centroid of the input polygon itself.

3.5 Implementation of the Rolling Algorithm

We are now ready to describe the complete rolling algorithm as a PASCAL program. The input polygon is represented in an array:

<u>type</u> Polygon = <u>array</u>[1..n] <u>of</u> Point

This type declaration is essentially the same as the declaration of Point Array in the first algorithm, however here, it is the users responsibility to make sure that the n points are vertices of a convex polygon given in clockwise order.

As in the first algorithm we defer certain parts of the program to subroutines, both of which are straightforward. The first computes the centroid and the area of a convex polygon.

```
procedure Centroid(
    P : Polygon; size : integer;
    var c : Point; var ar : real )
```

Here, P[1],...,P[size] are the vertices of the polygon (listed in clockwise order). The procedure returns the centroid in c and the area of the polygon in ar . Because the polygon is convex, it is very easily partitioned into triangles. The total area is the sum of the areas of each triangle and the centroid may be computed as a weighted sum of the centroids of the component triangles.

The second procedure computes the center of the circle of a given radius through two given points.

```
procedure circ(
    u , v : Point; r : real; var c : Point )
```

In general, there will be two such circles. In case of two different circles, the procedure returns the center which lies to the right of the line segment directed from u to v .

As the algorithm proceeds, certain vertices are deleted from the perimeter of the polygon. This suggests a linked-list structure to store the active set of vertices. The variable succ is defined as

```
array[0..n] of integer
```

and stores a circularly linked list of indices into P . In other words, for some k:P[k],P succ[k]],Psucc[succ[k]]],...,
P[kk]
are the active polygon vertices, where succ[kk] = k . Location succ[0] always stores such a value of k . The succ array is initialized by the following code segment:

The stack of contact points during the "rolling" operation is implemented quite conventionally:

```
s : array[1..n] of integer,
top : integer
```

where top is the top-of-stack pointer. The derived polygon is stored as the first top - 1 elements of the array:

D : Polygon

Since the "rolling" procedure has been described above, at least in "PIDGIN" PASCAL, the only major piece left is how to get an initial spanning circle. As in the first algorithm, a spanning circle through at least one vertex of the polygon is very easy to find. If c is our initial center, we search all the vertices to find the one which is furthest from c . If we let the initial radius, r , be the distance from c to this vertex, then the circle with center c and radius r is clearly a spanning circle through at least one point. Our "rolling" process, however, needs the initial circle touching two points.

The vertex which is furthest from c is clearly a contact point, call it A_1 . As the circle rolls about A_1 , which other point will it encounter first? In other words, for which other contact point A_2 does the circle have to swing through the smallest angle α about point A_1 ? Figure 7 shows two circles. One is centered at c and goes through A_1 only. The other circle, shown dotted, has the same radius but goes through both A_1 and A_2 . Let c_2 denote the center of the second circle. There is an angle, α formed by the two radii to A_1 . Our code segment avoids any trigonometric calculations by observing that the minimum value of α is achieved by that center c_2 which is closest to c . The code segments finishes with A_1 and A_2 as the first two points on the stack.

```
(*--Computer r--*)
    s[ 1 ]:=succ[ 0 ];
    r := dist( c , P[ s[ 1 ]]);
    k := succ[ succ [ 0 ]];
    while k<>succ[ 0 ] do begin
        if dist( c , P[ k])>r then begin
              s[ 1 ]:=k;
              r := dist( c , P[ k ])
        end;
```

```
        k := succ[ k ]
    end;
    (*Now r is max distance from c to a vertex*)
(*--Get initial spanning circle through two points--*)
    k := succ[ s[ 1 ]];
    alphamin := Inf,
    repeat
        circ( P[ s[ 1 ]] , P[ k ] , r , c1 );
        if dist( c , c1)<alphamin then begin
              alphamin := dist( c , c1 );
              s[ 2 ] := k
        end
        k := succ[ k ]
    until k = s[ 1 ];
    (*--Compute circle through P[s[1]] and P[s[2]]--*)
        circ( P[s[ 1 ]] , P[s[ 2 ]] , r , c )
```

An outline of the complete program is given below:

```
procedure MSC(
    P : Polygon, var c : Point; var r : real );
(*--Compute minimum spanning circle of convex polygon
 P[1],P[2],...,P[n] where n is a global variable.
 Return center in c and radius in r. --*)
const Inf = 1.0E10; (*Infinity*)
var
    succ : array [ 0..n ] of integer;
    (*Circularly linked list of non-discarded vertices.
       succ[ 0 ] always points to an entry in this list.
    *)
    D : Polygon;
    (*Derived polygon*)
    s : array [ 1..n1 ] of integer,(*n1 = n+1*)
    top : integer;
    (*Stack of contact points is s[1],...,s[top]*)
    ar : real;
    (*area of D*)
    (*Miscellaneous variables:*)
    c1 : Point;
```

```
    k , z : integer,
    alphamin : real;
begin
    (*Get initial center as centroid of P*)
       Centroid( P , n , c , ar );
    (*--Initialize succ--*)
        for k := 1 to n - 1 do succ [ k ] := k + 1;
        succ[ n ] := 1; succ[ 0 ] := 1;
    while ar>sqr( Epsilon ) do begin
        (*--Compute r--*)
        (*--Get initial spanning circle through two points--*)
        (*--Compute D_r--*)
        Centroid( D , top - 1 , c , ar )
    end
end(*MSC*)
```

4. Comments on the two programs

Both algorithms were brought up as working PASCAL programs on our local computer system. The shrinking algorithm was compiled under UNIX using the Berkeley compiler. The program is about 300 lines long, much of this being "support" code to perform basic geometric computations such as a rotation of the plane. The MSC procedure itself is 115 lines of code and is essentially a copy of what was described in the section 2. We experienced only one difficulty in getting the program. This involved comparisons of real numbers. For example, a point is considered a contact point of the circle if its distance from the current center is equal to the current radius. Our program was modified to accept a point as a contact point if its distance from the center is equal to the radius within a small tolerance (1.0E-8).

We ran the program on several small test cases to make sure that it was working, then tried it out on some larger set of points. Times for three of these sets are shown in below. In each case, only a few shrinking steps were needed. This is not surprising. The points were generated using a uniformly distributed random variable for both the x and y coordinates. A large number of points generated in this fashion nearly fills the square defined by the range of the random variable. The spanning circle bumps against two diagonally opposite corners of the square and then shrinks until this diagonal is a diameter of the spanning circle.

Point set size	time	number of shrinking steps
100	2.7 Sec	3
500	12.4	2
700	17.3	2

Sample times for Shrinking Algorithm

The second program was compiled under VMS. The complete program is about 250 lines. We also encountered a problem getting the second program to run. The problem arises in the special case that the minimum spanning circle is determined by a diameter. In this case the derived polygon may degenerate to a line segment when r_i gets close enough to r^*. Our Centroid procedure caused a divide by 0 since the area of a line segment is 0. This situation is recognized when the stack has exactly three points on it after completion of the rolling procedure. Instead of calling Centroid, the algorithm finds the midpoint of the segment and takes this as the final value for c. As mentioned above, it is not quite fair to compare the two programs in their present form because the second program requires a convex polygon as input. We ran the second program on the input shown in Figure 7. Fifty points were placed evenly on one half of the unit circle. The fifty-first point was placed at the origin. This is a difficult case for the second algorithm, because it is not able to discard any points. The centroid of this polygon is on the x-axis a bit to the right of the origin. This means that the initial radius computed by the algorithm is too large. The rolling circle must, therefore, touch each consecutive pair of points around the perimeter of the polygon, excepting only the point at the origin. Each new center moves closer to the origin, which is the center of the minimum spanning circle. Figure 9 shows the sequence of radii and areas of the corresponding D_r as the algorithm converges.

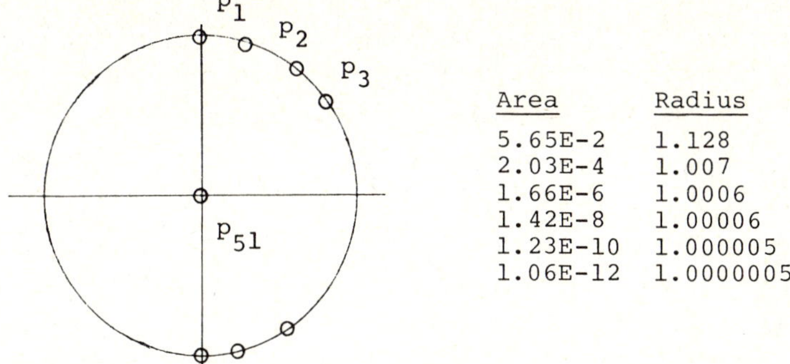

Area	Radius
5.65E-2	1.128
2.03E-4	1.007
1.66E-6	1.0006
1.42E-8	1.00006
1.23E-10	1.000005
1.06E-12	1.0000005

Difficult case for the Rolling Algorithm

5. References

[1] Hearn, D. W., Vijay, J., Efficient Algorithms for the (Weighted) Minimum Circle Problem. Operations Research 30; pp. 777-795; 1982.

[2] Shamos, M., Problems in Computational Geometry, Carnegie-Mellon University, 1977.

[3] Castells, C., Melville, R., An Unusual Algorithm for the Minimum Spanning Circle Problem. Technical Report, Electrical Engineering and Computer Science, Johns Hopkins University, Baltimore, MD.

[4] Chrystal, G., On the Problem to Construct the Minimum Circle Enclosing n Given Points in the Plane. Proceedings of the Edinburgh Mathematical Society, Vol.3, pp.30-33; 1885.

[5] Preparata, F. P., Hong, S. J., Convex Hulls of Finite Sets of Points in Two and Three Dimensions. CACM 20; pp.87-93; 1977.

[6] Graham, R., An Efficient Algorithm for Determining the Convex Hull of a Finite Planar Set. Information Processing Letters 1; pp. 132-133; 1972.

[7] Convexity. Proceedings of Symposia in Pure Mathematics, Vol. VII. American Mathematical Society, 1963. Edited by V. Klee.

[8] Graham, R., Yao, F., Finding the Convex Hull of a Simple Polygon. To appear in Journal of Algorithms.

[9] Rademacher, H. and Toeplitz, O., The Enjoyment of Mathematics, Princeton University Press, 1957.

[10] L. J. Bass and S. R. Schubert, "On finding the disc of minimum radius containing a given set of points," Mathematics of Computation, Vol. 21, 1967, pp.712-714.

[11] Toussaint, G. and Bhattacharya, B., On Geometric Algorithms that use the Furthest-Point Voronoi Diagram. Technical Report No. SOCS-81.3, McGill University School of Computer Science, 1981.

[12] Megiddo, N., Linear Time Algorithms for Linear Programming in R^3 and Related Problems, SIAM J. of Computing, Vol.12, No.4, pp.759-776, 1983.

COMPUTATIONAL GEOMETRY, G.T. Toussaint (Editor)
© Elsevier Science Publishers B.V. (North-Holland), 1985

Curve Similarity via Signatures

Joseph O'Rourke
Richard Washington

Department of Electrical Engineering and Computer Science
Johns Hopkins University
Baltimore, MD 21218

ABSTRACT

The signature is a function that can be associated with any continuous plane curve. Its definition combines local variations in the curve's tangent with the global layout of its length on the plane. The signature is invariant under translation, rotation, and scale changes to its curve, and nearly invariant under slanting transformations. These properties make it useful as the basis of a curve similarity measure. It is demonstrated in several experiments that the *signature distance* between two curves has appealing behavior, especially when applied to slanted handwriting samples.

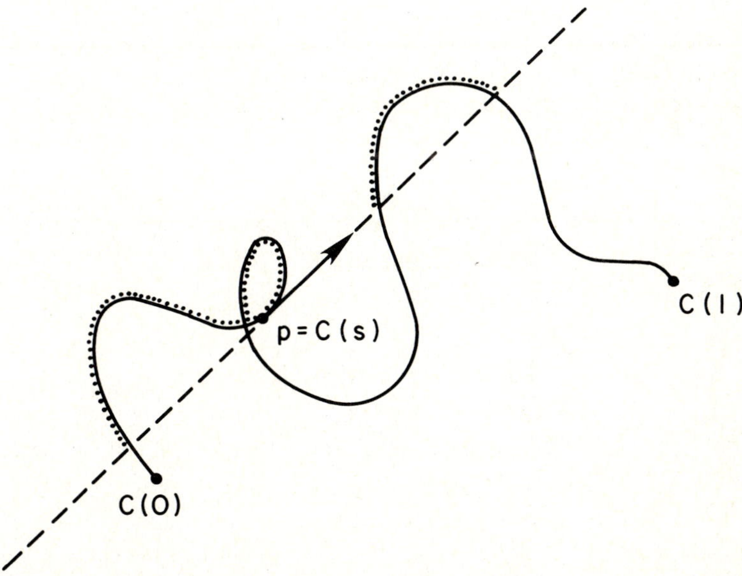

Figure 1: The signature value at arc length s is the length of the curve to the left of the directed tangent to the curve at the point $\mathbf{p}$ determined by s; here the portion to the left is shown dotted.

1. Introduction

The *signature* of a plane curve is a function in $[0,1]\times[0,1]$ that is defined in terms of both local shape variations in the curve and its global distribution on the plane. Let C be a curve normalized to have unit length. Assign it a traversal direction so that each arc length $s\in[0,1]$ identifies a point $\mathbf{p}=C(s)$ on the curve. The signature associates with each arc length s the total length of C on or to the left of the directed tangent to C at $\mathbf{p}=C(s)$; see Figure 1. The signature is undefined wherever the tangent is undefined.

The signature has several interesting properties that suggest its use for measuring shape similarity. These properties have been investigated theoretically in [8], and are described briefly in the next section. Succeeding sections test the efficacy of the signature as a shape measure in several comparison experiments. It is shown that the signature can form the basis of a rather robust curve similarity measure; it is especially effective at recognizing slanted handwriting samples.

2. Properties of the Signature

2.1. Polygonal Paths

The signature of a polygonal path is undefined at the vertices and constant along each straight edge. Thus a path of n edges has a signature that is a step function of n steps, as in Figure 2. All the experiments in this paper are conducted with polygonal paths.

The signature can be computed in $O(n^2)$ time by traversing the the entire path of n edges, noting the length to the left or on the tangent, for each of the n tangents determined by the edges. It is an open question whether this computation time can be reduced [8].

2.2. Convexity

The signatures of all smooth convex curves that are a subset of their convex hull, when traversed counterclockwise, are identical: the constant function 1. Thus the signature ignores differences in convexity: a circle, an ellipse, and every convex polygon (if undefined points are ignored), all have the same signature. Clockwise traversal of these curves results in signatures constant at zero if the curves have no flat sections; otherwise the asymmetry in the definition ("*on* or to the left of") results in non-zero signatures.

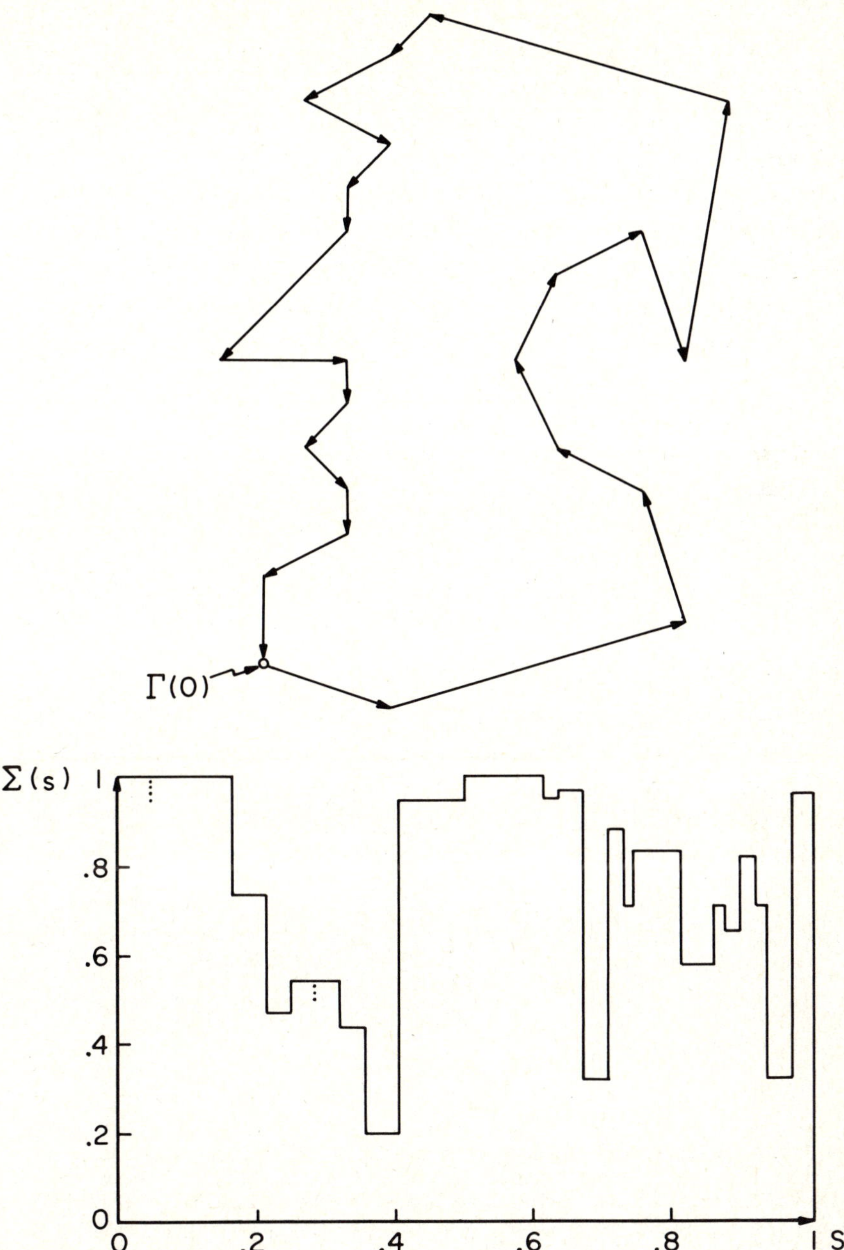

Figure 2: A simple polygon and its signature. Although the signature is undefined at every vertex, the signature is depicted as a continuous function by connected adjacent steps with vertical lines.

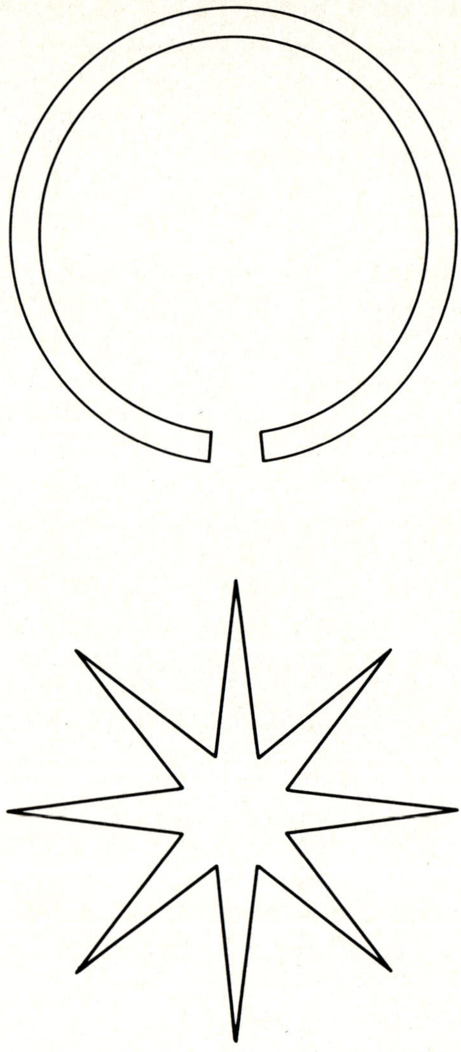

Figure 3: Two curves whose signature integrals approach ½. In the upper curve, the signature is 1 on the outer circle, and near 0 on the inner circle; as the thickness of the annulus approaches 0, the integral of the signature approaches ½. In the lower curve, the signature is constant at $½+\varepsilon$; as the number of arms of the star approaches ∞, ε approaches 0.

The integral of the signature is a measure of the convexity of a curve: a counterclockwise curve is convex iff its signature integral is 1, and small integrals represent highly non-convex shapes. For example, the two curves show in Figure 3 have signature integrals of ½. The smallest known signature integral for a simple closed curve is 1/4 [8].

2.3. Invariances

The signature is invariant under translation, rotation, and scale changes of the curve from which it is derived (the scale change invariance is due to length normalization). It preserves symmetries in the sense that symmetric curves have symmetric signatures; the converse is not necessarily true. One of the most useful properties of the signature is its "near" invariance with respect to "slanting" transformations. A curve is slanted by redrawing it on an oblique coordinate system without stretch: each point (x,y) is mapped to (x',y') according to the equations (see Figure 4)

$$x' = x + y\cos\vartheta$$

$$y' = y\sin\vartheta$$

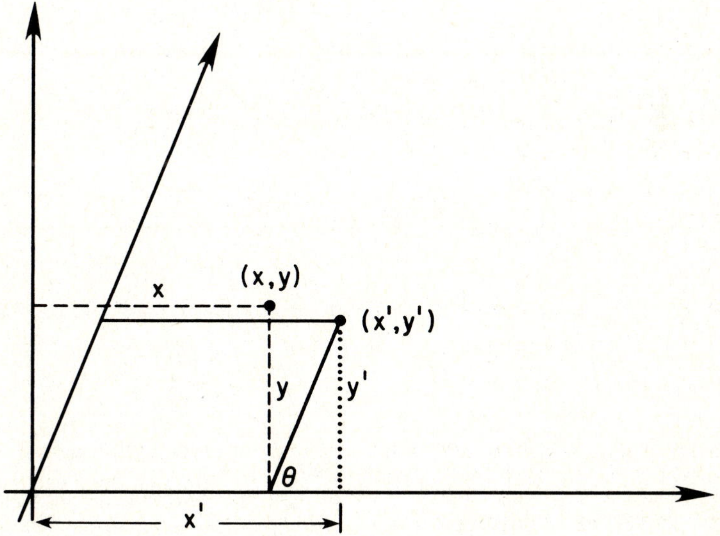

Figure 4: The slanting transformation. The point (x,y) is transformed to (x',y') by slanting the y-axis by an angle ϑ.

The signature of a rectilinear curve (one whose edges are aligned with the axes of an orthogonal coordinate system) is invariant under slanting transformations, as all the tangents slant in concert and no length changes occur. Arbitrary curves are not invariant, however, because of the possibility of differential shrinking of lengths; see Figure 5.

2.4. Inversion

The signature discards some information inherent in the curve -- orientation, convexity differences, slants, and so on. It is then natural to ask how much information is retained. It is rather surprising that in the case of closed rectilinear curves, it retains all the information: such a curve can be reconstructed (modulo a 90° rotation) from its signature [8]. Thus the signature transformation has a unique inverse for this class of curves.

3. Distance Measures

Three curve-to-curve distance measures are defined in this section, and then compared in a series of experiments in the two following sections.

3.1. Signature Distance

The properties described in the previous section suggest that two curves are "similar" if their signatures are similar, and "dissimilar" otherwise. We therefore define the *signature distance* between two curves as the distance between their signatures. The distance between two signatures is in turn defined by a standard dynamic warping correlation between discrete samples from the signature functions. As this warping technique is employed by all three distance measures used in this paper, we will describe its operation briefly.

Many applications require a measure of similarity between two finite strings of elements, and sophisticated metrics have been devised [6]. We use a dynamic programming algorithm that permits local "warping" to achieve an optimal global match. The optimality of the match is defined in terms of a local element-to-element distance: if $D(i,j)$ is the distance between element i of sequence 1 and element j of sequence 2, then the sum of $D(i,j)$ over all pairings (i,j) in the matching is minimized by the algorithm. The pairing does not have to be one-to-one, but may be warped to the extent that the pairing (i,j) may be followed by any of the three pairings $(i,j+1)$, $(i+1,j)$, or $(i+1,j+1)$. In addition, a bound was placed in our experiments on the total

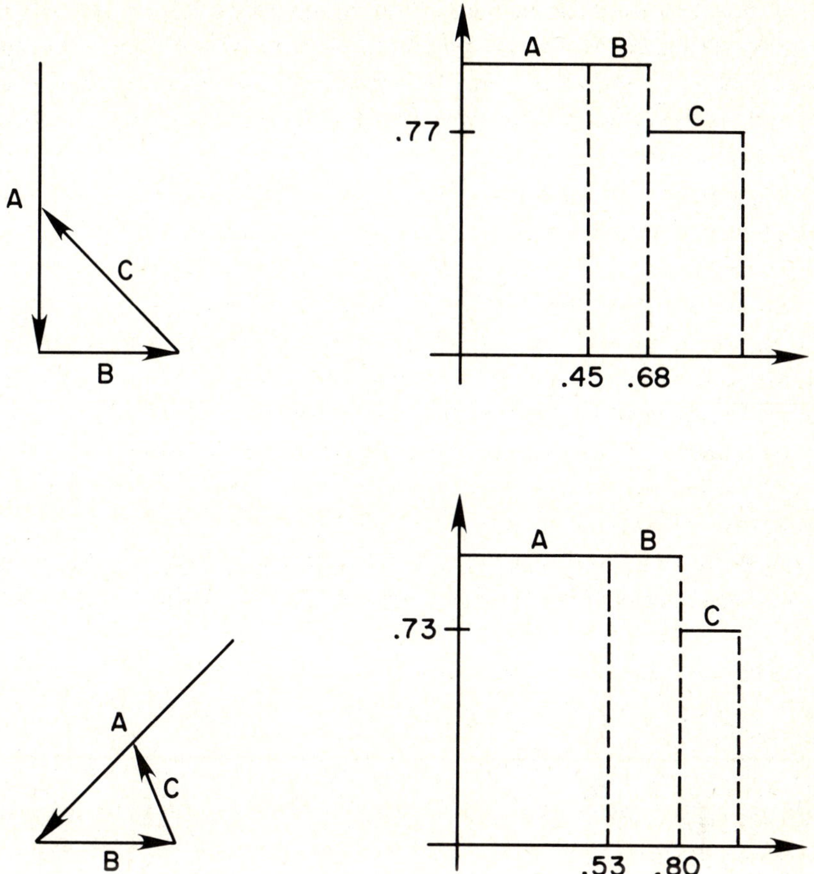

Figure 5: Not all curves are invariant under slant. Edges *A* and *B* retain their length, but *C* is shortened by the transformation.

amount of warp by restricting matches to be separated by no more than half the total string length. The dynamic programming algorithm can be implemented in $O(nm)$ time for strings of n and m elements. In all our experiments, $n \approx m$, so the computation time is $O(n^2)$. Further details on the dynamic warping algorithm may be found in [6][9][7][12][1].

For signature functions $y(x)$ and $z(x)$, the elements of the sequence are the ordinate values of the functions, and we use the element distance

$$D(i,j) = | y_i - z_j |$$

Note that this distance does not depend on the amount of warp along the x-axis: warping is "free" under this measure. The main justification for this measure is that most natural curve distortions (e.g., affine transformations[1]) warp the x-axis of the signature, but modify the ordinate values only slightly. The reason is that the x-axis of the signature records local information -- location along the curve specified by arc length -- and the ordinate records the global distribution of the curve -- how much is left of a particular line. Distortions that leave the "shape" recognizable tend to affect local sections more than global arrangement; if in addition the distortions do not greatly affect the curve's tangents (and affine transformations do not), then the signature is mainly modified along the x-axis. This is demonstrated for the slant transformation both in Figure 5 and more clearly in Figure 6. The latter figure displays two signatures, one derived from a curve back-slanted 45°, the other from the same curve forward-slanted 45°. The optimal match computed by the warping algorithm is shown as a series of connecting lines between the two signatures. It is clear that the warping algorithm has correctly found the "obvious" correspondence between the most conspicuous peaks and valleys of the two signatures. Despite a 90° difference in slant, the signature values do not differ greatly, and so the computed signature difference is rather small. This example will be revisited in Section 5.1.

Two other curve distances were implemented to serve as comparisons for the performance of the signature distance. The first is based entirely on the angles at the vertices of the polygonal curve, and the second is primarily a positional measure.

---

1. Affine transformations are linear transformations of coordinates that map straight lines into straight lines; stretch, slant, and perspective distortion are all affine transformations.

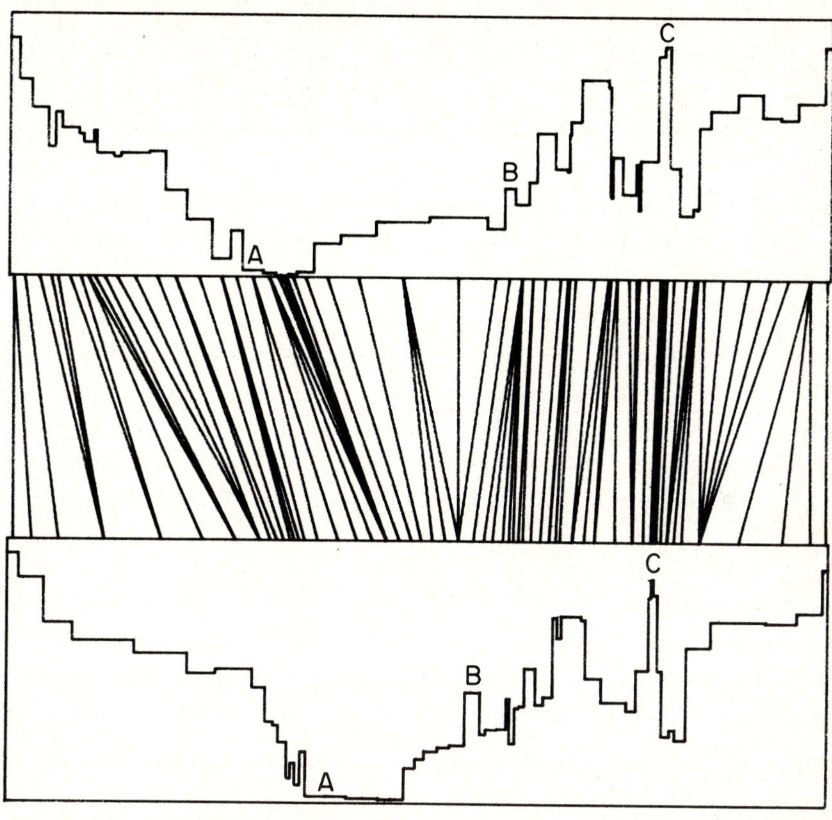

Figure 6: The optimal match computed by the dynamic warping algorithm. The signatures are derived from the two slanted curves shown in Figure 14: they are $-45°$ and $+45°$ slanted versions of the original Joe_1. Note that the sections marked A, B, and C in the two curves are matched by the algorithm.

3.2. Angular Distance

Angular disparity in curves is a common component in measures of dissimilarity [3][4][11], but it has serious limitations when used in isolation [1]. Because a measure based solely on angles ignores the lengths of edges, it must consider the two curves shown in Figure 7 to be identical, which hardly agrees with perceptual shape intuition. Nevertheless, when curves are uniformly digitized (as they are in our experiments), an angular distance can be quite effective.

We define the *angular distance* between two plane curves using the dynamic warping technique described above. In this case, however, the elements matched by the algorithm are the angles at each vertex of the polygonal path, measured from the edge preceding each vertex to the edge following; thus an angle of 0 represents no turn, and a positive angle represents a left turn. If α_i from curve 1 is matched with β_j of curve 2, the element-to-element distance measure is

$$D(i,j) = |\sin\alpha_i - \sin\beta_j| + |\cos\alpha_i - \cos\beta_j| .$$

This distance is maximized when the angles differ by 180°. The angular distance measure is invariant under translation, rotation, and scale changes of the curves.

3.3. Positional Distance

An effective curve distance measure that is neither translation, rotation, nor scale invariant has been proposed by Burr [1][2]. It incorporates both angular and positional terms in a heuristic mix; we call it a *positional distance* here to contrast with the angular distance defined above. The measure again uses the dynamic warping technique. The elements matched by the algorithm are edge vectors on the two curves. If vector A_i from curve 1 is matched to vector B_j from curve 2, then the element-element distance is defined to be

$$D(i,j) = D_p(i,j) + 2D_a(i,j) ,$$

where

$$D_a(i,j) = |B_j \times A_i| / |A_i| + 3U(-B_j \cdot A_i) |B_j \cdot A_i| / |A_i|$$

$$D_p(i,j) = |C \times A_i| / |A_i| + U(C \cdot A_i / |A_i| - |A_i|) |C \cdot A_i / |A_i| - |A_i||$$

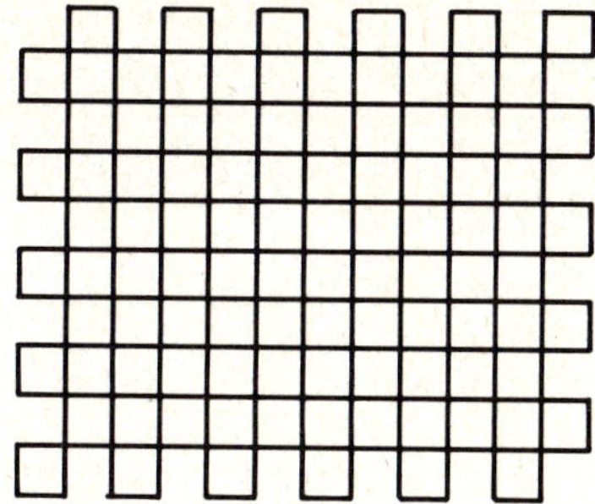

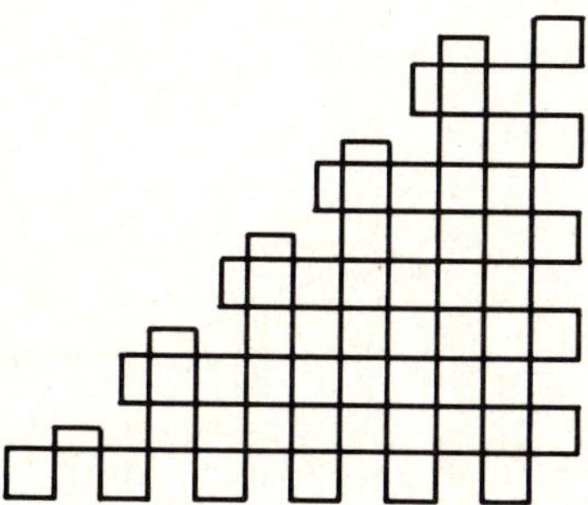

Figure 7: Two curves separated by an angular distance of 0: the sequence of angles in the two curves is identical, but their edge lengths are different.

$$+ U(-C\cdot A_i/|A_i|)|C\cdot A_i/|A_i|$$

with $U(t)=0$ when $t<0$ and 1 when $t\geq 0$, and where C is a vector from the tail of A_i to the midpoint of B_j (if $|A_i|>|B_j|$). The justification for these rather complex formulas is contained in [1].

Because of the dependence on absolute position, orientation, and scale, the two curves must be "registered" before the distance is computed. Burr has used various techniques suited to particular pattern recognition tasks, including a "deslanting" transformation [2]. Here we use a simple registration procedure: the curves are scaled to have roughly the same size, the center of gravities are overlayed, and their principle axes of inertia are aligned. We emphasize that this registration technique differs from Burr's, making this measure something of a "strawman." Nevertheless, it serves as a useful contrast with the signature and angular measures.

3.4. Sampling the Curves

When comparing distorted versions of the same curve (as we do in Section 5), it is important that the samples used by the dynamic warping algorithm be chosen independently in the original and the distorted curves. Otherwise the warping algorithm is given an unrealistic advantage by the existence of a natural one-to-one correspondence between the samples. Most of our curves have several hundred vertices. Each was sampled using a polygonal approximation technique [10] to attain 30-80 points. The sampling was always performed *after* any transformations applied to the curve, ensuring independence.

3.5. Computation Time

All three distance measures require $O(n^2)$ computation time for two curves of n edges each. In our implementations, the positional distance requires approximately 4 times the computation time as the signature and angular distances (whose times are nearly equal), due to the complexity of its element-element distance formula.

4. Similar/Dissimilar Separation

There is no one "correct" curve distance measure, and evaluations ultimately must depend on subjective criteria. This section examines the behavior of the signature, angular, and positional distance measures on three different classes of curves. Three curves of each class are used in each comparison: two that are subjectively similar, and a third that is subjectively dissimilar from the other two. To agree with intuition, the distance measures should cleanly separate similar from dissimilar curves. The three inter-curve distances are displayed as a triangle, where the lengths of the sides are proportional to the distance measures. Thus a near-equilateral triangle indicates little discrimination, whereas a triangle strongly acute at the dissimilar vertex reflects good separation of similar from dissimilar. As the three distance metrics are expressed in incompatible units, the distance between the subjectively similar vertices is normalized to 1 in all cases to facilitate comparision.

4.1. Printed Letters

The first class consists of the three printed letters *b*, *h*, and *a*; see Figure 8. The letters *b* and *h* are similar in appearance, differing only in a short piece closing the *b*. The letter *a* is rather different from both. The results of the three distance measures are displayed in Figure 9. All three measures successfully separate similar from dissimilar, although the signature makes the largest separation.

We should note that the signature alone would not be sufficient for recognition of individual hand-printed characters: its various invariances, which can be advantageous in certain circumstances, here limit its usefulness. Due to their convexity, the signature considers o and c identical; the signature's rotational invariance could cause the pairs *M* and *W*, *p* and *d*, and *b* and *q* to have a separation distance of zero if the traversal directions are opposed in the pair elements. Also, letters such as *t* are not easily represented as a single continuous curve, a necessary condition for the signature to be applicable. Burr, on the other hand, has achieved some success with his measure (which in our experiment is the least discriminating) for character recognition on the entire alphabet [1].

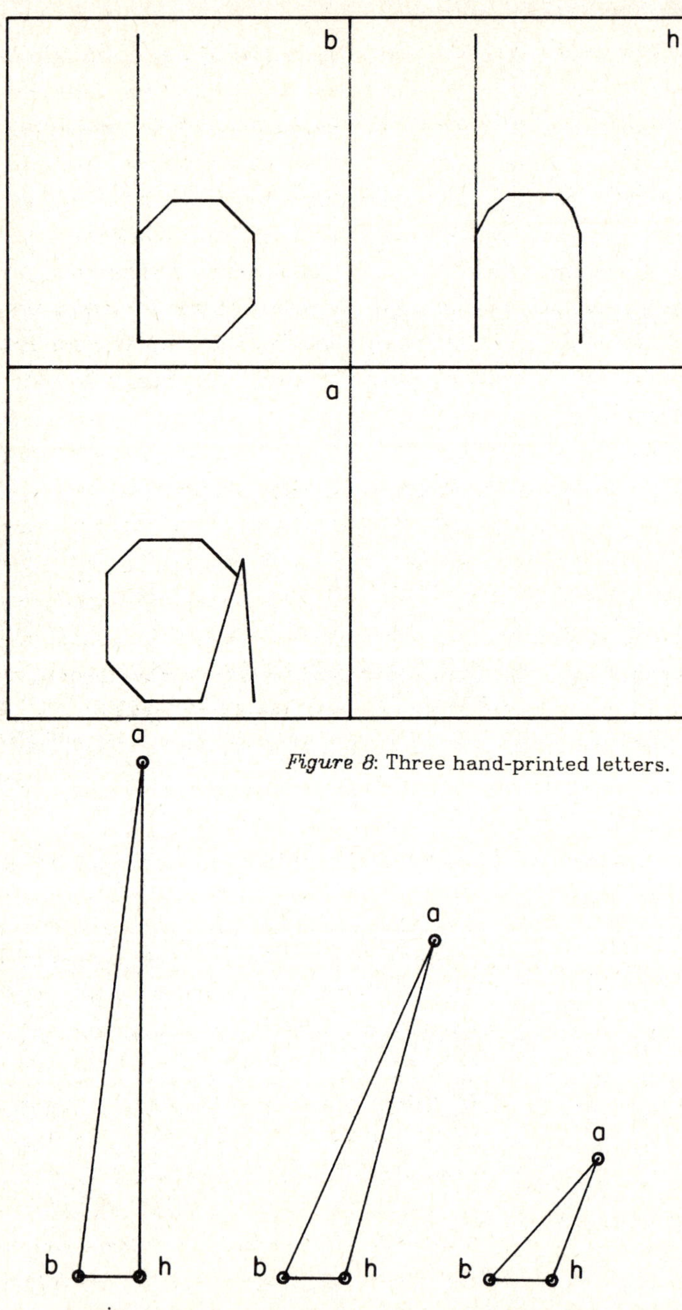

Figure 8: Three hand-printed letters.

Figure 9: The distances between the curves of Figure 8 according to the signature, angular, and positional metrics.

4.2. Handwritten Script

The second class of curves includes samples of extended (multiple-letter) cursive handwriting. Two of the samples are of the same word written in a different style (Joe_1 and Joe_2), and the third is a different word (*Job*) which is nevertheless rather similar for the majority of its length to Joe_1; see Figure 10. The comparison results are shown in Figure 11. All three measures consider Joe_1 at least as close to *Job* as to Joe_2, but none of the three measures distinguishes strongly among the samples. The obvious lesson is that there can be more shape variation within different versions of the same word than there is between different words.

4.3. Continents

Three closed shapes form the last class: the boundaries of *Africa*, *Australia*, and *South America*; see Figure 12. Although these shapes are closed, we treat them the same as open curves: each has a definite starting and ending point (marked in Figure 12), corresponding to arc lengths 0 and 1. If any of the three measures were to be applied exclusively to closed shapes, it would be appropriate to search for the optimal "shift" by running the dynamic programming algorithm for each of the n possible starting point correspondences.[2] This was not done in our experiments because of the computational cost ($O(n^3)$ versus $O(n^2)$, where $n \approx 50$); the starting points were chosen, however, to correspond in a "natural" way.

It is difficult to say what could be hoped for in this comparison; the three shapes are all different, but *Africa* and *South America* share several features. The signature considers the shapes to be indistinguishable, but the angular, and to a greater extent, the positional measure surprisingly consider *Africa* and *Australia* to be similar; see Figure 13.

5. Robustness under Distortion

Although some of the preceding experiments are difficult to interpret because our intuitions are not strong, we usually have strong intuitions when matching distorted shapes with undistorted originals. This section explores three types of distortion: slanting, perspective, and noise.

2. This approach is taken by e.g. Fuchs *et al* for finding optimal triangulations between two contours [5].

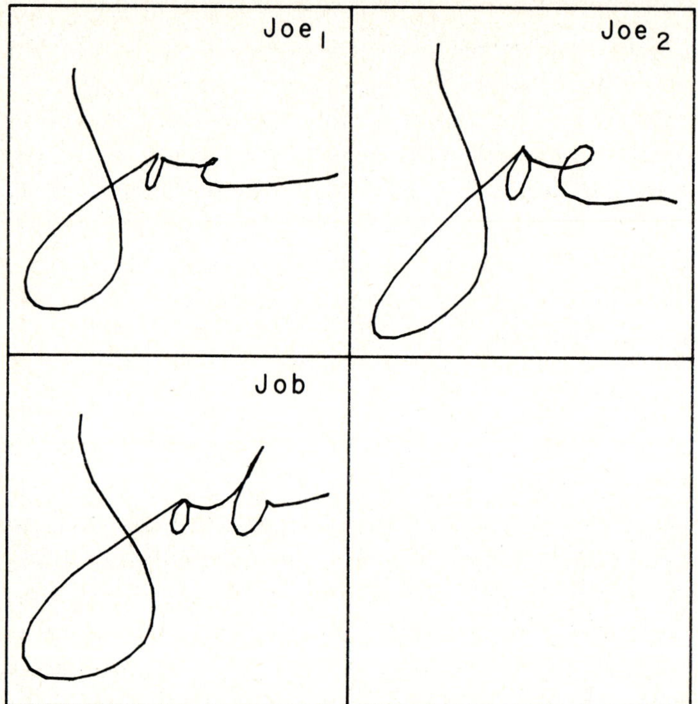

Figure 10: Three handwriting samples.

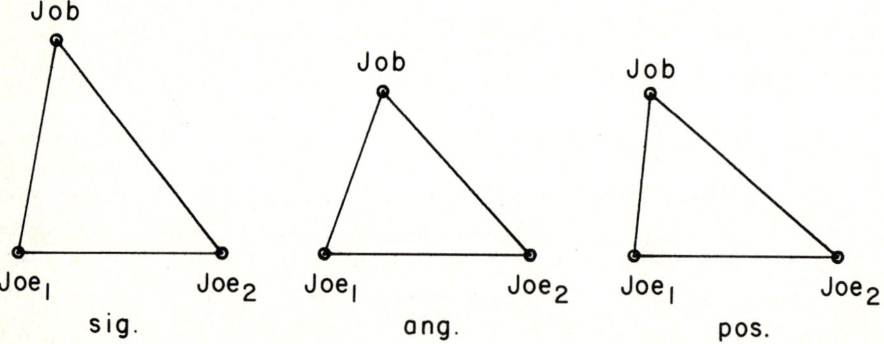

Figure 11: The distances between the curves of Figure 10 according to the three metrics.

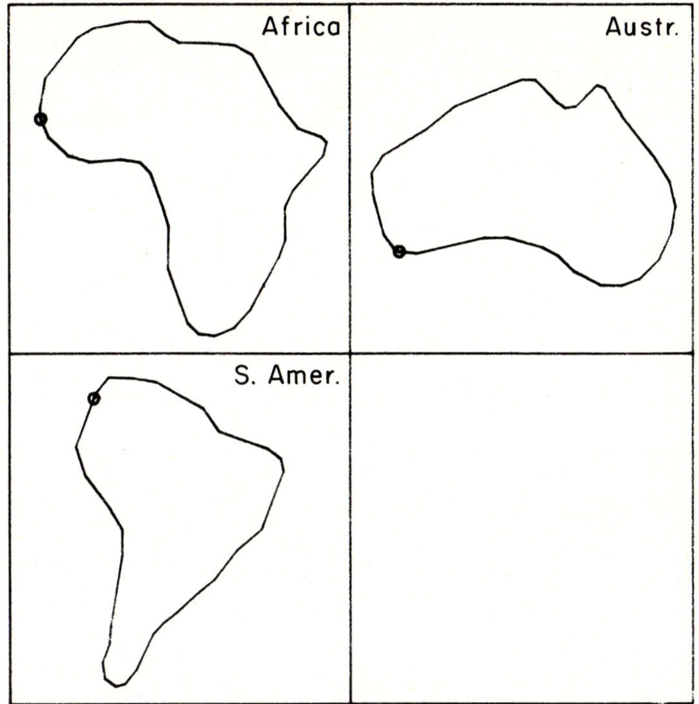

Figure 12: Three closed shapes. The starting point is marked on each curve.

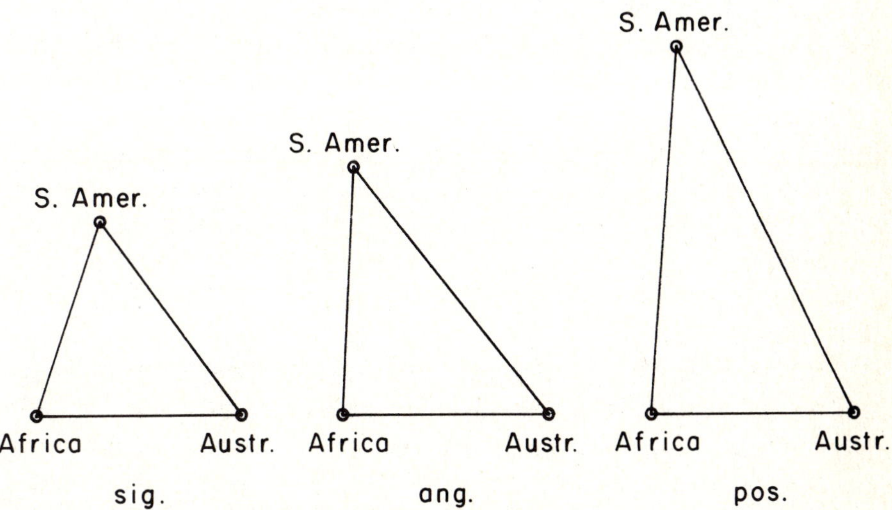

Figure 13: The distances between the curves of Figure 12 according to the three metrics.

5.1. Slant

Figure 14 shows two slanted versions of the script Joe_1 together with the original, and *Job*, which serves as a foil. The version of Joe_1 slanted −45° was compared with the other three curves, using all three distance metrics. The results are presented in the barchart of Figure 15; for all three measures, the distance to the undistorted Joe_1 is normalized to 1. In this experiment, the signature fares well, which is hardly surprising given its near invariance under slant: it considers the back-slanted sample closest to the original, just slightly further from the forward-slanted version, and quite distinct from *Job*. (Figure 6 shows the optimal warping computed for the −45° to +45° comparison.) The angular distance considers *Job* closer than the forward-slanted sample, and the positional measure does not significantly distinguish among any of the curves.

5.2. Perspective

The shape *Africa* was distorted by drawing it on a plane in space and tilting it with respect to a central projection point to varying angles to achieve different perspective effects. Three distortions of various degrees were made as shown in Figure 16. The *Africa-20/30* shape was then compared against the other distortions, the original, and the *Australia* and *South America* shapes. The results are presented in Figure 17. None of the three measures achieve clean separation of the *Africa*-derived shapes from the others. The angular measure finds *Africa-20/30* closer to *Australia* than to *Africa-40/50*. The positional measure finds that *Africa-40/50* is the most dissimilar from *Africa-20/30* among all the shapes. The signature measure is more consistent, finding the distances amongst the distorted *Africas* less than to either *Australia* or *South America*, but the difference between e.g. the *Africa* and *South America* distances is too small to be considered a successful separation.

5.3. Noise

The final comparison considers noisy, jagged curves. Uniformly distributed noise was added to the vertex coordinates of the script Joe_1. Figure 18 shows two noisy versions of Joe_1; they differ only in the choice of the seed for the random number generator. One of these noisy curves was then compared against the other noisy version, the original, and *Job*. The results for all three distance measures, normalized to the distance to the original Joe_1, is

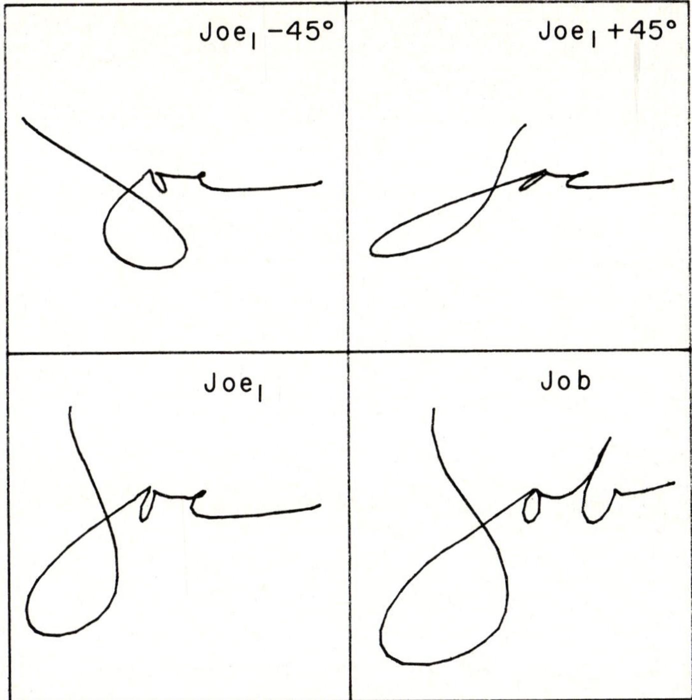

Figure 14: Slanted handwriting samples. The top two boxes show −45° and +45° slanted versions of Joe_1, shown in the lower left box.

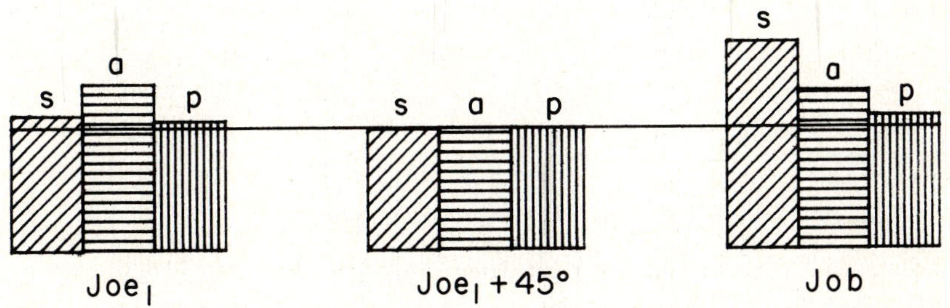

Figure 15: Results of a comparison of Joe_1−45° against Joe_1, Joe_1+45°, and *Job*, for all three distance metrics. The signature distances are marked with diagonal shading, and the angular and positional with horizontal and vertical shading respectively.

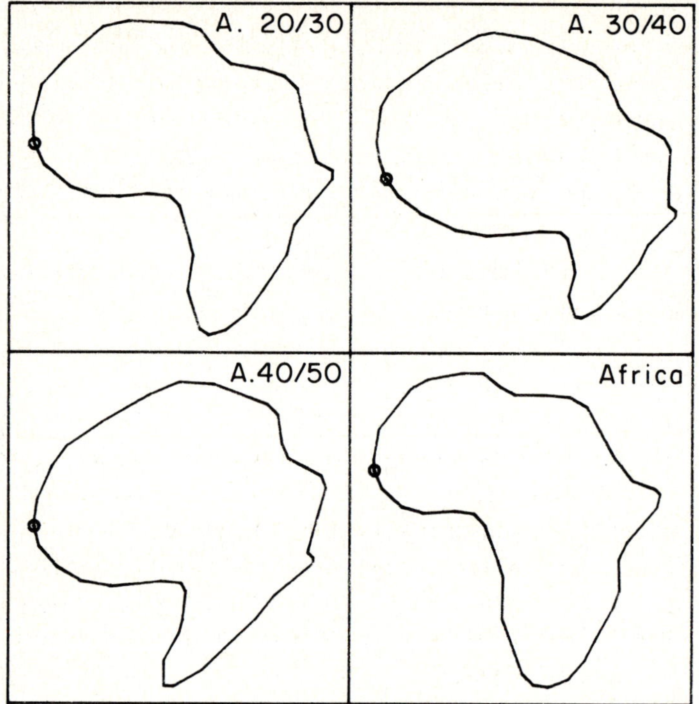

Figure 16: Versions of *Africa* distorted by perspective. The notation "20/30" means a 20° turn about the x-axis followed by a 30° rotation about the y-axis.

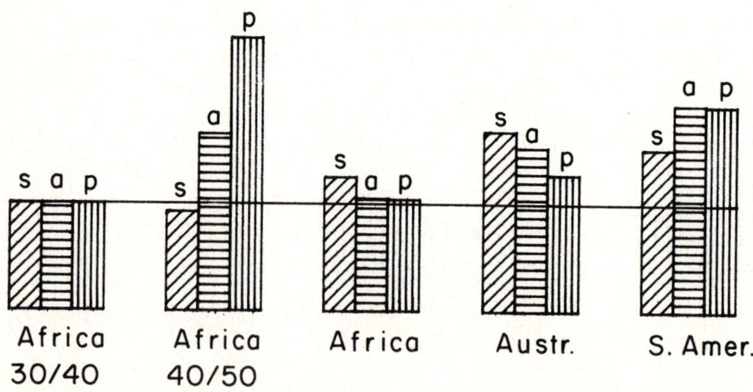

Figure 17: Comparison of *Africa-20/30* against the other shapes shown in Figure 16, plus two shapes from Figure 12.

displayed in Figure 19.

As might be expected, both the signature and the angular measures cannot "see through" the noise to the underlying shape: both consider all three candidates curves nearly equal matches to a noisy Joe_1. Noise has made these measures lose their discriminating power. The positional measure, surprisingly, cleanly identifies the original Joe_1, but cannot distinguish between the other noisy version and *Job*.

Thus, curves must first be smoothed for any of the metrics to be effective measures of similarity.

6. Conclusion

It is clear that the signature, in and of itself, cannot serve as a shape similarity panacea: it ignores some shape variations that are significant in some applications (e.g., different varieties of convexity), it does not degrade gracefully under noise, and it only applies to connected, continuous curves. But it is especially good at recognizing slanted versions of handwritten script, and it responds favorably to other types of distortion as well. These properties make it an attractive alternative in some applications to more conventional curve similarity measures.

Acknowledgments

We thank Alok Aggarwal, Heather Booth, Kathleen Callaway, and Prasoon Tiwari for their comments on this work. This research was partially supported by NSF grant MCS83-04780.

REFERENCES

[1] Burr, D. J., A Technique for Comparing Curves, *Proceedings of IEEE Conference on Pattern Recognition and Image Processing*, Chicago (1979) 271-277.

[2] Burr, D. J., Designing a Handwriting Reader, *Proceedings of the 5th International Conference on Pattern Recognition*, Miami (1981) 715-722.

[3] Davis, L. S., Understanding Shape, I: Angles and Sides, *IEEE Trans. on Computers*, **C-26** (1978) 236-242.

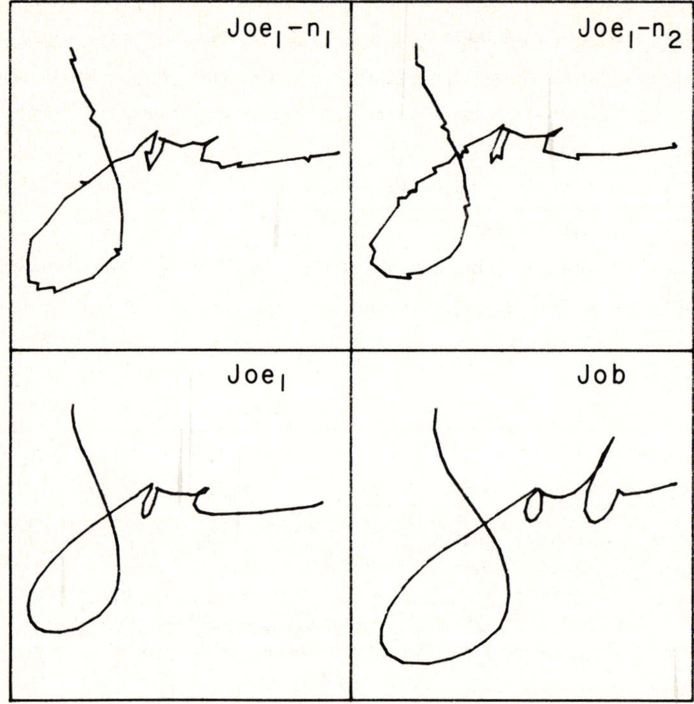

Figure 18: Noisy curves. The top two boxes were constructed using identical noise statistics, but with different random number generator seeds.

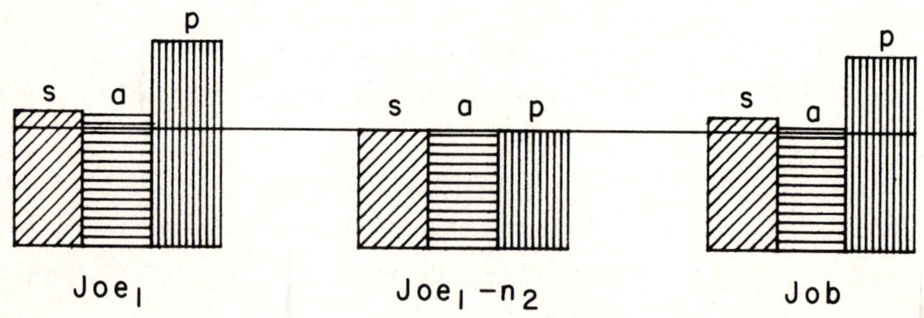

Figure 19: Comparison of one noisy curve with the original, the second noisy version, and *Job*.

[4] Davis, L. S., Shape Matching Using Relaxation Techniques, *IEEE Trans. on Pattern Analysis and Machine Intelligence*, **PAMI-1** (1979) 60-72.

[5] Fuchs, H., Kedem, Z. M., and Uselton, S. P., Optimal Surface Reconstruction from Planar Contours, *Communications of the ACM*, **20** (1977) 693-702.

[6] Kruskal, J. B., An Overview of Sequence Comparison: Time Warps, String Edits, and Macromolecules, *SIAM Review*, **25** (1983) 201-237.

[7] Meyers, C., Rabiner, L. R., and Rosenberg, A. E., Performance Tradeoffs in Dynamic Time Warping Algorithms for Isolated Word Recognition, *IEEE Trans. on Acoustics, Speech, and Signal Processing*, **ASSP-28** (1980) 623-633.

[8] O'Rourke, J., The Signature of a Plane Curve, *SIAM J. on Computing*, to appear, 1984.

[9] Sakoe, H., and Chiba, S., Dynamic Programming Algorithm Optimization for Spoken Word Recognition, *IEEE Trans. Acoustics, Speech, and Signal Processing*, **ASSP-26** (1978) 43-49.

[10] Sklansky, J., and Gonzales, V., Fast Polygon Approximation of Digitized Curves, *Proceedings of IEEE Conference on Pattern Recognition and Image Processing*, Chicago (1979) 604-609.

[11] Ullman, J. R., Subset Methods for Recognizing Distorted Patterns, *IEEE Trans. on Systems, Man, and Cybernetics*, **SMC-7** (1977) 180-191.

[12] Yasuhara, M., and Oka, M., Signature Verification Experiment Based on Nonlinear Time Alignment: A Feasibility Study, *IEEE Trans. on Systems, Man, and Cybernetics*, **SMC-7** (1977) 212-216.

COMPUTATIONAL GEOMETRY, G.T. Toussaint (Editor)
© Elsevier Science Publishers B.V. (North-Holland), 1985

A METHOD FOR PROVING LOWER BOUNDS FOR CERTAIN GEOMETRIC PROBLEMS

Raimund Seidel

Department of Computer Science
Cornell University
Ithaca NY. 14853

Abstract:

We prove lower bounds for a number of geometric problems. Our results show that certain types of additional input information cannot possibly permit faster solutions for these problems. The main idea in all our lower bound proofs is the use of the index of a point as an additional coordinate.

I. Introduction

Most of the research effort in the field of computational geometry goes into the design of algorithms for the efficient solution of geometric problems. Often several algorithms are known for particular problems and the question of comparing their respective efficiency arises. The so-called asymptotic worst case time complexity is one of the favorite measures for this comparison.

Whenever algorithms for a particular problem are compared and one of them is found to be best, the question naturally arises whether this algorithm can still be improved upon or whether the problem is intrinsically so difficult that there is no substantially better way of solving it in the given computational framework. This is the point where the computational geometer has to change his viewpoint entirely because instead of looking for (geometric) properties of the given problem that facilitate an easy and efficient solution he must search for "bad" properties that will force a certain amount of inefficiency on any algorithm that purports to solve the problem.

This lower bound question has been answered for a number of geometric problems such as the point uniqueness problem [1], different versions of the convex hull problem [22],[24], [14], the Voronoi Diagram [22],[17],[21], certain intersection problems [23], and a number of searching problems [6],[7],[8]. The lower bounds given in this paper are of a special kind. We look at problems for which in the general case a lower bound is known that is also matched by an optimal algorithm and we ask whether this lower bound is still applicable, if additional input information is included in the problem definition. It is well known, for instance, that the construction of the convex hull of a set S of n points in $\mathbf{R}^3$ requires $\Omega(n \log n)$ time in the worst case. But it is not clear whether this problem becomes easier if the order of the points in S with respect to each of the three coordinate directions is known. The answer to this question turns out to be negative. Similarly we show in this paper that if is not easier to find the convex hull of S even if the points of S are the vertices of a star-shaped polytope, or we prove that having n points in the plane arranged as a triangulation does in general not aid in locating every one of these points in a planar subdivision.

All the lower bounds in this paper are proved by reduction, i.e. we show that a fast solution for the problem of interest would yield a fast solution for some other problem and would thus contradict a known lower bound for this other problem. The unifying tool in all these proofs is the use of the index of a point as an additional coordinate.

As our underlying model of computation we assume the RAM with real arithmetic and unit cost per operation. Throughout the paper $x(p)$ and $y(p)$ denote the x and y-coordinate of a point p, respectively, and $\langle a, b \rangle$ denotes the ordinary scalar product between two vectors a and b.

II. Sorting Triangulated Points

It is well known that any algorithm that sorts a set $A = \{x_1, \ldots, x_n\}$ of n reals by the use of comparisons needs $\Omega(n \log n)$ time for some input set A. This implies immediately that the problem of sorting a set $S = \{p_1, \ldots, p_n\}$ of n points in the plane with respect to their x-coordinate has an $\Omega(n \log n)$ lower bound also. After all, any algorithm that sorts S in such a way could be used to sort the set of points $\{(x_1,0), \ldots, (x_n,0)\}$ and would thus produce the order of A. Now the question arises, what if additional information about the set S is available. What happens if we know a triangulation of S? By a *triangulation* $D(S)$ of S we mean a maximal set E of unordered pairs $\{p_i, p_j\} \subset S$, $i \neq j$, called *edges*, such that for each pair of edges $\{p_i, p_j\}$, $\{p_k, p_m\} \in E$ the open straight line segments joining p_i and p_j, and p_k and p_m, respectively, do not intersect. Figure 2.1 shows a point set and two of its triangulations.

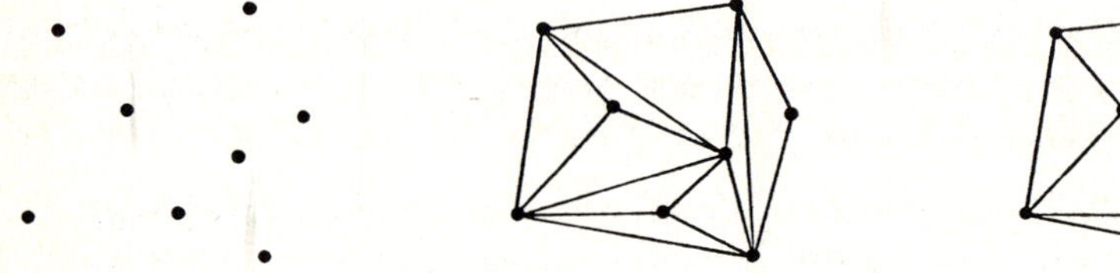

Figure 2.1

Thus a triangulation can be viewed as a planar graph whose vertices are the points of the given set. In the following we will therefore assume that triangulations are described by typical planar graph representations, such as cyclically ordered adjacency lists, or the quad edge data structure of [10].

A triangulation of a point set S obviously contains a fair amount of information about the structure of S. It has also been shown [22] that this information is sufficiently complex that in general $\Omega(n \log n)$ time is required to construct a triangulation for an n-set S in the plane. On the other hand, in the following we show that in general a triangulation contains so little information about S that sorting S by, say, the x-coordinate still takes time $\Omega(n \log n)$.

In order to prove this result we need the following lemma that will be important throughout the paper.

Lemma 1:

Let $A = \{x_1, \ldots, x_n\}$ be an (unordered) set of n real numbers.

In $O(n)$ time one can construct an ordered n-set $S = \{p_1, \ldots, p_n\} \subset \mathbf{R}^2$ along with a triangulation D of S, where for $1 \leq i \leq n$, $x(p_i) = x_i$, and for $1 \leq i < n$, $y(p_i) < y(p_{i+1})$.

Proof:

For $1 \leq i \leq n$ let $p_i = (x_i, i)$. Now clearly the p_i's have the two properties stated in the lemma. Thus it only remains to show that one can construct a triangulation D of S in linear time. We will prove by induction that S can be triangulated in time proportional to $n+e$, where e is the number of edges in the triangulation. Note that for planar graphs with n vertices always $e \leq 3n-6$. Thus $n+e$ is $O(n)$.

The basis for our induction is obviously true since a triangulation of a constant number of points can be constructed in constant time.

So let $S' = \{p_1, \ldots, p_{n-1}\}$, let D$'$ be a triangulation of S', and let e' be the number of edges in D$'$. Assume inductively that D$'$ has been constructed in time proportional to $n-1+e'$.

As $y(p_n) > y(p)$ for $p \in S'$, p_n lies outside the convex hull of S' and thus there is only one way of extending D$'$ to a triangulation D of S, namely by introducing edges between p_n and a number of points in S' that lie on the convex hull (see Figure 2.2). Assume there are $h \geq 2$ such new edges. By the y-maximality of p_{n-1} in S' one of these edges must be incident to p_{n-1}. From this fact and from D$'$ the remaining edges can be determined at constant cost per new edge. Thus D$'$ can be extended to D in time proportional to a constant plus h. Using the inductive assumption this means that D was constructed in time proportional to $n+e'+h$. But as clearly $e = e'+h$, D was therefore constructed in time proportional to $n+e$, i.e. it was constructed in $O(n)$ time.

Q.E.D.

Now we can state the first lower bound result of this paper.

Theorem 1:

Let S be a set of n points in the plane and let D be a triangulation of S.

Any algorithm that, given S and D as input, sorts the points of S with respect to their x-coordinates takes time $\Omega(n \log n)$ in the worst case.

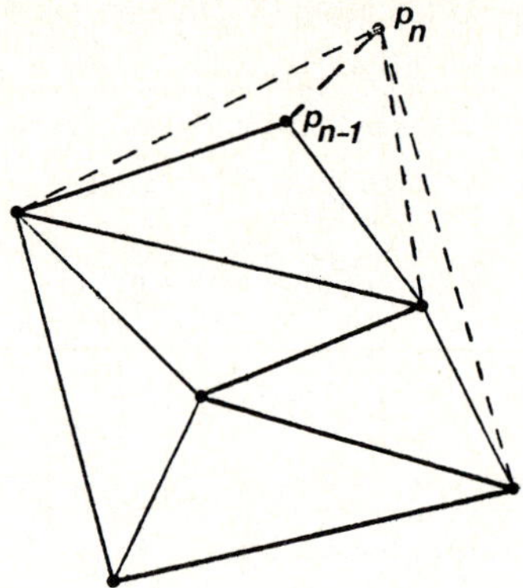

Figure 2.2

Proof: (reduction to sorting)

Consider the following algorithm for sorting a set $T = \{x_1, \ldots, x_n\}$ of reals.

1. Use Lemma 1 to construct the set $S = \{(x_1,1), \ldots, (x_n, n)\}$ and a triangulation D thereof.
2. Use any algorithm to sort S by x-coordinate that also makes use of D. As for $1 \le i \le n$, $x((x_i, i)) = x_i$, this effectively also yields the sorted order of T.

As by Lemma 1 step 1 takes time $O(n)$ and since sorting T requires time $\Omega(n \log n)$ in the worst case it is clear that step 2 must take time $\Omega(n \log n)$ in the worst case.

Q.E.D.

It is now worth asking whether some special classes of triangulations yield more information. One popular class of triangulations are the so-called Delaunay triangulations, which are the duals of Voronoi Diagrams (see [22]). A necessary and sufficient condition for a triangulation D of a set S to be Delaunay is that for any triangle T in D the open disc bounded by the circumscribed circle of T contains no point of S (see Figure 2.3). For a finite set S in the plane the Delaunay triangulation $\Delta(S)$ is unique[1]. We now show a result analogous to Lemma 1.

Lemma 2:

Let $A = \{x_1, \ldots, x_n\}$ be a set of n reals.

In $O(n)$ time one can construct an n-set $S = \{p_1, \ldots, p_n\} \subset \mathbf{R}^2$ along with its Delauny triangulation $\Delta(S)$, where S has the property that for $1 \le i \le n$, $x(p_i) = x_i$, and for $1 \le i < n$, $y(p_i) < y(p_{i+1})$.

[1]Strictly speaking, $\Delta(S)$ is only guaranteed to be a triangulation if no four points in S are co-circular. However, the "degenerate" cases where this is not true are not of interest here.

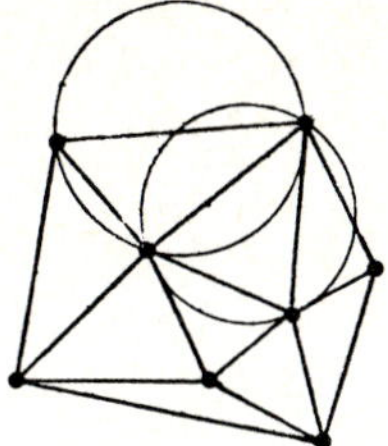

Figure 2.3

Proof:

The inductive proof for this lemma is almost identical to the proof of Lemma 1. One only needs to be more judicious about the choice of y-coordinate for a point p_i.

For the base case of the induction observe that because of the distinctness assumption for the points in A it is possible to choose in constant time reals $y_1 < y_2 < y_3$ such that $p_i = (x_i, y_i)$, $i = 1,2,3$, forms a triangle, which of course is the Delaunay triangulation of $\{p_1, p_2, p_3\}$.

Now assume inductively that we have found $y_1 < y_2 < \cdots < y_{n-1}$ and the Delaunay triangulation of $S' = \{p_1, \ldots, p_{n-1}\}$, where $p_i = (x_i, y_i)$, in time proportional to $n-1$ plus e', the number of edges in $\Delta(S')$. Also assume inductively that the real number

$$m' = \max\{\, y(q) \mid q \in C, \text{ where } C \text{ is the circumscribed disc of any triangle in } \Delta(S')\,\}$$

has been determined within that time bound.

Observe that $m' \geq y_{n-1}$. Now choose $y_n = m' + 1$, let $p_n = (x_n, y_n)$, and, as in the proof of Lemma 1, complete $\Delta(S')$ to a triangulation D of S in time proportional to h, the number of edges added. We claim that D $= \Delta(S)$.

To prove this we have to show that the circumscribed open disc of every triangle in D contains no points of S. This is obviously the case for all triangles that were already present in $\Delta(S')$, since the circumscribed disc of such a triangle cannot contain a point of S', as $\Delta(S')$ is a Delaunay triangulation, and since it cannot contain p_n by the choice of y_n.

Thus it remains to show that the circumscribed open disc of the $h-1$ "new" triangles contain no point of S'. Let T be such a triangle, let s be the edge of the convex hull of S' that is a side of T, and let T' be the unique triangle in $\Delta(S')$ that has s as a side (see Fig. 2.4). Furthermore, let C and C' be the circumscribed discs of T and T' respectively.

To see that C indeed contains no point of S' note that s cuts C into two parts. The part outside the convex hull of S' clearly cannot contain a point of S'. The other part must be contained in C' (since p_n not in C' !). But as T' is a triangle of the Delaunay triangulation $\Delta(S')$, C' does not contain a point of S', and hence C does not.

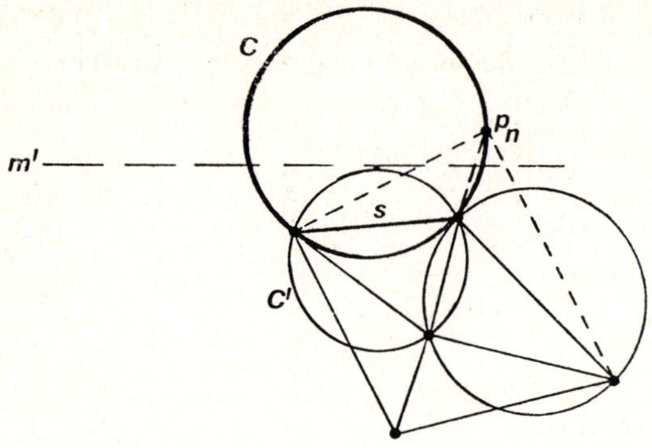

Figure 2.4

Finally note that clearly m' can be "updated" to a new y-maximum m in time proportional to h.

Q.E.D.

This lemma immediately yields a stronger version of Theorem 1.

Theorem 2:

Let S be a set of n points in the plane and let $\Delta(S)$ be its Delaunay triangulation.

Any algorithm that, when given S and $\Delta(S)$ as input, sorts the points of S with respect to their x-coordinates takes time $\Omega(n \log n)$ in the worst case.

Proof:

Identical to the proof of Theorem 1; just use Lemma 2 instead of Lemma 1.

We want to point out, however, that this proof takes advantage of the unit cost assumption in our RAM model of computation since there is no obvious upper bound for the y-coordinates of the points obtained in Lemma 2.

Q.E.D.

III. Preprocessed Planar Subdivision Searching

A classical 1-dimensional problem is the so-called dictionary problem: let $A = \{x_1, \ldots, x_n\}$ be a set in **R** with $x_1 < x_2 < \cdots < x_n$; for an arbitrary query number $z \in \mathbf{R}$ one is to determine whether there exists an i, $1 \leq i < n$, such that $x_i \leq z < x_{i+1}$, and in case it does, one is to report it. This problem is usually solved by binary search with $O(\log n)$ comparisons, and it also has been shown that $\Omega(\log n)$ comparisons are necessary even on fairly elaborate models of computation [8]. It is fairly easy to prove that solving the dictionary problem for an arbitrary set $B = \{z_1, \ldots, z_n\} \subset \mathbf{R}$ of n numbers requires time $\Omega(n \log n)$. The situation changes however if some additional information about B is available. If, for instance, B is sorted, i.e. $z_1 \leq z_2 \leq \cdots \leq z_n$, then these "batched" dictionary problems can be solved in $O(n)$ time by a simple merge procedure.

The dictionary problem can be reformulated as follows: for a given contiguous set of intervals on the real line and an arbitrary query number determine which interval (if any) contains that number. The 2-dimensional generalization of this problem is quite straightforward: for a given contiguous set of triangles in the plane and an arbitrary query point determine which triangle (if any) contains that point. This is of course the classical planar subdivision search problem in computational geometry and practical $O(\log n)$ time and $O(n)$ space solutions have been found by Kirkpatrick [13] and by Edelsbrunner et al. [5]. It should be clear that the $\Omega(\log n)$ lower bound for a single query and the $\Omega(n \log n)$ lower bound for multiple queries of the 1-dimensional dictionary problem carry over to the 2-dimensional subdivision search problem since it is easy to express the 1-dimensional problem as an instance of the 2-dimensional one.

Now the question arises whether, like in the 1-dimensional case, the multiple query subdivision search problem can be solved in $o(n \log n)$ time if additional information is available about the set of query points Q. Is, for instance, an improvement possible, if Q is sorted by one coordinate, or if a triangulation of Q is available?

The following theorems anwer these questions in the negative.

Theorem 3:

Let $S \subset \mathbf{R}^2$ be a set of n points and let Δ be a triangulation of S. Let $Q \subset \mathbf{R}^2$ be a set of n query points and let D be a triangulation of Q.

The problem of determining for each point in Q which triangle (if any) of Δ contains it requires time $\Omega(n \log n)$ in the worst case.

Proof:

Let us assume we have some algorithm that solves the problem stated in the theorem in time $f(n)$ in the worst case. We will show that with an additional time expenditure of $O(n)$ this algorithm could be used to solve the multiple query dictionary problem. However, for that problem there is an $\Omega(n \log n)$ lower bound. Hence $f(n)$ must be $\Omega(n \log n)$.

So let $A = \{x_1, \ldots, x_n\}$ with $x_1 < x_2 < \cdots < x_n$ be a dictionary and let $B = \{z_1, \ldots, z_n\}$ be an arbitrary set of query points. Using Lemma 1 obtain from B in linear time a set $Q = \{q_1, \ldots, q_n\} \subset \mathbf{R}^2$ and a triangulation D of Q, where for $1 \leq j \leq n$, $x(q_j) = z_j$, and for $1 \leq j < n$, $y(q_j) < y(q_{j+1})$.

Obtain in linear time from A a set $S = \{p_1, \ldots, p_n, p_1', \ldots, p_n'\}$, with $p_i = (x_i, y_{min})$ and $p_i' = (x_i, y_{max})$ for $1 \le i \le n$, where $y_{min} = y(q_1) - 1$ and $y_{max} = y(q_n) + 1$. Next construct the triangulation Δ of S as depicted in Figure 3.1. This can be done in linear time since the set A is sorted. Now run the assumed algorithm to locate all points of Q in the triangulation Δ in time $f(n)$. It should be clear that some $z_j \in B$ is "contained" in the dictionary A if and only if the corresponding point $q_j \in Q$ lies in some triangle of Δ. Furthermore, knowing the triangle that contains q_j allows to determine in constant time where z_j is to be found in the dictionary, i.e. for which i, $x_i \le z_j < x_{i+1}$. Thus we have solved the multiple query dictionary problem in time $O(n)+f(n)$ and this completes the proof. Q.E.D.

The preceding theorem can be strengthened in the following way.

Theorem 4:

Let $S \subset \mathbf{R}^2$ be a set of n points and let $\Delta = \Delta(S)$ be its Delaunay triangulation. Let $Q = \{q_1, \ldots, q_n\} \subset \mathbf{R}^2$ and let $D = \Delta(Q)$ be its Delaunay triangulation.

The problem of determining for each point in Q which triangle (if any) of Δ contains it requires time $\Omega(n \log n)$ in the worst case.

Proof:

The proof is basically identical to the proof of Theorem 3. Use Lemma 2 instead of Lemma 1 to obtain the set Q and its Delaunay triangulation.

The set S used in the proof of Theorem 3 is degenerate in the sense that for all i, $1 \le i < n$, the four points p_i, p_{i+1}, p_i', and p_{i+1}' are co-circular, and the edges connecting p_i' and p_{i+1} should not really be in $\Delta(S)$. However, this problem can be circumvented by determining from A in linear time $\delta = \min_{1 \le i < n} (x_{i+1} - x_i)$ and letting $p_i' = (x_i + \frac{1}{4}\delta, y_{max})$ for $1 \le i \le n$.

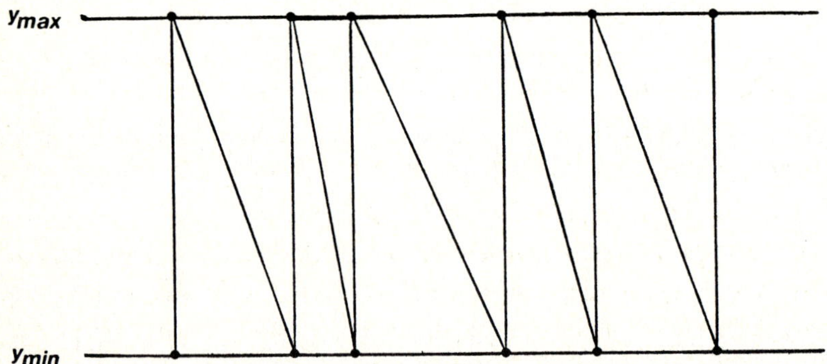

Figure 3.1

It can then be shown that the Delaunay triangulation of the set S defined this way has the same structure as the one of the original S is depicted in Figure 3.1. Furthermore, if $q_j \in Q$ is located in some particular triangle of $\Delta(S)$, then only one of two possible intervals of the dictionary A can contain the corresponding $z_j \in B$, and it can be determined in constant time which one does.

Q.E.D.

Before closing this section we want to point out something rather peculiar. We just showed that given two n-sets S and Q and their Delaunay triangulations $\Delta(S)$ and $\Delta(Q)$, generally $\Omega(n \log n)$ time is necessary to locate all points of Q in $\Delta(S)$. On the other hand Kirkpatrick [12] has shown that given $\Delta(S)$ and $\Delta(Q)$ one can construct the Delaunay triangulation of $S \cup Q$ in *linear* time.

IV. Finding Convex Hulls of 3-Dimensional Sorted Sets

The convex hull problem is one of the best studied problems in computational geometry. For the 2-dimensional case, finding the smallest convex polygon that contains a given set of n points in $\mathbf{R}^2$, a number of different $O(n \log n)$ algorithms have been proposed ([9],[22], [19],[18],[14]), and in [22] it is shown that this problem[2] has an $\Omega(n \log n)$ lower bound matching the upper bound achieved by these algorithms.

For the 3-dimensional case, finding the smallest convex polytope that contains a given set of n points in $\mathbf{R}^3$, only four non-trivial algorithms are known ([4],[19],[20],[11]). However, only the algorithm in [19] has $O(n \log n)$ worst case time complexity and is worst case optimal, whereas the other three algorithms have $O(n^2)$ worst case time complexity.

The fact that a fair number of optimal algorithms for the 2-dimensional problem are known, whereas there is only one single optimal algorithm for the 3-dimensional case indicates that intuitively the 3-dimensional problem is harder than the 2-dimensional one, although this difference in difficulty is not quantifiable by the usual means of asymptotic complexity measures. In this and the following section we give evidence that the 3-dimensional problem is indeed "harder" by showing that certain additional input information that allows linear time algorithms in the 2-dimensional case does not reduce the $\Omega(n \log n)$ complexity in the 3-dimensional case.

The first kind of additional input information we consider is the ordering of the input points with respect to some coordinate. If for a set $S = \{p_1, \ldots, p_n\} \subset \mathbf{R}^2$ it is guaranteed that $x(p_1) < x(p_2) < \cdots < x(p_n)$, then the planar convex hull of S can be constructed in $O(n)$ time by an incremental construction akin to the one used in the proof of Lemma 1. In contrast to this we show in the following that for a set $S \subset \mathbf{R}^3$ that is ordered in such a way there is still an $\Omega(n \log n)$ lower

[2]To be exact, the bound given in [22] is for the version of the convex hull problem, where one is to find the ordered sequence of vertices around the convex hull. A stronger lower bound for the weaker problem of just identifying the vertices of the convex hull is given in [14].

bound for constructing its 3-dimensional convex hull.

At first, though, we have to make more specific what we mean by constructing the convex hull of a point set $S \subset \mathbf{R}^3$. Such a convex hull is a 3-dimensional polytope whose vertices are points in S. Its boundary complex, i.e. the system of its vertices, edges, and faces, can be viewed as a planar graph. Thus for our purposes, an adequate description of the convex hull C of S has to identify which points of S are vertices of P and it also has to provide a good representation of the planar graph formed by the boundary complex of C, e.g. in form of cyclically ordered adjacency lists. This is the description that is to be produced by a 3-dimensional convex hull algorithm. All four algorithms mentioned above can be made to yield such a representation.

For the theorems in this section we need the following

Lemma 3:

Let S be a set of n points in $\mathbf{R}^3$ and assume an adequate representation of its convex hull C is available.

Let H be a plane in $\mathbf{R}^3$ and let S' be the orthogonal projection of S into H.

With this information the 2-dimensional convex hull C' of S' in H can be determined in $O(n)$ time.

Proof: (Outline)

Let H be the plane $\{ x \in \mathbf{R}^3 \mid \langle x, a \rangle = c \}$ for some $c \in \mathbf{R}$ and some non-zero $a \in \mathbf{R}^3$.
It should be clear that the boundary of C' is the projetion of some closed path P of edges and vertices of C. The criterion for an edge e of C to be in this path P is simple and can be tested in constant time: let N_e and N_e' be the outward normal vectors of the two faces of C that contain e, and let $s_e = \langle N_e, a \rangle$ and $s_e' = \langle N_e', a \rangle$; the edge e is in P iff one of s_e and s_e' is greater than zero and the other one is not.

Thus all the edges of P can be identified in time proportional to the number of edges of C which is linear in n. Furthermore, it is easy to see that the cyclic order in which these edges appear on P can be determined in linear time also.

Finally let $V = \{v_0, \ldots, v_{k-1}\}$ be the cyclic sequence of vertices that appear on P and let $V' = \{v_0', \ldots, v_{k-1}'\}$ be their orthogonal projections into H. Remove in linear time from V' all v_i' that are collinear with $v'_{(i-1) \bmod k}$ and $v'_{(i+1) \bmod k}$. (These collinearities can appear if a face of C is normal to H.) The resulting sequence is the sequence of vertices around C'.

Q.E.D.

With this lemma the following result is easy:

Theorem 5:

Let $S = \{p_1, \ldots, p_n\}$ be a set of n points in $\mathbf{R}^3$ with $x(p_1) < x(p_2) < \cdots < x(p_n)$.

Any algorithm that finds the convex hull of such an ordered set requires time $\Omega(n \log n)$ in the worst case.

Proof:

Assume some algorithm that solves this problem in worst case time $f(n)$. We will use this algorithm to solve a 2-dimensional convex hull problem.

Let $T = \{(y_1, z_1), \ldots, (y_n, z_n)\}$ be a set of n points in $\mathbf{R}^2$. We construct the convex hull of T in the following way.

Form the set $S = \{ (i, y_i, z_i) \mid 1 \leq i \leq n \} \subset \mathbf{R}^3$ in linear time. Clearly S is sorted by x-coordinate. Thus we can apply our assumed algorithm to find C, the 3-dimensional convex hull of S in time $f(n)$.

Now let H be the y-z-plane and use the construction of Lemma 3 to find in linear time the 2-dimensional convex hull of the set $S' = \{ (0, y_i, z_i) \mid 1 \leq i \leq n \}$ which is the orthogonal projection of S into H.

Clearly S' can be identified with T and thus we have in effect constructed the 2-dimensional convex hull of T in time $O(n) + f(n)$. However, because of the $\Omega(n \log n)$ lower bound for the 2-dimensional convex hull problem $f(n)$ must therefore be $\Omega(n \log n)$.

Q.E.D.

Theorem 5 says that having a set in $\mathbf{R}^3$ sorted with respect to the x-coordinate does not really help in finding its convex hull. David Kirkpatrick has suggested a strengthening of this result.

Let a be a non-zero vector in $\mathbf{R}^3$. For a set $S = \{p_1, \ldots, p_n\} \subset \mathbf{R}^3$ we say we know the *order of S with respect to the direction a,* if we know a permutation π_a of the integers from 1 to n, such that $\langle a, p_{\pi_a(i)} \rangle \leq \langle a, p_{\pi_a(i+1)} \rangle$ for $1 \leq i < n$. The intuitive meaning of this definition is straightforward: if for some line L with direction a, l_i is the orthogonal projection of p_i onto L for $1 \leq i \leq n$, then when traversing L in direction a one encounters $l_{\pi_a(1)}, l_{\pi_a(2)}, \ldots, l_{\pi_a(n)}$ in this sequence.

Theorem 6:

Let k be a fixed positive integer, and let $A \subset \mathbf{R}^3$ be a set of k non-zero vectors in $\mathbf{R}^3$.

Let $S = \{p_1, \ldots, p_n\}$ be a set of n points in $\mathbf{R}^3$ for which the ordering is known with respect to each direction $a \in A$.

Any algorithm that finds the convex hull of S requires time $\Omega(n \log n)$ in the worst case, even if it has all these orderings available.

Proof:

Assume some algorithm that solves this problem in worst case time $f(n)$. The proof that $f(n) = \Omega(n \log n)$ is again by reduction to the 2-dimensional convex hull problem.

Let b be a vector in $\mathbf{R}^3$ such that $\langle b,a \rangle \neq 0$ for all $a \in A$. It is easy to see that such a b must always exist and that it can be found in constant time if k is assumed to be a constant. Let $A^+ = \{ a \in A \mid \langle a,b \rangle > 0 \}$ and let $A^- = \{ a \in A \mid \langle a,b \rangle < 0 \}$.

Now let $T = \{q_1, \ldots, q_n\}$ be a set of n points in the plane $H = \{ x \in \mathbf{R}^3 \mid \langle x,b \rangle = 0 \}$. We want to construct the 2-dimensional convex hull C of T in H.

We first construct a set $S = \{p_1, \ldots, p_n\} \subset \mathbf{R}^3$, where $p_i = q_i + \lambda_i b$ for $1 \leq i \leq n$. Obviously for any choice of the reals λ_i the orthogonal projection of S into H is T. Furthermore, we can choose the λ_i's such that S is sorted with respect to each direction $a \in A$. To be more precise, we will choose the λ_i's such that the index permutation π_a that gives the ordering of S with respect to some direction $a \in A$ is the identity permutation π_I if $a \in A^+$, and the reversal permutation π_R if $a \in A^-$, where π_I and π_R are defined by $\pi_I(i) = i$ and $\pi_R(i) = n+1-i$, for $1 \leq i \leq n$.

The reader may verify that this can be achieved by setting λ_1 to 1 and letting

$$\lambda_i = \max \left(\left\{ \frac{\langle p_{i-1} - q_i, a \rangle}{\langle b,a \rangle} \mid a \in A^+ \right\} \cup \left\{ \frac{\langle q_i - p_{i-1}, a \rangle}{\langle b,a \rangle} \mid a \in A^- \right\} \right)$$

for $i = 2, \ldots, n$.

As $A = A^+ \cup A^-$ contains only a constant number of elements all the λ_i's can be determined in $O(n)$ time and we have constructed the desired set S, whose projection into H is T and which is sorted with respect to each of the k directions in A.

Now apply the assumed algorithm to find the 3-dimensional convex hull of S in time $f(n)$. Subsequently, using the construction of Lemma 3, recover from this hull the 2-dimensional convex hull C of T in the plane H in time $O(n)$.

Thus we constructed C in time $O(n) + f(n)$. But because of the $\Omega(n \log n)$ lower bound for the 2-dimensional convex hull problem $f(n)$ must therefore be $\Omega(n \log n)$.

Q.E.D.

V. Convex Hulls of Simple and Star-Shaped Polytopes

Let us informally define a *simple polytope* to be a 3-dimensional solid which is not necessarily convex but does not have any "holes" and whose boundary consists of a finite number of planar pieces. A *star-shaped polytope* is a simple polytope P that contains a point z (called a *center* of P) such that for every $x \in$ P the straight line segment joining z and x is totally contained in P.

The results of this section were motivated by the attempt of several researchers to find linear time algorithms for constructing the convex hull of 3-dimensional star-shaped and simple polytopes. There was some encouraging evidence that such algorithms exist since the corresponding 2-dimensional problems do have linear time solutions. The second phase of Graham's algorithm [9] finds the convex hull of a star-shaped polygon in linear time and algorithms in [2] and [16] do so for simple polygons. (Several other algorithms have been published but their correctness has been challenged.)

A linear time convex hull algorithm for star-shaped polytopes would have had some nice applications (e.g. in the work presented in [15]) and furthermore it would have implied the existence of a linear time procedure for constructing the convex hull of the union of two convex polytopes.[3]

Before proving the main theorem of this section we need to introduce the notion of an upper hull. For a non-zero vector b and a finite set T in the plane we define the *upper hull* of T with respect to the direction b to be the ordered sequence of all the edges of the convex hull of T for which the outward normals form a positive inner product with b. This definition is made more intuitive in Figure 5.1.

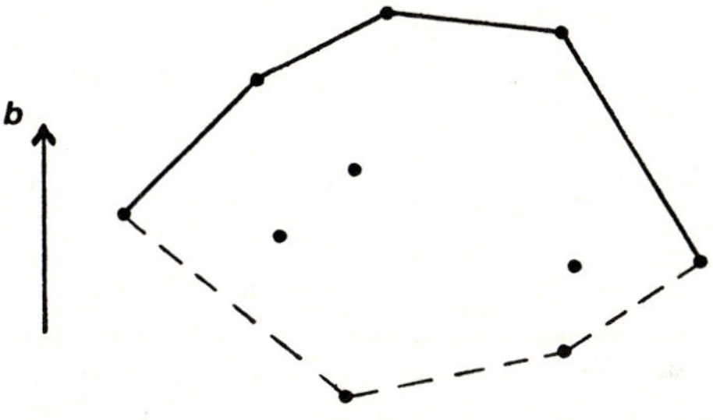

Figure 5.1

It should be clear that constructing the upper hull of a set T of n points with respect to some direction b requires $\Omega(n \log n)$ time in the worst case since this upper hull together with the upper hull with respect to $-b$ yields a complete description of the convex hull of T.

Finally, let us note that, like in the case of convex polytopes, we assume that a simple or star-shaped polytope is represented by a planar graph description of its boundary together with the coordinates of its vertices.

Theorem 7:

Any algorithm that constructs the convex hull of a star-shaped polytope with n vertices requires $\Omega(n \log n)$ time in the worst case.

---

[3]The problem whether there exists such a linear time procedure is still open in general and only solved for the case where the two polytopes are separated by a plane [19]. The claim in [3] that the solution in [19] generalizes to the non-separable case is as yet completely unsupported.

Proof:

Assume some algorithm that solves this problem in time $f(n)$. We prove that $f(n) = \Omega(n \log n)$ by reduction to the 2-dimensional upper hull problem.

Let $T = \{q_1, \ldots, q_{n-1}\}$ be a set of points in the x-z-plane with $q_i = (x_i, 0, z_i)$. We want to construct the upper hull of T with respect to the direction $b = (0,0,1)$.

Using the construction of Lemma 1 determine in $O(n)$ time a triangulation D of the set $T' = \{q_1', \ldots, q_{n-1}'\}$ in the x-y-plane, where $q_i' = (x_i, i, 0)$ for $1 \le i < n$. Re-endow every point in T' with its "original" z-coordinate to yield the set $S = \{p_1, \ldots, p_{n-1}\}$, where $p_i = (x_i, i, z_i)$ for $1 \le i < n$, and let S "inherit" the triangulation D of T'. This inherited triangulation can be interpreted as a piecewise planar surface F with the property that no planar piece is orthogonal to the x-y-plane, and every "vertical" line, i.e. a set of the form $\{ a + \lambda b \mid \lambda \in \mathbf{R} \}$, $a \in \mathbf{R}^3$, intersects F at most once.

This property ensures the existence of a sufficiently large μ such that the point $p_n = p_1 - \mu b$ lies "below" all the planes that contain a planar piece of F. Furthermore, this μ can be determined from D in $O(n)$ time since D has no more than $2n-7$ triangles and hence F consists of at most $2n-7$ planar pieces.

Next determine in $O(n)$ time the edges of D that correspond to the boundary of F (i.e. essentially the convex hull edges of T') and for each such edge e add the triangle spanned by e and p_n to the surface F. This yields a closed piecewise planar surface bounding a simple polytope P with vertex set $S' = \{p_1, \ldots, p_n\}$. As a matter of fact, P is star-shaped with center p_n.

Now use the assumed algorithm to find the 3-dimensional convex hull of P in time $f(n)$ and use Lemma 3 to obtain in linear time the projection of this hull onto the x-z-plane.

It is easy to see that this projection consists exactly of p_n and the vertices of the upper hull of the initial set T with respect to b. Thus in time $O(n)+f(n)$ we effectively constructed the upper hull of T. The $\Omega(n \log n)$ lower bound for the upper hull problem implies therefore that $f(n) = \Omega(n \log n)$.

Q.E.D.

Theorem 8:

Any algorithm that constructs the convex hull of a simple 3-dimensional polytope with n vertices requires $\Omega(n \log n)$ time in the worst case.

Proof:

This follows immediately from Theorem 7 since a star-shaped polytope is a special case of a simple polytope.

Q.E.D.

Acknowledgements:

I would like to thank John Gilbert for his careful reading of the manuscript and for his comments.

References:

[1] Ben-Or,M. "Lower Bounds for Algebraic Computation Trees." Proc. 15th Annual ACM Symposium on Theory of Computing, pp.80-86; 1983.

[2] Bhattacharya,B.K., El-Gindy,H. "A New Linear Convex Hull Algorithm for Simple Polygons." IEEE Transactions on Computers; pp.85-88; 1984.

[3] Bentley,J.L., Shamos,M.I. "Divide and Conquer for Linear Expected Time." Information Processing Letters 7; pp.87-91; 1978.

[4] Chand,D.R., Kapur,S.S. "An Algorithm for Convex Polytopes." Journal of the ACM 17; pp.78-86; 1970.

[5] Edelsbrunner,H., Guibas,L.J., Stolfi,J. "Optimal Point Location in a Monotone Subdivision." Manuscript; 1983.

[6] Fredman,M.L. "Inherent Complexity of Range Query Problems." Proc. of the 17th Annual Allerton Conf. on Communication, Control, and Computing; pp.231-240; 1979.

[7] Fredman,M.L. "A Lower Bound on the Complexity of Orthogonal Range Queries." Journal of the ACM 28; pp.696-705; 1981.

[8] Friedman,N. "Some Results on the Effect of Arithmetics on Comparison Problems." Proc. 13th IEEE Symp. Switching and Automata Theory, pp.139-143; 1972.

[9] Graham,R.L. "An Efficient Algorithm for Determining the Convex Hull of a Finite Planar Set." Information Processing Letters 1, pp.132-133; 1972.

[10] Guibas,L.J., Stolfi,J. "Primitives for the Manipulation of General Subdivisions and Computation of Voronoi Diagrams." Proc. 15th Annual ACM Symposium on Theory of Computing, pp.221-234; 1983.

[11] Kallay,M. "Convex Hull Algorithms for Higher Dimensions." Manuscript, Univ. of Oklahoma, Norman OK; 1981.

[12] Kirkpatrick,D.G. "Efficient Computation of Continuous Skeletons." Proc. of 20th Annual IEEE Symposium on Foundations of Computer Science, pp.18-27; 1979.

[13] Kirkpatrick,D.G. "Optimal Search in Planar Subdivisions." SIAM J. on Computing 12; pp.28-35; 1983.

[14] Kirkpatrick,D.G., Seidel,R. "The Ultimate Planar Convex Hull Algorithm?" Cornell Univ., CS Tech. Rep. 83-577; 1983.

[15] Little,J. "An Iterative Method for Reconstructing Extended Gaussian Images." Proc. of the 1983 National Conf. on Artificial Intelligence; pp.247-250; 1983.

[16] McCallum,D., Avis,D. "A Linear Algorithm for Finding th Convex Hull of a Simple Polygon." Information Processing Letters 9; pp.201-206; 1979.

[17] Paschinger,I. "Konvexe Polytope und Dirichletsche Zellenkomplexe." Univ. Salzburg, Austria; Math. Inst. Arbeitsbericht 1-2; 1982.

[18] Preparata,F.P. "An Optimal Real Time Algorithm for Planar Convex Hulls." Communications of the ACM 22; pp.402-405; 1979.

[19] Preparata,F.P. Hong,S.J. "Convex Hulls of Finite Sets in Two and Three Dimensions." Communications of the ACM 20; pp.87-93; 1977.

[20] Seidel,R. "A Convex Hull Algorithm Optimal for Point Sets in Even Dimensions." Univ. of British Columbia, Dept. of Computer Science, Rep. 81-14; 1981.

[21] Seidel,R. "On the Size of Closest Point Voronoi Diagrams." Tech. Univ. Graz, Austria; Inst. fuer Informationsverarbeitung, Rep. F94; 1982.

[22] Shamos,M.I. "Computational Geometry." Ph.D. Thesis; Yale Univ., New Haven CT; 1978.

[23] Swart,G. "An $\Omega(n^2)$ Lower Bound For Intersection Detection." Univ. of Washington, Dept. of Computer Science; Manuscript; 1983.

[24] Yao,A.C. "A Lower Bound to Finding Convex Hulls." Journal of the ACM 28; pp.780-787; 1981.

COMPUTATIONAL GEOMETRY, G.T. Toussaint (Editor)
© Elsevier Science Publishers B.V. (North-Holland), 1985

MOVABLE SEPARABILITY OF SETS

Godfried T. Toussaint

School of Computer Science
McGill University
805 Sherbrooke St. West
Montreal, Canada H3A 2K6

Abstract

Spurred by developments in spatial planning in robotics, computer graphics, and VLSI layout, considerable attention has been devoted recently to the problem of moving sets of objects, such as line segments and polygons in the plane or polyhedra in three dimensions, without allowing collisions between the objects. One class of such problems considers the *separability* of sets of objects under different kinds of motions and various definitions of separation. This paper surveys this new area of research in a tutorial fashion, presents new results, and provides a list of open problems and suggestions for further research.

Key Words and Phrases:
sofa problem, polygons, polyhedra, movable separability, visibility hulls, hidden lines, hidden surfaces, algorithms, complexity, computational geometry, spatial planning, collision avoidance, robotics, artificial intelligence.

CR Categories: 3.36, 3.63, 5.25, 5.32, 5.5

Research supported by NSERC grant no. A9293 and FCAC grant no. EQ1678.

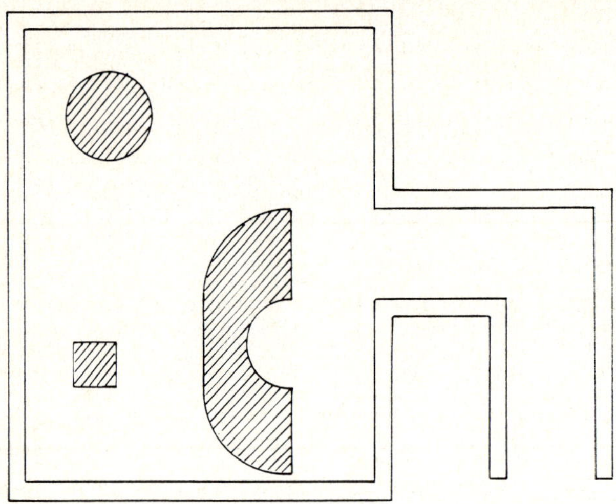

Figure 1: Moving furniture out of the apartment.

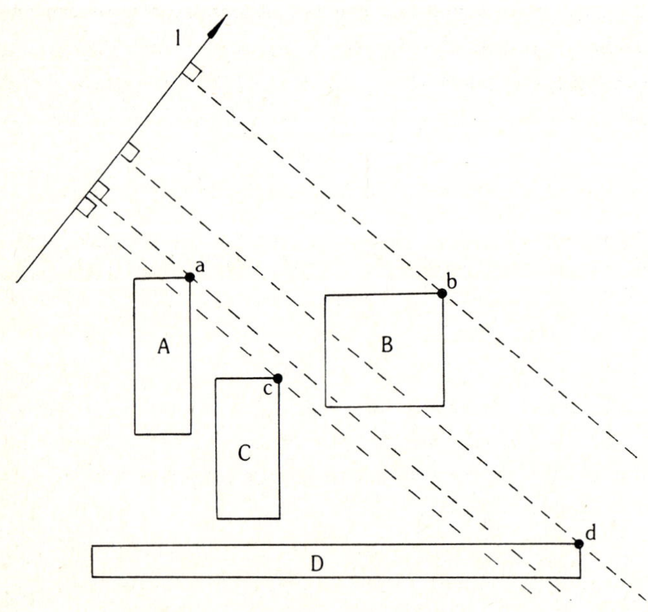

Figure 2: Rectangle D cannot be translated in direction l before rectangle C, and yet d occurs before c on line l.

I. Introduction

Consider a large one-room apartment with a corridor connecting it to the outside world, as illustrated in Figure 1. The corridor is assumed to have unit width and a right-angled corner. We can ask what is the figure of largest area that can be moved out of the apartment and into the street through the connecting corridor. This is known as the *sofa problem* and has received some attention in the mathematics and computing literature [1] - [5]. Note that the room is as large as desired and we assume it is no obstacle to position the desired object into the first "wing" of the corridor. Thus we are really concerned only with getting by the corner of the corridor. An obvious lower bound on the solution to this problem is *unity* since a square of unit area can be moved out with two translations. However, we can find non-convex figures with areas as large as $(\pi/2) + (2/\pi) \cong 2.2074$ that can be moved out with a sequence of translations and rotations [5]. A variety of such problems present themselves if we vary the shape of the hallway (such as allowing left-angled as well as right-angled corners) and restrict the class of objects considered [5]. For example, we may ask for the largest (in some sense) *convex*, or *star-shaped* figure. We can also ask whether a given object or figure can be moved out of a specified apartment. Furthermore, we can ask for a sequence of motions that will free the object if such an action is possible. All these problems belong to a large family of problems which in this paper are termed *movable separability of sets*. While movable separability can be investigated for quite general sets, in this paper we consider the sets to be simple objects such as *line segments*, *circles*, and *simple polygons* in the plane or *spheres* and *polyhedra* in three dimensions. We also distinguish between *movable separability* problems and *collision avoidance* problems in *robotics* such as the *findpath* problem [6]. For a survey of those problems and their relation to computational geometry the reader is referred to [7]. It is difficult to formally define the class of problems labeled "movable separability". Nevertheless the problems considered here are in some sense more concerned with the notion of *separability* than the typical collision avoidance problem found in robotics.

As another example of a class of *movable separability* problems let us look at the problem of *translating* rectangles. Consider a set of non-intersecting rectangles in the plane whose sides are parallel to the x and y axes, as illustrated in Figure 2. Such rectangles are termed *isothetic*.

A problem that arises in graphics and VLSI [8] is that of translating the entire collection by some common vector to a new location with two constraints: first, the rectangles can only be moved one at a time and, second, during the entire process no *collisions* are allowed between the rectangles. A *collision* occurs if at some instant in time the interiors of two rectangles intersect. One problem that arises immediately is whether such a translation ordering property holds for all sets of rectangles and all directions. Another is the efficient computation of such an ordering if it exists. This is clearly a separability problem since each rectangle being translated is also being separated.

Referring to Figure 2, let ℓ denote the desired direction of translation and construct that line of support to each rectangle perpendicular to ℓ that maximizes the displacement in direction ℓ. These lines of support intersect the rectangles A,B,C, and D at the *support vertices* a, b, c, and d, respectively. It is tempting, at first glance, to claim that a translation ordering can be obtained by sorting the projections of the *support vertices* on ℓ and that therefore B can be moved first. Note that this is not the case, however, since this method would require D to move second which is impossible since it is blocked by C. Guibas and Yao [8] have shown that given a set of n rectangles and a direction ℓ, a translation ordering always exists and can be computed in O(n log n) time. In fact they prove this also for a set of n convex polygons. Thus we will say that convex polygons in the plane exhibit the *translation-ordering property*. A simpler O(n log n) algorithm for solving this problem was later discovered by Ottman and Widmayer [9].

This problem was generalized to consider other types of polygons, and motions besides simple translations in [10] - [12]. One class of problems that results when convexity is relaxed concerns "interlocking polygons". Thus one class of problems considered in [10] - [12] deals with the determination of whether a given collection of polygons is "movably separable" in a specified sense. Some of the results of [10] - [12] were also independently obtained by Chazelle et al. [13] for the special case of *isothetic* polygons.

II. Movable Separability of Sets: Definitions

In this section we define some notions of *movable* separability and relate them to the well-known concepts of *linear* and *nonlinear* separability of sets [14] - [15], [19].

Definition: An object P can be *moved* from position A to position B if there exists a finite sequence of translations and rotations (possibly simultaneous) of P that carries P from A to B.

Definition: Let two objects P and Q be moved during a time interval starting at t_1 and ending at t_2. We say that a *collision* occurs between P and Q if there exists a time t such that $t_1 \leq t \leq t_2$ and the *interiors* of P and Q intersect.

Definition: Two objects are *movably separable* if one of them can be *moved* an arbitrary distance away from the other *without colliding* with it.

If two objects are *not* movably separable then they are said to be *interlocked*. Note that a variety of definitions of movable separability are immediately apparent by specifying the type of motion considered. The simplest might be a *single translation*. A more complicated case may allow many translations but no rotations, and going further, any number of both types of displacements may be allowed but not simultaneously.

Note that when we speak of *distance* between two objects or between one object and a collection of objects we are measuring the distance between two sets, say S_1 and S_2. The distance used here refers to the minimum euclidean distance between an element in S_1 and an element in S_2.

It is interesting to see what the relation is between *movable separability* and the more well-known concepts of separability in discriminant analysis [14], [15], [19].

Definition: Let P and Q be two sets in Euclidean d-space, $\mathbb{R}^d$. We say that P and Q are *linearly separable* if there exists a hyperplane H that partitions $\mathbb{R}^d$ into two half-spaces H(P) and H(Q) such that P is contained in H(P) and Q is contained in H(Q).

Definition: P and Q are *nonlinearly separable* if there exists a partition of $\mathbb{R}^d$ into two non-intersecting regions R(P) and R(Q) containing P and Q, respectively.

These concepts are illustrated in Figure 3. In Figure 3 (a) we see that linear separability implies movable separability. For example, either P or Q can be translated to infinity in a direction parallel to the line of separability. In Figure 3 (b) we note first, that movable separability does not imply linear separability. This follows from a well-known theorem in convexity theory [16] which states that "two sets

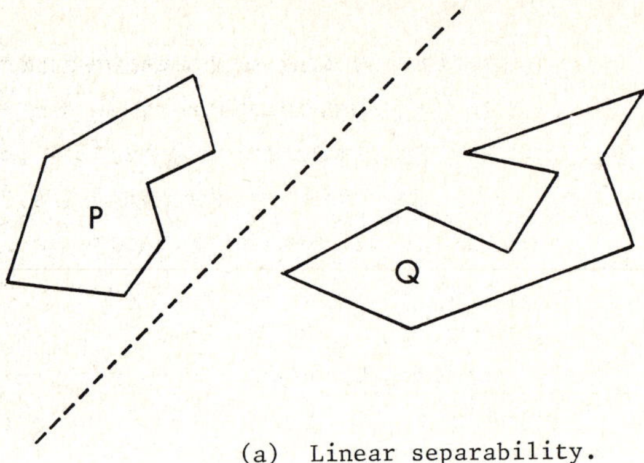

(a) Linear separability.

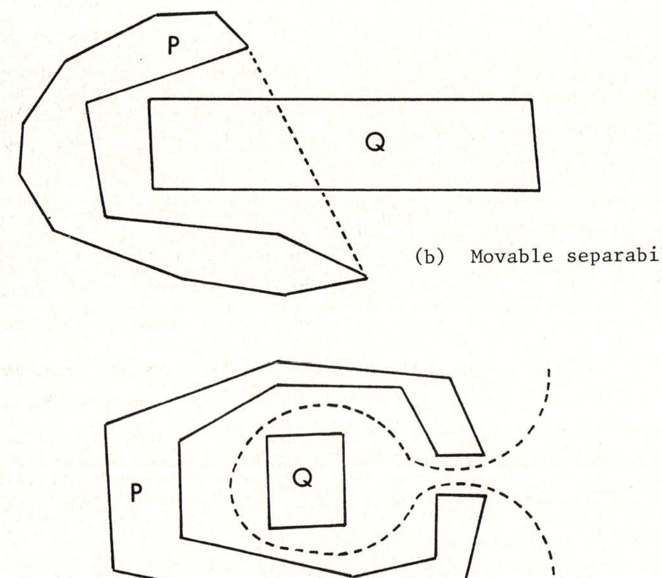

(b) Movable separability.

(c) Nonlinear separability.

Figure 3: Illustrating the relation between linear separability, movable separability, and nonlinear separability.

are *linearly separable* if, and only if, their convex hulls do not intersect" and the fact that convex hull intersection need not hinder movable separability. Secondly, Figure 3 (b) also illustrates that movable separability implies nonlinear separability. Finally, Figure 3 (c) shows that nonlinear separability need not imply movable separability. In fact it is obvious that even a low-order polynomial discriminant function [15] such as a *quadratic* does not imply movable separability. In summary we have:

linear separability	$\Longrightarrow$	movable separability	$\Longrightarrow$	nonlinear separability

If we are dealing with a collection of more than two objects then a variety of different types of movable separability is possible. Let $\mathbb{P} = \{P_1, P_2, \ldots, P_n\}$ be a collection of objects. Denote an ordering of $\mathbb{P}$ by $\mathbb{P}' = \{P_{(1)}, P_{(2)}, \ldots, P_{(n)}\}$, and let $\mathbb{P}'_{(i)} = \{P_{(i)}, P_{(i+1)}, \ldots, P_{(n)}\}$.

Definition: A set of objects $\mathbb{P}$ is *sequentially movably separable* if there exists an ordering $\mathbb{P}'$ such that for $i = 1,2,\ldots,n-1$, $P_{(i)}$ can be moved an arbitrary distance away from $\mathbb{P}'_{(i+1)}$ without colliding with any object in $\mathbb{P}'_{(i+1)}$.

Figure 4 illustrates a set of *sequentially movably separable* polygons. Here $\mathbb{P}' = \{P_2, P_3, P_4, P_5, P_6, P_1\}$ and P_2 through P_6 can be moved to infinity with two translations each.

Definition: If there exists a set of motions on a collection of objects $\mathbb{P}$ such that for each object P_i, $i = 1,2,\ldots,n$, P_i can attain an arbitrarily large distance from $\{\mathbb{P} - P_i\}$ without collisions, then $\mathbb{P}$ is said to be *movably separable*.

Figure 5, a variation of an example due to Chazelle, et al., [13], illustrates a collection of four *edge-visible* polygons which are *not sequentially movably separable* and yet they are *movably separable*. Note that in Figure 5 no object can be moved to infinity without colliding with the remaining objects. Yet if we allow repeated alternating translations of P_1 and P_2 in the +x direction then eventually P_1 and P_2 lie outside the convex hull of P_4 and all four polygons can be translated to infinity. Note however that by making the separation δ, between P_1 and P_2, arbitrarily small we can require an arbitrarily large number of motions for achieving separation, independent of the number of objects. Chazelle, et al. [13] also give an example where the number of motions is *exponential* in the number of objects. (A polygon P is *edge-visible* if it contains an

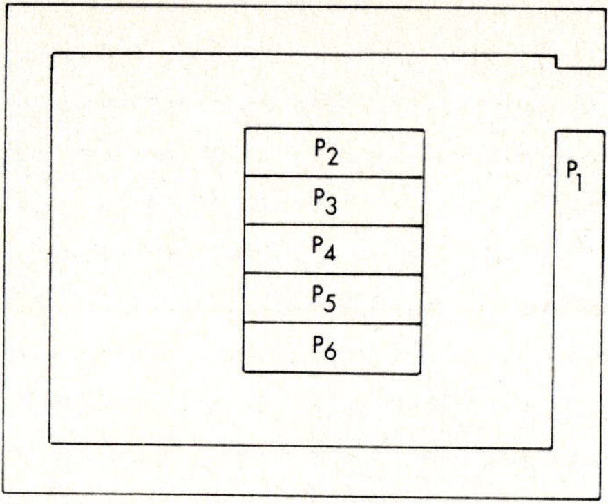

Figure 4: A set of sequentially movably separable polygons.

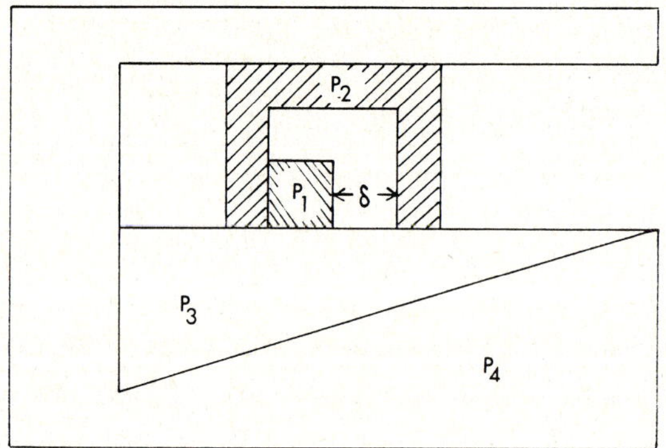

Figure 5: A set of four movably separable edge-visible polygons which is not sequentially separable.

edge e such that for every point x in P there exists a point y in e such that the line segment [x,y] lies in P. Here [x,y] denotes *closed* line segment. Similarly, an *open* line segment will be denoted by (x,y)).

Note that in the example of Figure 5 an object is moved as often as required. However, separability is still attained by moving only *one* object at a time. It is possible that a set of objects is *separable* but *only if simultaneous* motions of several objects are required. One such example is illustrated in Figure 6. This set is not *sequentially movably separable*. Neither is it separable with repeated individual motions, and yet it is separable under simultaneous motions. In fact there exists an infinite number of non-trivial sets of motions that will separate this set. One example is to first translate P_1 and P_2 as a block in the + y direction until the convex hull of P_1 and P_2 does not intersect the convex hull of P_3 and P_4. Subsequently P_1 and P_4 can be translated in the - x direction and P_2 and P_3 can be translated in the + x direction. This example leads us to the following definition.

Definition: A set of objects $\mathbb{P}$ is *simultaneously movably separable* if $\mathbb{P}$ is separable *only* by moving a subset of $\mathbb{P}$, of cardinality greater than one, *simultaneously*. (Note that a motion with velocity zero clearly cannot count as a motion.)

Definition: A set of objects $\mathbb{P}$ which is *not* movably separable is said to be *interlocked*.

III. A Hierarchy of Simple Polygons

Let $P = (p_1, p_2, \ldots, p_n)$ be a simple polygon, i.e., we are given a list of its vertices, in clockwise order, along with their cartesian coordinates. We assume the polygon is in *standard form*, i.e., the vertices are distinct and no three consecutive vertices are collinear. A pair of vertices, say $p_i\ p_{i+1}$, defines the ith edge of P. The sequence of vertices and edges forming the boundary of a polygon P, and denoted by bd(P), partitions the plane into two open regions: one bounded, termed the *interior* of P and denoted by int(P), and the other the unbounded *exterior* of P, denoted by ext(P). When we consider a collection of polygons, each polygon is assumed to contain n vertices to simplify notation.

We saw with the example of Figure 2 that any finite collection of *rectangles*, no matter how large, is movably separable under translations. In fact, all fixed directions admit a translation ordering. On the other

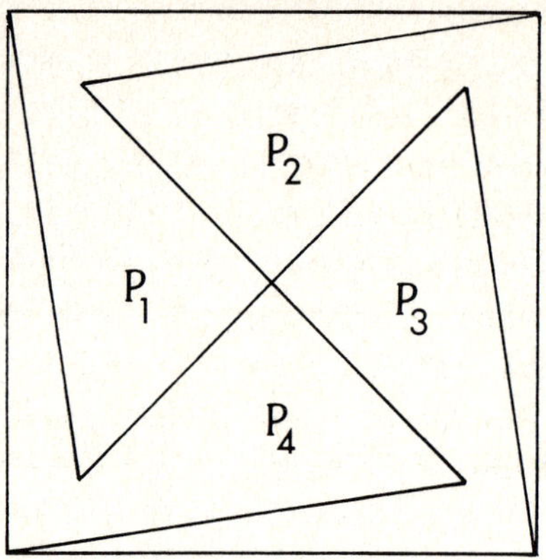

(a) Initial configuration.

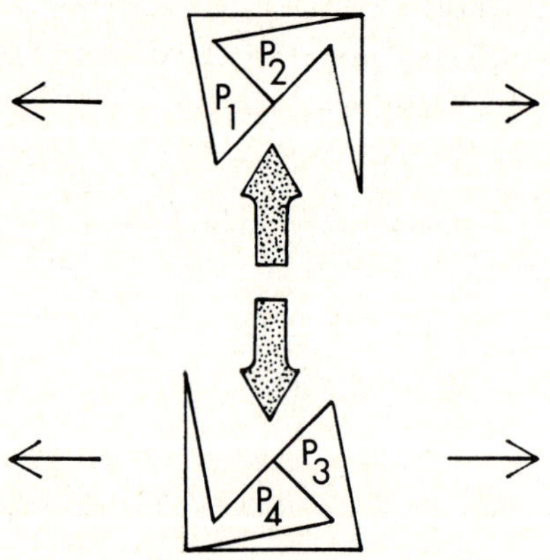

(b) After simultaneous translation of P_1 and P_2 in the +y direction.

Figure 6: A set of simultaneously separable polygons.

hand, from Figure 3 (c) we see that two *arbitrary simple* polygons are sufficient to form an *interlocked* set. Now, *rectangles* and *simple polygons* are extremes among a hierarchy of families of simple polygons of varying degrees of shape complexity. An example of a hierarchy of nine families of polygons is given by the directed graph in Figure 7. A node in this graph represents a family of polygons. A directed path connects node A to node B (A → B) if, and only if, polygons belonging to family A also belong to family B. For example, *convex* polygons are *star-shaped* but not vice-versa. It is thus interesting to follow different paths along this graph and to determine the places along the path where *movability properties* change. Actually, the graph in Figure 7 can be enlarged by the inclusion of a score of additional families of polygons [17] but this set will suffice to illustrate the point. Before considering movability we provide some definitions.

Definition: The *convex deficiency* of a polygon P is a set of *deficiency polygons* $D_1, D_2, \ldots, D_k$ obtained by subtracting int(P) from the convex hull of P and deleting the edges of P that are also edges of the convex hull of P. (Deficiency polygons are also more affectionately termed *pockets*.)

Definition: The edge of a pocket of P which is not also an edge of P is the *lid* of the pocket.

Definition: A polygon P is *completely visible* from edge e if for every point x in e and every point y in P, the line segment [x,y] lies in P.

Definition: A polygon P is *completely externally visible* if every deficiency polygon of P is completely visible from its *lid*.

Definition: A polygon P is *strongly visible* from edge e if there exists a point x in e such that for all y in P, the line segment [x,y] lies in P.

Definition: A polygon P is *strongly externally visible* if every deficiency polygon of P is strongly visible from its lid.

Definition: A polygon P is *weakly externally visible* if every deficiency polygon of P is *edge-visible* from its lid.

Definition: A polygon P is said to be *star-shaped* if there exists a region K in P, termed the *kernel* of P, such that for all $x \in K$ and all $y \in P$ the line segment [x,y] lies in P.

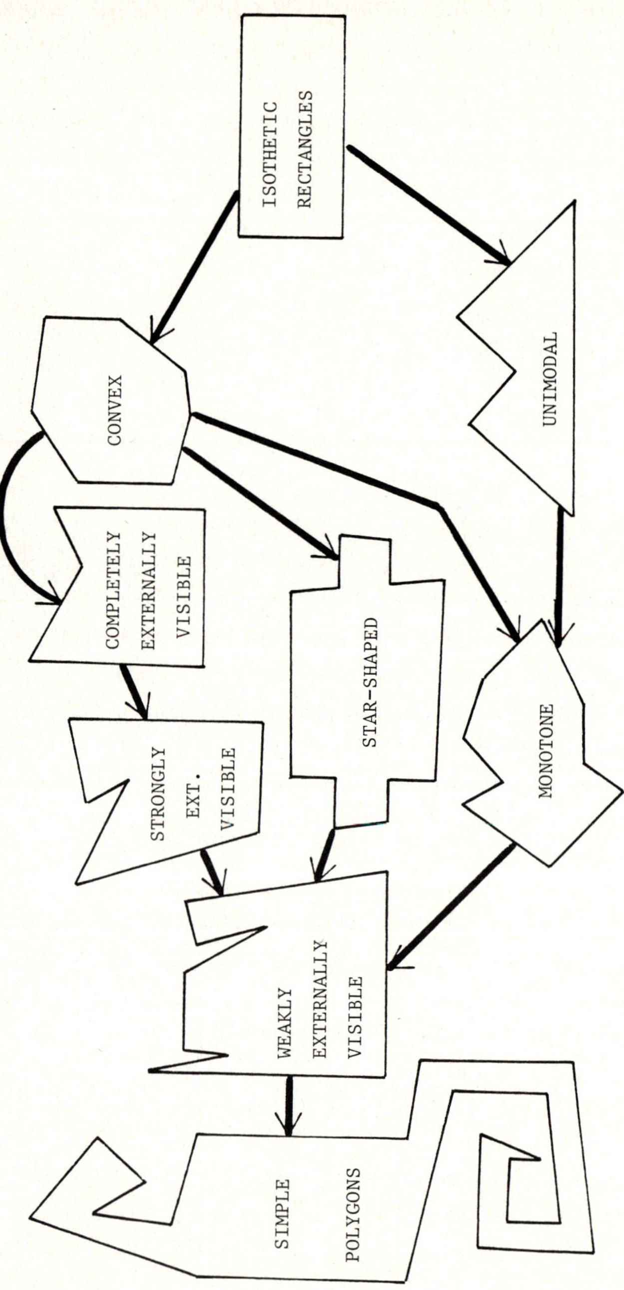

Figure 7: A hierarchy of simple polygons.

Definition: A vertex p_i of a polygon P is said to be *unimodal* if the euclidean distance function $d(p_i, p_{i+1}), d(p_i, p_{i+2}), \ldots, d(p_i, p_{i-1})$ has only one local maximum.

Definition: A polygon P is *unimodal* if every vertex of P is unimodal.

Note that at first glance there appears to be a close relation between *unimodal* and *convex* polygons. In fact there is no relationship whatsoever. For a detailed treatment of this topic and other definitions of unimodality the reader is referred to [18].

Definition: A polygon P is *monotone* if there exists a direction Θ such that the two opposite extreme vertices in direction Θ partition the polygon into two polygonal chains each of which, when traversed, yields a monotonically increasing projection onto a line in direction Θ.

IV. Some Movable Separability Problems in the Plane

4.1 Isothetic rectangles

Consider following the uppermost path in the graph of Figure 7 starting with *isothetic rectangles*. As mentioned earlier, Guibas and Yao [8] have shown that for a set of n isothetic rectangles all directions Θ admit a *translation ordering* and such an ordering can be computed in $O(n \log n)$ time. This property was shown by Guibas and Yao [8] to remain true for sets of convex polygons. Thus there appears to be no basic difference in the separability properties between these two families of polygons. Actually, there is a slight difference worth mentioning.

Referring to Figure 2 we note that sorting the projections of the *support vertices* on ℓ to obtain a translation ordering in direction ℓ can be viewed as *sweeping* a line in the opposite direction and identifying the support vertices in the order in which the "sweep line" traverses them. This idea will be referred to as the *line-sweep heuristic* and it is, in fact, the first idea that a person invariably proposes for obtaining a translation ordering of convex polygons. Although the *line-sweep* heuristic can fail even for rectangles it is interesting to ask whether there exist classes of objects for which it is guaranteed to produce a valid ordering. In fact, one such class is precisely the set of *isothetic rectangles* when ℓ is restricted to the $\pm$ x and $\pm$ y directions. It is trivial to show that in this particular case the first rectangle traversed by the "sweep line" can always be moved first. It follows that for these four directions a translation ordering can be easily obtained

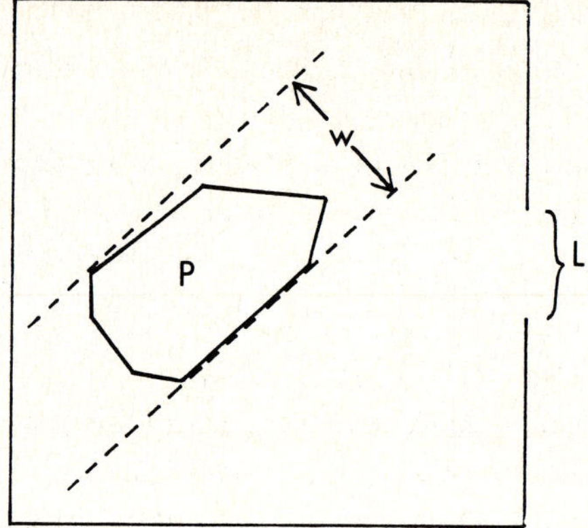

Figure 8: Passing a convex polygon through a slit.

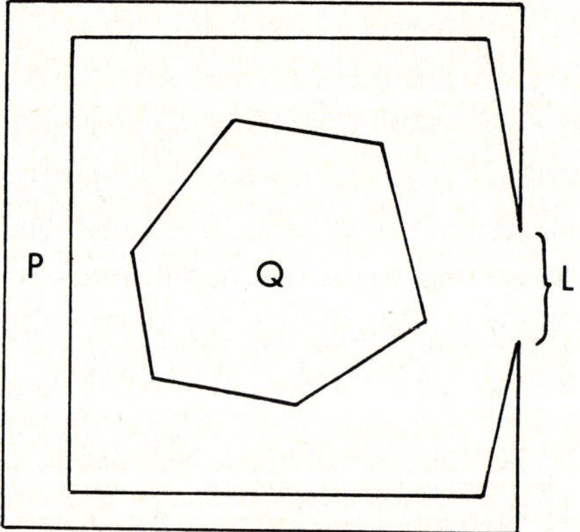

Figure 9: Two completely externally visible polygons can interlock.

by *sorting* the corresponding edges of the rectangles. Thus, although the order of the complexity is not changed, the algorithm (pure and simple sorting) is much simpler than the O(n log n) algorithms of either Guibas and Yao [8] or Ottmann and Widmayer [9]. Other examples where the "sweep line" technique works successfully are given in [20].

4.2 Completely externally visible polygons

Referring to Figure 7 and moving along the graph from *convex* to *completely externally visible* polygons we encounter a dramatic change in movability properties. First we consider the problem of passing a convex polygon through a slit (see Figure 8). Let $P = (p_1, p_2, \ldots, p_n)$ be a convex polygon. Let L be the length of the slit.

Definition: The *width* w of a convex polygon P is the *minimum* distance between parallel lines of support of P.

Theorem 1: (G. Strang [21])
A convex polygon P can pass through a slit of length L if, and only if, $w \leq L$.

We remark here that the width of a convex n-gon can be computed in O(n) time using the "rotating calipers" [22], [23] and thus we obtain the following theorem.

Theorem 2: Given a convex n-gon P and a slit I of length L, whether or not P can pass through I can be determined in O(n) time.

We are now ready to show that two *completely externally visible* polygons can interlock. One such example is illustrated in Figure 9. First we construct a polygon P with a large deficiency polgyon compared to its lid of length L. The lid plays the role of the slit in Figure 8. Next we construct a convex polygon of width $w > L$ in the interior of the deficiency polygon. From Theorem 1 it follows that P and Q are interlocked. Thus we see that in terms of movable separability *completely externally visible* polygons offer no additional freedom over arbitrary simple polygons in this sense.

4.3 Star-shaped polygons

We can ask the same question as in section 4.2 for star-shaped polygons. It turns out that *two star-shaped* polygons are always separable with a *single translation*, which we now prove.

The following lemma is proved in [12].

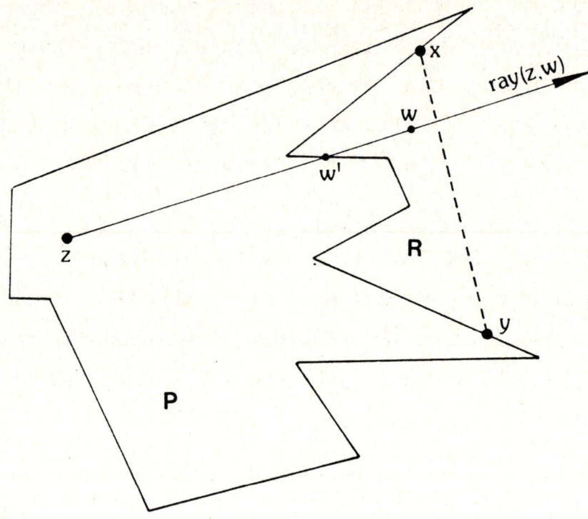

Figure 10: Illustrating the proof of lemma 2.

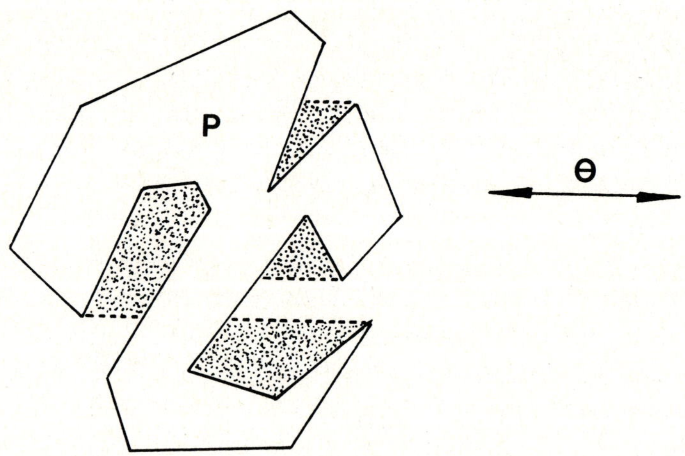

Figure 11: The visibility hull of P in direction theta.

Lemma 1: Two polygons *monotonic* in a common direction Θ, are movably separable with a single translation in a direction *orthogonal* to Θ.

Lemma 2: If P is *star-shaped* with respect to a point z and if x,y are two points on bd(P) such that (x,y) ε ext(P), then the polygon P* = P ∪ R, where R is the bounded exterior of P is also *star-shaped* with respect to z.

Proof: (Refer to Figure 10) Let w be a point in R and construct the half line emanating from z and passing through w, denoted by ray (z,w). Since z lies in the *kernel* K(P), ray (z,w) can intersect bd(P) only at one point, say w'. Now, the creation of P* involves substituting the boundary of P between x and y by [x,y]. It is easy to see that for all w ε R, the corresponding w' must lie on this section of bd(P). This follows from the fact that x and y are visible from z. Since w' is removed in the creation of P* it follows that w is visible from z in P*. Since this is true for all w, P* remains star-shaped with respect to z. Q.E.D.

Definition: Given a simple polygon P and a direction Θ, the *visibility hull of P in direction* Θ, denoted by VH(P,Θ), is the set obtained by taking the union of P with all line segments [a,b] parallel to Θ such that a,b ε P. Note that VH(P,Θ) is monotonic with respect to a direction orthogonal to Θ. The visibility hull of P is the union of P with some "pockets" as illustrated in Figure 11, and can be interpreted as the region bounded by the portions of P visible from ± ∞ in direction Θ. Note that VH(P,Θ) may have new vertices which are not vertices of P.

Definition: The *visibility deficiency* of P, denoted by VD(P,Θ), in direction Θ is the closure of the set-difference between P and the *visibility hull* of P.

The *visibility defficiency* polygons or "pockets" are of two types: *lower pockets* and *upper pockets*. Let p_i be a vertex of P that determines some pocket. If p_{i-1} and p_{i+1} do not lie above the line collinear with the lid of the pocket then the pocket is a *lower* pocket. Similarly if p_{i-1} and p_{i+1} do not lie below the lid line then we obtain an *upper* pocket. (We assume here without loss of generality that Θ is in the direction of the x axis.)

Lemma 3: If L_z is a straight line in direction Θ through a point z ε K(P) then int[VD(P,Θ)] is *not visible* from any point on L_z ∩ ext[VH(P,Θ)].

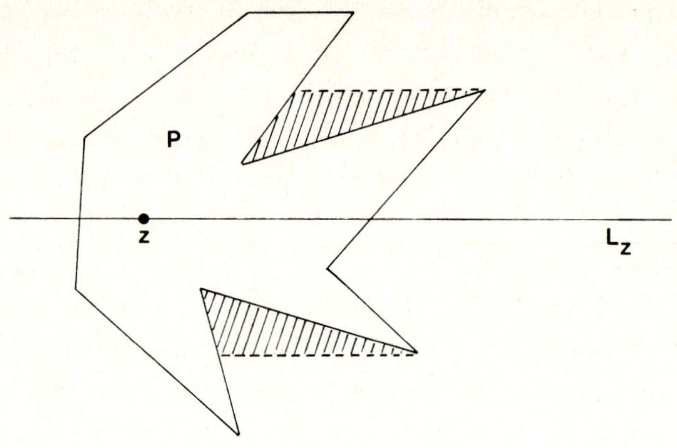

Figure 12: Illustrating the proof of lemma 3.

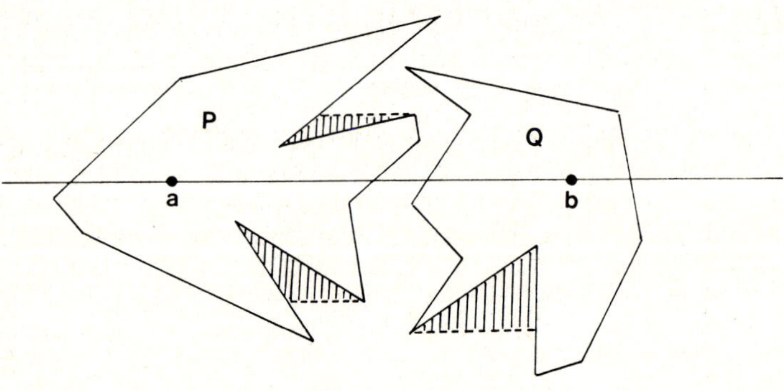

Figure 13: Two star-shaped polygons are separable with a single translation.

Proof: (Refer to Figure 12) First we note that a pocket whose lid is above L_z is a lower pocket. For if it were an upper pocket it would imply, by the Jordan curve theorem that P was not star-shaped from z, a contradiction. Similarly, a pocket whose lid is below L_z must be an upper pocket. It follows that any point x in int(upper pocket) can only see points y lying below L_z. Similarly, any point x in int(lower pocket) can only see points y above L_z. Q.E.D.

Theorem 3: Two *star-shaped* polygons are movably separable with a single translation.

Proof: (Refer to Figure 13) Let P and Q be two non-intersecting star-shaped polygons. Let $a \in K(P)$ and $b \in K(Q)$ and construct a line L(a,b) through a,b. Let Θ be the direction of this line. Now construct $VH(P,\Theta)$ and $VH(Q,\Theta)$. Since $P \in VH(P,\Theta)$ and $Q \in VH(Q,\Theta)$, it is sufficient to show that the visibility hulls can be separated. First we note that $VH(P,\Theta)$ and $VH(Q,\Theta)$ are *star-shaped* with respect to a and b, respectively, by lemma 2. Next we show that the interiors of $VH(P,\Theta)$ and $VH(Q,\Theta)$ do not intersect. For, let $x \in int(Q)$. Then clearly x cannot lie in P. Furthermore, if x lies in a pocket of P then by lemma 3 x is not visible from $L(a,b) \cap ext[VH(P,\Theta)]$ and thus not visible from b which contradicts the star-shapedness of Q. On the other hand if x lies in a pocket of Q it cannot lie in P as it would contradict the star-shapedness of P with respect to a. Neither can it also lie in a pocket of P as it would contradict the star-shapedness of $VH(Q,\Theta)$ with respect to b. Similar arguments hold for the case where $x \in int(P)$. Thus, $VH(Q,\Theta)$ and $VH(P,\Theta)$ are two non-intersecting polygons monotonic in a common direction $\Theta+\Pi$, and by lemma 1, the result follows. Q.E.D.

Theorem 3 also provides us with an algorithm for finding a direction of translation. All we need to do is to find a point $a \in K(P)$ and a point $b \in K(Q)$. This can be done with the linear-programming algorithm of Dyer [24].

Theorem 4: Given two *star-shaped* n-gons, a direction for separating them can be determined in O(n) time.

Actually, *all* directions of separability by translation determined by points in the kernels can be found in O(n) time. Since the kernels are convex polygons, all directions determined by two points $a \in K(P)$ and $b \in K(Q)$ lie in a *cone* defined by the *critical lines of support* between

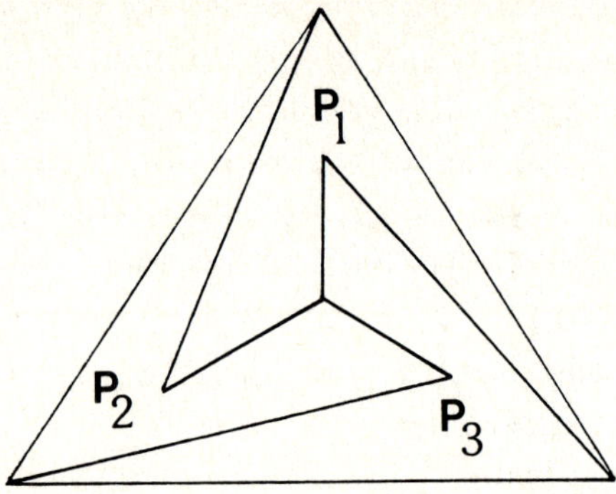

Figure 14: Three sequentially interlocking star-shaped polygons.

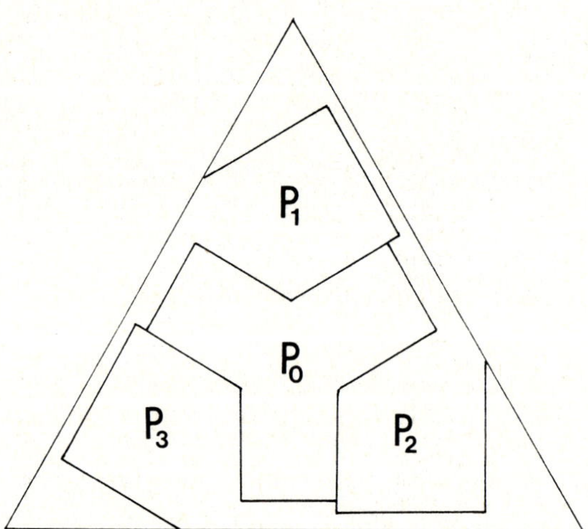

Figure 15: Four monotone polygons interlocked under simultaneous general motions (From Dawson /34/).

K(P) and K(Q). The kernels can be computed in O(n) time with the algorithm of Lee and Preparata [25], and the *critical lines of support* can be found in O(n) time using the "rotating calipers" [22]. It is obvious that the above set of directions for separability is not the complete set. *All* directions for separating two arbitrary simple polygons by a single translation can however be determined in O(n log n) time [26].

Note that theorem 3 provides additional ways of separating the configuration of Figure 6 by *simultaneous* translations. Since $P_1 \cup P_2$ and $P_3 \cup P_4$ make up two star-shaped polygons they can be separated with one translation. Since all four polygons are themselves star-shaped two more translations will separate P_2 from P_1 and P_3 from P_4.

Theorem 3 also implies that two star-shaped polygons can be separated by translating *both* P and Q *simultaneously* in some pairs of directions with respect to an arbitrary fixed point in the plane. It is sufficient to guarantee that the *relative* motion between P and Q is correct. Accordingly, let a* and b* be any pair of points in the plane such that L(a*,b*) intersects K(P) and K(Q). Let x be any reference point on the plane, and consider the vectors $\overrightarrow{xa^*}$, $\overrightarrow{xb^*}$, and $\overrightarrow{a^*b^*}$. We can see now that if we translate P and Q in directions $\overrightarrow{xa^*}$ and $\overrightarrow{xb^*}$ with velocities proportional to the magnitudes of $\overrightarrow{xa^*}$ and $\overrightarrow{xb^*}$, respectively, the correct relative motion between P and Q is maintained. Different pairs of points a*, b* only change the relative velocity of separation. An alternate, very elegant, proof of this result for the restricted case in which $a^* \in K(P)$ and $b^* \in K(Q)$ was given by Dawson [27].

So much for *two* polygons - what about *three* star-shaped polygons. It turns out that as few as three star-shaped polygons can be *sequentially interlocked*. One such example is shown in Figure 14. A surprising result, however, due to Dawson [27] is that any finite collection of star-shaped polygons can still be separated by *simultaneous translations*.

4.4 Monotone polygons

Examination of Figure 14 reveals that the three polygons are *monotonic* in addition to being star-shaped, and thus three monotonic polygons can be *sequentially interlocked*. Furthermore from lemma 1 we know that two monotone polygons are separable with the restriction that they share a common direction of monotonicity. This restriction is removed in [28], [35] with the following theorem.

Theorem 5: Given two polygons P and Q monotonic in directions Θ and ϕ, respectively, then P and Q are separable with a single translation in at least one of the two directions $\Theta + \frac{\Pi}{2}$, $\phi + \frac{\Pi}{2}$.

This theorem immediately suggests the following algorithm for determining a direction of separability for two monotone polygons.

Algorithm SEPARATE

Input. Two non-intersecting monotone polygons $P = (p_1, p_2, \ldots, p_n)$ and $Q = (q_1, q_2, \ldots, q_n)$.

Output. A direction ψ for separating P and Q.

BEGIN

Step 1: Compute the directions of monotonicity for P and Q.

Step 2: If P and Q have a common direction of monotonicity ζ, then EXIT with $\psi \leftarrow \zeta + \frac{\Pi}{2}$.

Step 3: Pick two directions of monotonicity for P and Q, say Θ and ϕ respectively.

Step 4: Compute $VH(Q, \Theta + \frac{\Pi}{2})$.

Step 5: If P intersects $VH(Q, \Theta + \frac{\Pi}{2})$

then EXIT with $\psi \leftarrow \phi + \frac{\Pi}{2}$

else EXIT with $\psi \leftarrow \Theta + \frac{\Pi}{2}$

END

Theorem 6: Algorithm SEPARATE determines a direction of separability for two monotone polygons P and Q in $O(n)$ time.

Proof: The correctness of the algorithm follows from theorem 5. Thus we turn to the complexity. Steps 1-3 can be performed in $O(n)$ time using the algorithm in [29] by Preparata and Supowit. Computing the visibility hull of Q in step 4 can be done in $O(n)$ time with a variety of hidden line algorithms [30]-[31]. Finally, step 5 can be performed in $O(n)$ time using a simple modification of the slab method of Shamos and Hoey [32] for intersecting two convex polygons. This follows from the fact that P and $VH(Q, \Theta + \frac{\Pi}{2})$ are two polygons monotonic in direction Θ and therefore their intersection can only contain a linear number of pieces. See for example Guibas and Stolfi [33]. Q.E.D.

Just as *star-shaped* polygons can be sequentially interlocked but any number of them are always separable under simultaneous translations, so we can ask this question for monotone polygons. Dawson [34] has shown that *three* monotone polygons are separable with simultaneous translations but *four* can interlock under simultaneous general motions (see Figure 15).

4.5 Simple polygons

Since arbitrary simple polygons may or may not interlock in a variety of senses, it is interesting to determine if a configuration of polygons does or does not interlock in any of these senses. We illustrate some results along these lines.

Theorem 7: Two simple polygons P and Q are movably separable by a single translation in direction Θ if, and only if,

$$\text{int}[VH(P,\Theta)] \cap \text{int}\,[VH(Q,\Theta)] = \Phi$$

where Φ is the null set.

Proof: (sufficiency) In this case we have two polygons, the interiors of which do not intersect, and which are monotonic in a common direction $\Theta + \frac{\Pi}{2}$. Thus the result follows from lemma 1. For the necessity four cases arise:

(a) A point $z \in \text{int}(Q)$ lies in the interior of a pocket of VH(P). In this case, if z is translated in direction Θ it cannot exit the pocket through its lid since the lid is parallel to Θ. Therefore z must collide with P and so must Q.

(b) A point $z \notin \text{int}(Q)$ but $z \in \text{int}VH[Q]$ lies in a pocket of VH(P). In this case if z is translated in direction Θ it must collide either with P or Q. In the former case we are done. In the latter case let z collide with Q at $z' \in Q$. But now case (a) applies with z'. The last two cases where $z \in \text{int}[VH(P)]$ are similar. Q.E.D.

Theorem 8: Given two simple n-gons P and Q, and a direction Θ, whether or not P and Q are *separable* by a single translation in direction Θ can be determined in O(n) time.

Proof: From theorem 7 it follows that it is sufficient to compute the visibility hulls of P and Q in direction Θ and then determine if the hulls intersect. The first step can be done in O(n) time with either of the algorithms of El Gindy and Avis [30] or Lee [31]. The second step can be done in O(n) time with a slight variation of the "slab" method of Shamos & Hoey [32] for intersecting convex polygons. Q.E.D.

Theorem 9: A set of polygons $\mathbb{P} = \{P_1, P_2, \ldots, P_M\}$ admits a *translation ordering* in direction Θ if, and only if, every pair of polygons, viewed in isolation, is separable with a single translation in direction Θ.

Proof: Without loss of generality let Θ be the x axis. First consider the case when a translation ordering exists. Let each polygon be translated in *order* by some fixed magnitude and assume we have just translated $P_{(i)}$ from position A to position B. Clearly $P_{(i)}$ is separable from each polygon in the set $\{P_{(1)}, P_{(2)}, \ldots, P_{(i-1)}\}$. Furthermore, $P_{(i)}$ can be translated back to A from B. Therefore $P_{(i)}$ is separable from each polygon in the set $\{P_{(i+1)}, P_{(i+2)}, \ldots, P_{(M)}\}$. Since this remains true for all i it follows that every pair of polygons is separable with a single translation in direction Θ. Next, consider the case where every pair of polygons, viewed in isolation, is separable with a single translation in direction Θ. We must show that $\mathbb{P}$ admits a translation ordering in direction Θ. From theorem 7 it follows that for all i and j $\text{int}[VH(P_i,\Theta)] \cap \text{int}[VH(P_j,\Theta)]=\Phi$. Thus it is sufficient to show that if we are given M non-intersecting polygons monotonic in a common direction $\Theta + \frac{\pi}{2}$, then they admit a translation ordering in direction Θ. Let $RCH(P_i)$ denote the *right chain* of P_i, i.e., $RCH(P_i) = (p_t^i, p_{t+1}^i, \ldots, p_b^i)$ where p_t^i and p_b^i are the vertices with maximum (top) and minimum (bottom) y coordinates, respectively. Clearly, a polygon can be translated in direction Θ if, and only if, its *right chain* can be so translated. Hence, we need only consider the right chains. Now imagine a light source at $x = +\infty$ and mark all the top vertices of the chains which are illuminated (see Figure 16). To establish that a translation ordering exists it is sufficient to show that one of these monotonic chains can always be translated to $x = +\infty$ without disturbing the others. It turns out that the chain with the lowest (minimum y coordinate) *marked top* vertex can always be moved out first. Let $RCH(P_i)$ be the desired chain and assume it cannot be translated to $x = +\infty$. This means that there must be another chain, say $RCH(P_j)$, blocking $RCH(P_i)$. Two cases arise depending on how $RCH(P_j)$ blocks $RCH(P_i)$.

Case 1: p_t^j lies above p_t^i. Because the monotonicity of the chains it follows that if $RCH(P_j)$ is to block $RCH(P_i)$ then p_t^i cannot be illuminated, a contradiction.

Case 2: p_t^j lies below p_t^i. Because of monotonicity it follows that if $RCH(P_j)$ is to block $RCH(P_i)$ then p_t^j must lie above p_b^i and to the right of $RCH(P_i)$. Now p_t^j cannot be illuminated since it is below p_t^i and would lead to a contradiction. Therefore p_t^j must be blocked by some third chain $RCH(P_k)$. Furthermore, using similar arguments to the above p_k^t must lie in between p_j^t

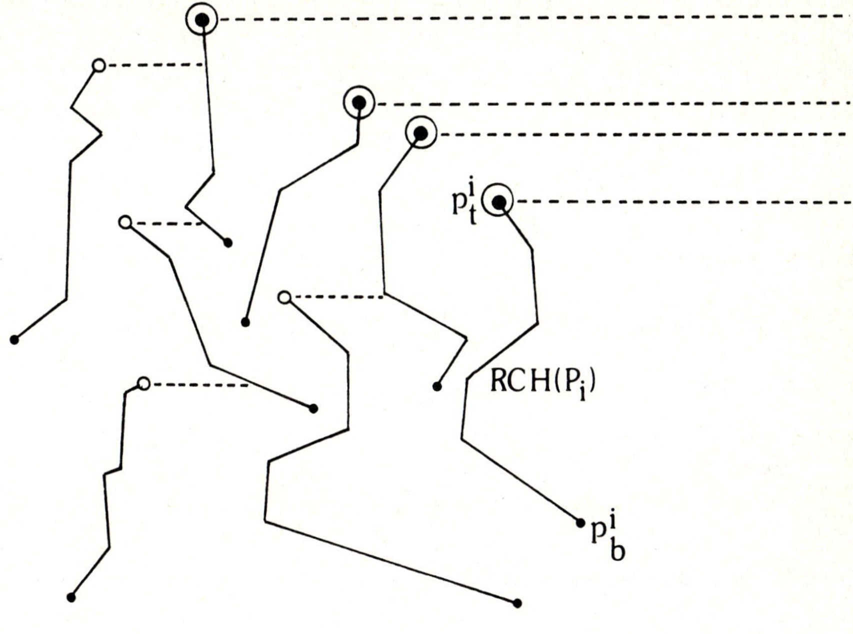

Figure 16: The monotone chain with the lowest illuminated top vertex can always be moved out first.

and p_i^t. But this in turn requires an unlimited number of additional blocking chains. Since we only have M available we eventually obtain an illuminated vertex lower than p_t^i, a contradiction. Q.E.D.

Theorems 7 and 9, together with existing solutions to certain geometric problems [30] - [32], [36] offer immediate algorithms for solving the problems of *detecting* and *computing* translation orderings for sets of simple polygons.

The Detection Problem

Theorem 10: Given a collection of M simple n-gons $\mathbb{P} = \{P_1, P_2, \ldots, P_M\}$ and a direction Θ, whether or not a *translation ordering exists* can be determined in time $O(\min(M^2n, Mn\log Mn))$.

Proof: From theorems 7 and 9 it follows that we can first compute the visibility hulls $VH(P_i, \Theta)$, $i = 1, 2, \ldots, M$ and subsequently determine if any pair intersects. The first step can be done in $O(Mn)$ time [30] - [31]. The second step can be done in two ways. Since the $VH(P_i, \Theta)$ are monotonic polygons with a common direction of monotonicity $\Theta + \frac{\pi}{2}$, it follows that whether any pair intersects can be determined in $O(n)$ time. Testing all such pairs yields $O(M^2n)$ time for this method. The second method is to consider the M n-gons as a set of Mn line segments and to use the method of Shamos & Hoey [32] to yield a complexity of $O(Mn\log Mn)$. Q.E.D.

Note that which method is faster depends on whether M or n dominates in a given problem. If M is a constant we have $O(n)$ versus $O(n\log n)$; if n is a constant we have $O(M^2)$ versus $O(M\log M)$.

The Computation Problem

Theorem 11: Given a collection of M simple n-gons $\mathbb{P} = \{P_1, P_2, \ldots, P_M\}$ and a direction Θ that admits a translation ordering of $\mathbb{P}$, then such a translation ordering can be computed in time $O(\min(M^3n, M^2n\log Mn))$.

Proof: Two approaches are immediately obvious.

Method 1: For every top vertex construct a half-line in direction Θ and determine if it intersects any of the remaining Mn-1 edges. This can be done in $O(M^2n)$ time. Now in $O(M)$ time we can determine which polygon is the first to move. We repeat this for the remaining polygons until none remain for a total complexity of $O(M^3n)$.

Method 2: Considering the M polygons as a set of Mn line segments, we can compute the *next-element subdivision* in direction Θ in $O(Mn\log Mn)$ time [36]. From this structure we can pick out the first polygon to move in $O(M)$ time. Repeating for the remaining polygons leads to a total complexity of $O(M^2n\log Mn)$. Q.E.D.

Note that, as before, which method is faster depends on whether M or n dominates. Hossam El Gindy has reduced the complexity of both of the above problems to O(Mn log M).

4.6 Isothetic polygons

In some applications areas such as VLSI [8] the collection of polygons is *isothetic* with respect to a *common* pair of orthogonal directions (say the x and y axes). One would expect movability to enjoy some additional freedom as compared to the case of arbitrary polygons. We illustrate this idea with the following example. Recall from Figure 14 that three star-shaped polygons can be *sequentially interlocked*. If on the other hand the polygons are *isothetic* then any number of them are sequentially separable.

Theorem 12: A collection of M *isothetic star-shaped* polygons admits a *translation ordering* in the x and y directions.

Proof: First we note that if an isothetic polygon is *star-shaped* then it is *monotonic* in the x and y directions [38]. The result then follows from lemma 1 and theorem 9. Q.E.D.

V. Separability in Three Dimensions

5.1 Translating spheres

In section 4.1 it was pointed out that the *line-sweep* heuristic is successful in determining a translation ordering in the x and y directions for a collection of *isothetic* rectangles. It is clear that the arguments carry over to three dimensions (x, y, z directions) for *isothetic* rectangular solids. It turns out that it also works correctly for sets of spheres of equal radii [20]. Furthermore, a variation of *line-sweep* will work for unequal spheres [20]. (Actually, a *plane-sweep* in 3-D)

Let $S = \{S_1, S_2, \ldots, S_n\}$ denote a set of n non-intersecting spheres in 3-space with centers $0 = \{o_1, o_2, \ldots, o_n\}$, where S_i is specified by the cartesian coordinates of o_i, namely (x_i, y_i, z_i), and the radius r_i. A *great circle* on a sphere S_i is a circle on the surface of S_i that partitions S_i into two *hemispheres*. Consider a sphere S_i and assume there is a light source at infinity casting rays of light in direction ℓ. The *shadow tunnel* of S_i in direction ℓ, denoted by $ST(S_i, \ell)$ is that subset of space not illuminated by the light source along with its boundary.

Lemma 4: Given $S = \{S_1, S_2, \ldots, S_n\}$, S_i can be translated to infinity in

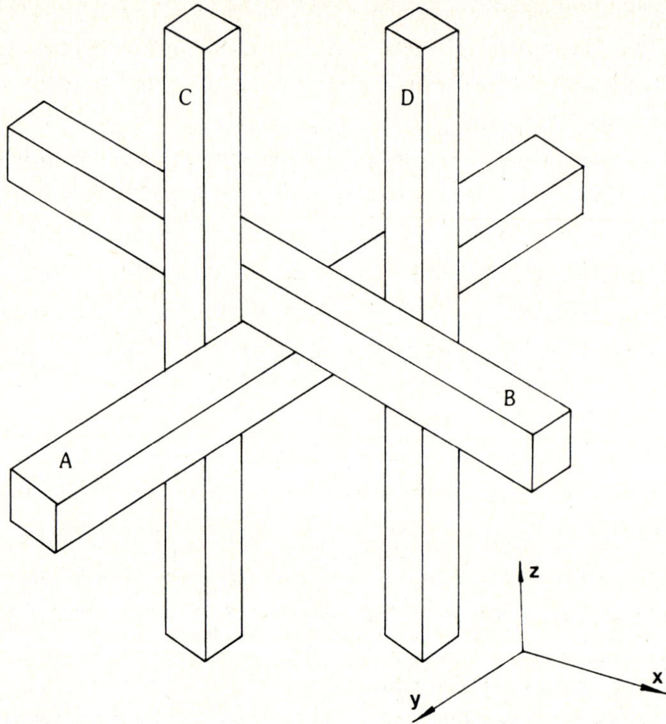

Figure 17: A set of isothetic convex polyhedra that does not allow a translation ordering in some directions such as x + y.

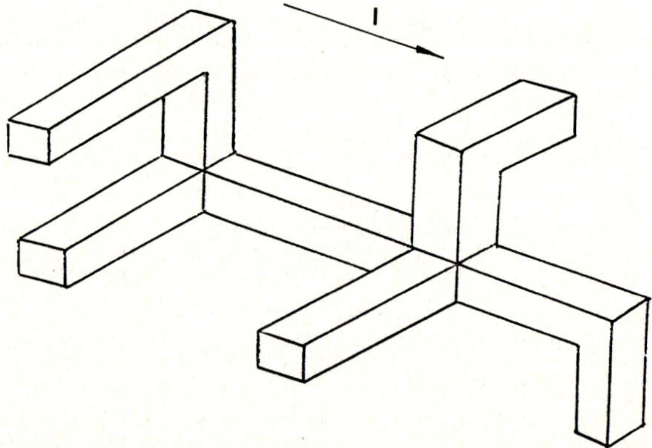

Figure 18: Illustrating a weakly monotonic polyhedron with monotonic intersections.

direction ℓ without disturbing the other spheres if, and only if, no point of S_j, j = 1, 2,..., n, j ≠ i lies in the interior of $ST(S_i, \ell)$.

Let $HS(S_i^+, \ell)$ denote the open half-space determined by the plane orthogonal to ℓ that cuts S_i at a *great circle* and contains $ST(S_i, \ell)$. Similarly let $HS(S_i^-, \ell)$ denote the complement of the union of $HS(S_i^+, \ell)$ and the cutting plane. These half-spaces will be referred to as closed when the cutting plane is included in the set.

Lemma 5: Given $S = \{S_1, S_2, \ldots, S_n\}$, if there exists a direction ℓ and a sphere S_i such that the o_j, j = 1, 2, ..., n, j ≠ i all lie in closed $HS(S_i^-, \ell)$, then S_i can be translated to infinity in direction ℓ without disturbing the other spheres.

Lemmas 4 and 5 imply that if we apply *plane-sweep* to the *centers* of the spheres rather than the spheres themselves we can obtain a valid translation ordering. We thus obtain the following two theorems proved in [20].

Theorem 13: Given $S = \{S_1, S_2, \ldots, S_n\}$, for *all* directions ℓ there exists an ordering on S such that the spheres can be translated by some common vector in direction ℓ *one at a time* without collision and such an ordering can be computed in O(n log n) time.

Dawson [27] has shown that given a collection of n spheres in 3-space at least min{n, 4} of them can be translated to infinity without disturbing the others. Lemmas 4 and 5 allow us to sharpen this result.

Let CH(O) denote the convex hull of $O = \{o_1, o_2, \ldots, o_n\}$ and let H be the number of sphere centers that lie on vertices, edges, and faces of CH(O). We then have the following theorem [20].

Theorem 14: Given $S = \{S_1, S_2, \ldots, S_n\}$, then the number of spheres that can be translated to infinity without disturbing the others is at least H and they can be identified in O(n log n) time.

5.2 Translating convex polyhedra

Consider first the *isothetic* case. It turns out that four isothetic rectangular polyhedra can be arranged such that for some directions in 3-space no translation ordering exists. The example due to Guibas and Yao [8] is reconstructed in Figure 17. A more surprising result is that if we relax the *isothetic* requirement we can find sets of rectangular polyhedra that do not admit a translation ordering in *any* direction. Therefore, convex polyhedra in 3-space do *not* exhibit the *translation ordering property*. An example of twelve convex polyhedra that do not admit an ordering in any direction is given in [39]. In the example in [39], two polyhedra can be moved to infinity for some chosen directions. An even more surprising result due to Dawson [27] is an example of twelve convex polyhedra none of which may be translated in any direction without disturbing the others. A similar example with only six objects was discovered by Post [40].

5.3 Passing a convex polyhedron through a window

A generalization of the problem illustrated in Figure 8 asks for whether a *convex* polyhedron P can be *passed* through a *convex window* W. Here W may be a circle, a rectangle, or an arbitrary convex polygon. Let H be a plane not intersecting P and assume a light source at infinity somewhere on the same side of H as P.

Definition: The *shadow* of P, denoted by SH(P), is a convex polygon determined by the projection of P onto H, i.e., the portion of H not illuminated. If Θ is the direction of the "light rays" we will also use $SH(P,\theta)$ to denote the shadow of P in direction Θ.

One would hope for a theorem analogous to theorem 1 relating SH(P) to the window W. Unfortunately this is not the case. While it is true that if there exists a direction of projection Θ such that SH(P) fits into W, then P can be passed through W (a single translation will do), there exist cases where no shadow of P on any plane H fits into W and yet P can still be passed through W by some sequence of displacements [21]. Nevertheless, the *shadow* concept is still relevant to the problem of separability, in particular if we are interested in separation with a single translation after initial positioning. We can also limit the initial positioning displacements to translations only. Let $P = (p_1, p_2, \ldots, p_n)$ be a convex polyhedron and let $W = (w_1, w_2, \ldots, w_m)$ be the convex window on a plane H. Chazelle [41] has shown that given two convex polygons with n and m vertices, respectively, whether the first can be fitted into the second with translations only can be determined in $O(n + m)$ time. Furthermore, given a polyhedron P in 3-space and a direction Θ orthogonal to a plane H, the $SH(P, \Theta)$ can easily be computed in $O(n)$ time. We thus have the following theorem.

Theorem 15: Given a convex polyhedron $P = (p_1, p_2, \ldots, p_n)$ and a convex window $W = (w_1, w_2, \ldots, w_m)$ on a plane H, whether P can pass through W by a *single translation* orthogonal to H, after prepositioning with only a single translation, can be determined in $O(n + m)$ time.

Note that if $SH(P, \theta)$, where Θ is orthogonal to H, cannot be contained in W it is still possible that P can pass through W by a single translation after pre-positioning with only a single translation.

Chazelle [41] has also shown that given two convex polygons $Q(n)$ and $R(m)$ whether Q can be contained in R by translations and rotations can be determined in $O(n\ m^2)$ time, and we therefore have.

<u>Theorem 16</u>: Given a convex polyhedron $P = (p_1, p_2, \ldots, p_n)$ and a convex window $W = (w_1, w_2, \ldots, w_m)$ on a plane H, whether P can pass through W by a *single translation* orthogonal to H, after pre-positioning with arbitrary translations but rotations only with respect to axes of rotation orthogonal to H, can be determined in $O(nm^2)$ time.

A related problem asks for the smallest, in some sense, window W on a plane H through which a given convex polyhedron P can be passed with a single translation orthogonal to H after pre-positioning P with arbitrary translations and rotations. This problem is directly related to finding smallest shadows. McKenna and Seidel [37] give an $O(n^2)$ algorithm for computing the minimum-area shadow of a convex polyhedron with n vertices.

As a final example of this type of problem, we can ask whether there exists a direction for a polyhedron to pass through a hole by a single translation *without* pre-positioning. Let $P = (p_1, p_2, \ldots, p_n)$ be a convex polyhedron arbitrarily positioned, on one side of a plane H, with respect to a convex window $W = (w_1, w_2, \ldots, w_m)$ in H. The following lemma can be easily proved.

<u>Lemma 6</u>: P can pass through W with a single translation in direction Θ if, and only if, each vertex of P can be passed through W with a single translation in direction Θ.

Note that this no longer holds true for non-convex holes in H.

Lemma 6 allows us to solve the above problem. In fact we can do more; we can compute *all* directions for passing P through W with a single translation and *no* pre-positioning. Alternately, we can view this problem as computing *all* the shadows $SH(P, \Theta)$ on H that can be contained in W.

<u>Theorem 17</u>: Given a convex polyhedron $P = (p_1, p_2, \ldots, p_n)$ arbitrarily positioned on one side of a plane H, with respect to a convex window $W = (w_1, w_2, \ldots, w_m)$ in H, all directions for translating P through W

can be computed in $O(mn \log mn)$ time.

Proof: By lemma 6 we can restrict ourselves to computing all directions for simultaneously translating the *vertices* of P through W. Consider vertex p_i. All directions for translating p_i from its initial configuration through W are defined by all the vectors emanating at p_i and intersecting H in W. Therefore the *cone* determined by the half-lines from p_i through w_j, $j = 1, \ldots, m$ specifies all such directions for p_i. Denote such a cone by CONE (p_i, W). Construct a 3-dimensional euclidean *direction* space D and translate all the cones CONE(p_i, W), $i = 1, 2, \ldots, m$ in D such that the p_i all overlap with the origin of D. Then the intersection of all the cones in D gives another cone which is the set of directions for simultaneous translation of all the p_i, and hence of P. Each cone can be computed in $O(m)$ time and thus all cones can be found in $O(mn)$ time. All the cones can be translated to D in $O(mn)$ time. All that remains is to compute the intersection of all the cones. Now, each cone can be viewed as the intersection of m half-spaces determined by the planes co-planar with the m faces of each cone. The *interior* half-space contains the cone. Therefore the solution cone is the intersection of all the interior half-spaces determined by all the cones in D. Now, the intersection of k half-spaces in 3-dimensional space can be computed in $O(k \log k)$ time using an algorithm of Preparata and Muller [45]. Therefore the solution cone can be obtained in $O(mn \log mn)$ time. Q.E.D.

5.4 Separating star-shaped polyhedra

The results of section 4.3 on star-shaped polygons in the plane extend to three dimensions. Let $\mathbb{P} = \{P_1, P_2, \ldots, P_M\}$ be a collection of M non-intersecting star-shaped polyhedra with n vertices each. Dawson [27] has shown that $\mathbb{P}$ can always be separated with simultaneous translations. Let K_i be the *kernel* of P_i and let k_i be a point in K_i. Let x be any point in 3-space. The vector $\overrightarrow{xk_i}$ determines the velocity and direction of translation for P_i in a valid set of simultaneous translations. Since the linear programming algorithm of Dyer [24] runs in linear time in 3 dimensions as well, we have the following theorem.

Theorem 18: Let $\mathbb{P}$ be a collection of star-shaped polyhedra. A set of translations for simultaneous separation of $\mathbb{P}$ can be determined in $O(nM)$ time.

5.5 Interlocking monotone polyhedra

In the previous section we observed that some of the movability properties of star-shaped polygons in the plane carry over exactly to star-shaped polyhedra in 3-space. The definition of monotone polygons, on the other hand, does not generalize straightforwardly or uniquely to three dimensions. In this section we explore several families of "monotone" polyhedra and consider some of their separability properties.

One common definition or characterization of monotone polygons is as follows. A polygon P is monotonic in direction ℓ if for every line L orthogonal to ℓ that intersects P, the intersection $L \cap P$ is a line segment (or point). We generalize this definition to 3-dimensional space to obtain a "large" family of monotone polyhedra we call *weakly monotonic*.

Definition 1: A polyhedron P is *weakly monotonic* in direction ℓ if there exists a direction ℓ such that the intersection, of each plane orthogonal to ℓ that intersects P, is a *simple polygon* (or a line segment or point).

Note that there exist a score of rather well-known special classes of simple polygons [17], [42] - [43]. By substituting these for the word *simple* in the above definition we obtain a score of families of *weakly monotonic* polyhedra. Thus we say that if all the intersections are *convex*, we have a *weakly monotonic* polyhedron in the *convex sense*. Figure 18 illustrates a *weakly monotonic* polyhedron in the *monotonic* sense, i.e., with monotonic polygons as intersections.

In [35] it was shown that two monotone polygons in the plane, even if they do *not* share a *common* direction of monotonicity, can always be separated with a single translation. In three dimensions, on the other hand, *weakly monotonic* polyhedra lose all their freedom of motion, even if they share a common direction of monotonicity, as we now demonstrate with an example.

Fact: Two polyhedra P, Q *weakly monotonic* with respect to a common direction ℓ can *interlock* under *all motions*.

Consider the two polyhedra in Figure 19; P is an "isothetic coil" and Q an "isothetic cross". They are weakly monotonic with respect to a direction parallel to the x axis. By thickening the four "arms" of the "cross" until they touch the "coil" we cannot subsequently move one without the other. This holds true for all possible *translations*, *rotations*, and *screw* motions, i.e., simultaneous translations and rotation.

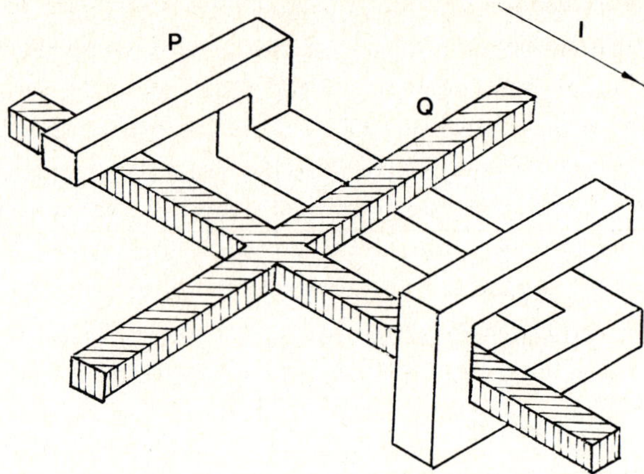

Figure 19: Two polyhedra weakly monotonic with respect to a common direction can interlock under all motions.

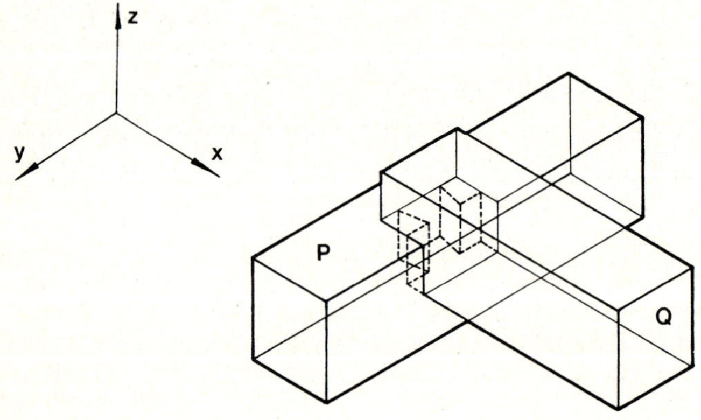

Figure 20: Two polyhedra strongly monotonic with respect to PL(z). This example is taken from /44/ and illustrates the "Ari-kake" joint in Japanese carpentry. The only way to separate this pair of polyhedra is to translate either P or Q in the z direction.

Another characterization of monotone polygons in the plane follows from the definition of *directional convexity*. A polygon P is monotonic in direction ℓ if for every pair of points a, b ε P, such that the line L(a,b) through a and b is orthogonal to ℓ, the *line segment* [a,b] lies in P. Alternately, we say the polygon is *directionally convex* with respect to $\ell + \frac{\Pi}{2}$. Several possibilities exist for generalizing this notion to higher dimensions.

Definition: A polyhedron P is *directionally convex* with respect to ℓ, if there exists a direction ℓ such that for every pair of points a, b ε P with L(a,b) *parallel* to ℓ, the line segment [a,b] lies in P.

Fact: Two polyhedra P, Q *directionally convex* with respect to a common direction ℓ can *interlock* under *all motions*.

The example of Figure 19 illustrates the point. Both P and Q are directionally convex with respect to the y axis.

Note that if a polyhedron P is directionally convex with respect to ℓ it does not necessarily follow that P is weakly monotonic with respect to all directions orthogonal to ℓ. Polyhedron P in Figure 19 is directionally convex with respect to the y axis but it is not weakly monotonic in the z direction. In the above definition only one direction ℓ was used for convexity. We can obtain different families of directionally convex polyhedra by increasing the number of directions used. One way of doing this is to consider all directions lying on a plane, as in the following definition.

Definition: A polyhedron P is *directionally convex orthogonal* to ℓ, if there exists a direction ℓ such that for every pair of points a, b ε P lying on a *plane orthogonal* to ℓ, the line segment [a,b] lies in P.

Note that such a polyhedron must be *weakly monotonic* with respect to ℓ in the *convex* sense.

Under all three definitions above we have seen that two polyhedra can interlock under all motions. Now we introduce a class of polyhedra which we call *strongly monotonic* which possesses some movability properties analogous to monotone polygons in the plane. Let PL(ℓ) denote a plane orthogonal to a direction ℓ.

Definition: A polyhedron P is *strongly monotonic with respect to* PL(ℓ) if there exists a direction ℓ such that all planes *parallel* to ℓ that intersect P form as their intersection with P a *simple* polygon that is *mono-*

tonic in a direction *orthogonal* to ℓ.

Figure 20 illustrates two polyhedra P, Q *strongly monotonic* with respect to PL(z). It is taken from [44] and is known in Japanese carpentry as the "ari-kake" joint. Note that because of the "dove-tail" P and Q can only be separated with a translation in the z direction. It is proven in [39] that such polyhedra are always separable if they share a common direction ℓ.

Theorem 19: Two polyhedra P, Q both of which are *strongly monotonic* with respect to PL(ℓ) are separable with a single translation in direction ℓ.

VI. Satins and Twills: The Computational Geometry of Weaving

Another class of movable separability problems occurs in the design of fabrics, i.e., weaving [47]. We give a brief description of one such problem but for details the reader is referred to [47] - [49]. We borrow some definitions from Grünbaum and Shephard [47].

The word *fabric* will be used in a mathematical sense to mean, roughly speaking, two layers of congruent strands in the same plane E such that the strands of different layers are nonparallel and they "weave" over and under each other in such a way that the fabric "hangs together". To be precise, "weaving" means that at any point P of E which does not lie on the boundary of a strand, the two strands containing P have a stated *ranking*, that is to say, one strand is taken to have precedence over the other, and this ranking is the same for each point P contained in both strands. This concept may be conveniently expressed by saying that one strand *passes over* the other, in accordance with the obvious practical interpretation. By saying that the fabric *hangs together* we mean that it is impossible to partition the set of all strands into two nonempty subsets so that each strand of the first subset passes over every strand of the second subset.

It is convenient to let the plane E be the xy plane in 3-space and to consider translations of the strands in the z direction. Note that it is not necessary that the strands be ideal in the sense of having zero or negligible thickness. See, for instance, [46] for examples of "polyhedral fabrics" such as tabbies and twills. Here the *strands* are *convex-weakly-monotonic* polyhedra. If the set of all *strands* admits a translation ordering in the z direction the fabric "falls completely apart". Thus is our terminology determining whether a fabric *hangs together* is equivalent to determining if there exists a translation ordering of strict subsets of strands in the z direction.

A fabric can be represented by a matrix of 1's and 0's and for this representation efficient algorithms for determining whether a fabric hangs together are given in [48] and [49].

If we are given a "polyhedral fabric" we can first obtain the matrix representation of the fabric by solving the hidden surface problem for all strands taken together in the z direction.

VII. Conclusion

We conclude by discussing some open problems in this area.

It was mentioned in section III that a score of different families of polygons have made their appearance in the computational geometry literature. As we see in section IV, movability properties are known only for for a few classes of polygons; convex, star-shaped, and a monotone being noteworthy examples. Consider the case of *unimodal* polygons. One would expect unimodal polygons to experience a greater degree of freedom of motion than monotone polygons since unimodal polygons are closer to rectangles in the hierarchy of Figure 7. Recall from Figure 14 that three *monotone* polygons can *sequentially interlock* and four can interlock under simultaneous general motions. It is conjectured that three *unimodal* polygons are *sequentially separable* and four are always separable under simultaneous general motions.

Consider the problem of detecting and computing a translation ordering in a specified direction Θ of a given set of polygons $\mathbb{P} = \{P_1, P_2, \ldots, P_M\}$, each of which contains n vertices. It is shown in theorems 10 and 11 that the detection and computation problems, respectively, can be solved in $O(\min(M^2n, Mn \log Mn))$ and $O(\min(M^3n, M^2n \log Mn))$ time. These upper bounds are obtained by a simple combination of existing algorithms. Thus it is reasonable to expect that these bounds can be improved.

Consider the problem of translating circles and spheres. In reference [20] it is shown that given a collection of circles $C = \{C_1, C_2, \ldots, C_n\}$, all the circles in C that can be translated to infinity individually without disturbing the others can be identified by taking each circle C_i in turn and applying the circle-reachability algorithm described therein. This produces O(n) sets of circles for which the

the Laguerre Voronoi diagram is computed O(n) times leading to the $O(n^2 \log n)$ complexity. Can this complexity be reduced? Perhaps this problem can be solved in $O(n \log n)$ time if the Laguerre Voronoi diagram [50] is computed only once on C and thereafter each circle is treated in $O(\log n)$ time in a *query* type mode. Another open problem is of course determining all the spheres that can be so translated in a collection of spheres $S = \{S_1, S_2, \ldots, S_n\}$ in 3 dimensions. There is no conceptual difficulty in applying the two-dimensional technique to this case. In fact, fast hidden sphere algorithms for intersecting spheres in 3 dimensions exist [51]. However, it is an open question how fast we can compute the Laguerre Voronoi diagram in 3 dimensions, or otherwise compute the *contour-surface* of the union of n spheres.

Theorem 19 states that two polyhedra P and Q *strongly monotonic* with respect to $PL(\ell_1)$ and $PL(\ell_2)$, respectively, are separable with a single translation when $\ell_1 = \ell_2$. It is an open problem whether this is true for $\ell_1 \neq \ell_2$. More generally, Dawson [27] has shown that any finite collection of *star-shaped* polyhedra are separable under simultaneous translations. It is not known whether this also holds true for *strongly monotonic* polyhedra in 3-space.

Section 5.3 illustrates some solutions to problems concerning the passing of a convex polyhedron P through a convex hole W in a "thin" wall. Theorem 17 gives a solution to the problem of determining whether P can be passed through W with a single translation and *no* pre-positioning. It is an open problem how fast this can be determined if *one translation* is also allowed for pre-positioning.

Finally, it is not yet known how fast we can determine if a convex polyhedron can be passed through a convex hole when arbitrary motions are allowed.

References

[1] L. Moser, "Problem 66-11: Moving furniture through a hallway", *SIAM Review*, vol. 8, 1966, p. 381.

[2] M. Goldberg, "A solution of problem 66-11: Moving furniture through a hallway", *SIAM Review*, vol. 11, 1969, pp. 75-78.

[3] J. Sebastian, "A solution to problem 66-11: Moving furniture through a hallway", *SIAM Review*, vol. 12, 1970, pp. 582-586.

[4] W. E. Howden, "The sofa problem", *Computer Journal*, vol. 11, 1968, pp. 299-301.

[5] K. Maruyama, "An approximation method for solving the sofa problem", *International Journal of Computer and Information Sciences*, vol. 2, 1973, pp. 29-48.

[6] T. Lozano-Perez and M. Wesley, "An algorithm for planning collision-free paths among polyhedral obstacles", *Communications of the ACM*, vol. 22, 1979, pp. 560-570.

[7] S. Whitesides, "Computational geometry and motion planning", in *Computational Geometry*, Ed., G. Toussaint, North-Holland, 1985.

[8] L. J. Guibas and F. F. Yao, "On translating a set of rectangles", *Proc. Twelfth Annual ACM Symposium on Theory of Computing*, 1980, pp. 154-160.

[9] T. Ottman and P. Widmayer, "On translating a set of line segments", *Computer Vision, Graphics, and Image Processing*, vol. 24, 1983, pp. 382-389.

[10] G. T. Toussaint, "The complexity of movement", *IEEE International Symposium on Information Theory*, St. Jovite, Canada, September 1983.

[11] J.-R. Sack and G. T. Toussaint, "Movability of objects", *IEEE International Symposium on Information Theory*, St. Jovite, Canada, September 1983.

[12] G. T. Toussaint and J.-R. Sack, "Some new results on moving polygons in the plane", *Proc. Robotic Intelligence and Productivity Conference*, Detroit, Michigan, Nov. 18-19, 1983, pp. 158-163.

[13] B. Chazelle, et al., "The complexity and decidability of SEPARATION", Technical Report CS-83-34, University of Waterloo, November 1983.

[14] G. T. Toussaint, "On a simple Minkowski metric classifier", *IEEE Trans. Systems Science and Cybernetics*, vol. SSC-6, October 1970, pp. 360-362.

[15] G. T. Toussaint, "Polynomial representation of classifiers with independent discrete-valued features", *IEEE Trans. Computers*, vol. C-21, February 1972, pp. 205-208.

[16] J. Stoer and C. Witzgall, *Convexity and Optimization in Finite Dimensions I*, Springer-Verlag, 1970.

[17] G. T. Toussaint, "A hierarchy of simple polygons", manuscript in preparation.

[18] G. T. Toussaint, "Complexity, convexity, and unimodality", *Int. Journal of Computer and Information Sciences*, in press.

[19] T. M. Cover, "Geometrical and statistical properties of systems of linear inequalities with applications in pattern recognition", *IEEE Trans. Electronic Computers*, vol. EC-14, June 1965, pp. 326-334.

[20] G. T. Toussaint, "On translating a set of spheres", Tech. Report, SOCS-84.4, School of Computer Science, McGill University, March 1984.

[21] G. Strang, "The width of a chair", *The American Mathematical Monthly*, vol. 89, no. 8, October 1982, pp. 529-534.

[22] G. T. Toussaint, "Solving geometric problems with the rotating calipers", *Proc. IEEE MELECON '83*, Athens, Greece, May 1983.

[23] M. Houle and G. T. Toussaint, "Computing the width of a set", manuscript in preparation.

[24] M. E. Dyer, "Linear time algorithms for two- and three-variable linear programs", *SIAM Journal of Computing*, vol. 13, no. 1, February 1984, pp. 31-45.

[25] D. T. Lee and F. P. Preparata, "An optimal algorithm for finding the kernel of a polygon", *Journal Ass. Comput. Mach.*, vol. 26, no. 3, 1979, pp. 415-421.

[26] J. R. Sack and G. T. Toussaint, "Movability of polygons in the plane", *Proc. STACS*, Saarbrücken, January, 1985.

[27] R. J. Dawson, "On removing a ball without disturbing the others", *Mathematics Magazine*, vol. 57, no. 1, January 1984, pp. 27-30.

[28] G. T. Toussaint and H. A. El Gindy, "Two monotone polygons can be separated in linear time", Tech. Rept. SOCS-84.7, School of Computer Science, McGill University, May 1984.

[29] F. P. Preparata and K. Supowit, "Testing a simple polygon for monotonicity", *Information Processing Letters*, vol. 12, 1981, pp. 161-164.

[30] H. El Gindy and D. Avis, "A linear algorithm for computing the visibility polygon from a point", *Journal of Algorithms*, vol. 2, 1981, pp. 186-197.

[31] D. T. Lee, "Visibility of a simple polygon" *Computer Vision, Graphics, and Image Processing*, vol. 22, 1983, pp. 207-221.

[32] M. I. Shamos and D. Hoey, "Geometric intersections problems", *Seventeenth Annual IEEE Symposiums on Foundations of Computer Science*, October 1976, pp. 208-215.

[33] L. J. Guibas and J. Stolfi, *Notes on Computational Geometry*, XEROX PARK and Stanford University, 1982.

[34] R. J. Dawson, "On the mobility of bodies in $\mathbb{R}^n$", manuscript, Cambridge University, England.

[35] G. T. Toussaint and H. A. El Gindy, "Separation of two monotone polygons in linear time", *Robotica*, in press.

[36] H. Edelsbrunner, M. H. Overmars, and R. Seidel, "Some methods of computational geometry applied to computer graphics", Tech. Rept. F117, Technical University of Graz, June 1983.

[37] M. McKenna and R. Seidel, "Shadows of convex polyhedra", *Proc. Symposium on Computational Geometry*, Baltimore, June 1985.

[38] J. R. Sack, "Rectilinear computational geometry", Ph.D. thesis, School of Computer Science, McGill University, Montreal, 1984.

[39] G. T. Toussaint, "On translating a set of polyhedra", *Conference on Polyhedra*, Smith College, Northampton, Mass., April 6-8, 1984.

[40] K. Post, "Six interlocking cylinders with respect to all directions", internal manuscript, University of Eindhoven, The Netherlands, December 1983.

[41] B. Chazelle, "The polygon containment problem", in *Computational Geometry*, Ed., F. P. Preparata, *Advances in Computing Research*, vol. 1, JAI Press, Inc., 1983, pp. 1-33.

[42] G. T. Toussaint, "Pattern recognition and geometrical complexity", *Proc. Fifth International Conference on Pattern Recognition*, vol. 2, Miami Beach, December 1980, pp. 1324-1347.

[43] G. T. Toussaint, "Computational geometric problems in pattern recognition", in *Pattern Recognition Theory and Applications*, Ed., J. Kittler, NATO ASI, Oxford University, April 1981, pp. 73-91.

[44] K. Seike, *The Art of Japanese Joinery*, Weatherhill/Tankosha, New York, Tokyo, Kyoto, 1977.

[45] F. P. Preparata and D. E. Muller, "Finding the intersection of n half-spaces in time O(n log n)", *Theoretical Computer Science*, vol. 8, 1979, pp. 45-55.

[46] A. Holden, *Orderly Tangles: Cloverleafs, Gordian Knots, and Regular Polylinks*, Columbia University Press, New York, 1983.

[47] B. Grunbaum and G. C. Shephard, "Satins and twills: an introduction to the geometry of fabrics", *Mathematics Magazine*, vol. 53, no. 3, May 1980, pp. 139-161.

[48] C. R. J. Clapham, "When a fabric hangs together", *Bull. London Math. Soc.*, vol. 12, 1980, pp. 161-164.

[49] T. C. Enns, "An efficient algorithm determining when a fabric hangs together", *Geometriae Dedicata*, vol. 15, 1984, pp. 259-260.

[50] H. Imai, M. Iri, and K. Murota, "The Voronoi diagram in the Laguerre geometry and its applications", Tech. Report # RMI 83-02, Department of Mathematical Engineering, University of Tokyo, March 1983.

[51] W. R. Franklin, "An exact hidden sphere algorithm that operates in linear time", *Computer Graphics and Image Processing*, vol. 15, 1981, pp. 364-379.

[52] H. El Gindy, personal communication.

COMPUTATIONAL GEOMETRY, G.T. Toussaint (Editor)
© Elsevier Science Publishers B.V. (North-Holland), 1985

Computational Geometry and Motion Planning

S. H. Whitesides

School of Computer Science
McGill University
Montréal, Québec
Canada

Abstract

Research and development work in robotics and industrial automation has created a need for good motion planning algorithms for collision avoidance; this in turn has stimulated research on the inherent computational geometry of motion planning. The purpose of this paper is to survey some recent algorithms and complexity results for motion planning, placing particular emphasis on computational geometry issues.

A brief discussion of a variety of the many different types of problems that arise in the context of motion planning is included, as many of these problems involve geometric questions. The main focus of the paper, however, is on the following problem and its variants: Given descriptions of an object and a collection of obstacles in 2 or 3-dimensional space, and given the initial and desired final configuration for the object, find a collision-free trajectory that moves the object to its goal or determine that no such trajectory exists.

Partially supported by NSERC A0368 and NSF MCS-8302024

1. Introduction

Research in robotics has stimulated considerable interest in finding algorithms for solving collision avoidance problems; this in turn has created interest in the computational complexity of motion planning problems.

Robotics research takes many directions and is prompted by a wide variety of goals. These goals include both the development of mobile automatons capable of understanding and responding to new environments and also, the development of sophisticated flexible, reprogrammable, computer-controlled machine tools. This variety of goals and intended applications creates a corresponding variety of motion related problem areas. For example, a mobile automaton cannot reasonably be expected to have an accurate, detailed model of its environment. On the other hand, a robot designed to be used on an automated assembly-line may have very detailed information from a CAD/CAM database about the objects it manipulates. The general nature of motion planning problems in these two situations is very different, although some subproblems may be shared by both situations.

The main purpose of this paper is to describe some representative examples of recent research on the following specific motion problem, which is often called the *Mover's Problem* or the *FINDPATH Problem.*

Definition: The Mover's Problem, or FINDPATH: Given the initial and desired final configurations of an object in 2- or 3-dimensional space, and given descriptions of the obstacles in the space, determine whether there is a continuous motion of the object from the one configuration to the other, and find such a motion if one exists. (The objects and the motion are typically required to satisfy certain specified properties.)

Various more specific forms of this general definition have appeared in the literature. For example, the objects are sometimes assumed to be polygons or polyhedra and the motions sought might be sequences of pure rotations and pure translations. A variety of words have been used to evoke an image of the object being moved. It is has been called a chair, a piano, and a sofa, for example. These words suggest the assumption that the object is inflexible, but the general definition just given allows the possibility, which has been considered in the literature, that the object consists of more than one part and that these parts are attached to one another in some flexible way, say by revolving or sliding joints. Additional possibilities are that the "parts" of the object are not attached to one another and can function independently, or that they can change shape as well as configuration as they move. In case the moving object consists of several independent pieces, the Mover's Problem is generally called a *Motion Coordination Problem.* In case the only objects are the ones to be moved (i.e., there are no obstacles) and the desired final configuration is only specified by requiring that the objects be spread out, the problem is called a *Separability Problem.* This type of problem is surveyed by Toussaint '85.

A fundamental assumption made in Mover's Problems is that a detailed description of the objects is available as input data for any algorithmic solution. Consequently, perceptual issues of great interest in some areas of robotics research are deliberately ignored in this paper. Also ignored are the control problems involved in achieving a specified path of motion. Nevertheless, it is important to be aware of these and other issues and to view the Mover's Problem in the more general context of a variety of other motion problems. Future progress in robotics will require the simultaneous solution and coordination of many kinds of motion problems.

While robotics has been responsible for stimulating much interest in motion planning problems, one can also view these problems not simply as part of robotics but rather as part of a discipline of object design, editing, representation, transformation, and manipulation that is emerging from several areas including computational geometry, VLSI, image processing, artificial intelligence, computer graphics, and CAD/CAM as well as robotics per se.

The above considerations motivate the inclusion of a brief overview of several other motion problems at the end of this survey.

The rest of the paper is organized as follows. Section 2 focuses on solutions to the Mover's Problem. It discusses general approaches and techniques and then describes the contents of a sampling of recent papers on this topic. Section 3 reviews recent work on the Motion Coordination variant of the Mover's Problem. Section 4 discusses computational complexity results and describes some motion problems that have been found to be NP-hard or complete or PSPACE-hard or complete. Section 5 puts the Mover's Problem into a larger context by presenting a variety of different motion problems areas, and Section 6 concludes the paper with a review of some issues of interest to computational geometry that occur in the context of motion planning.

2. The Mover's Problem, or FINDPATH

An important source of inspiration for research on algorithms for motion planning has come from the need to solve path finding problems for mobile automatons

(Nilsson '69; Thompson '77; Giralt, Sobek, Chatila '79; Moravec '83) and computer-controlled manipulators (Pieper '68; Lewis '74; Widdoes '74; Udupa '77a,b; Campbell and Luh '80). The need for motion planning algorithms has also arisen in computer graphics. (See Guibas and Yao '80, for example.) It is interesting to note that there has long been a fanciful interest in geometric motion problems (See Moser '66, Howden '68, Goldberg '69, Sebastian '70 for the "sofa" problem, which is to move a maximum area planar figure around a right angle bend in a hallway, and Strang '82 for the problem of passing a convex set through an interval.)

Recent research work on motion planning includes concern on the one hand with the computational complexity of motion planning and on the other hand with the search for fast algorithms. This section begins with a discussion of algorithmic approaches to the Mover's Problem and continues with a discussion of specific algorithmic results. Complexity results are discussed in Section 3.

2.1 General strategies for algorithmic motion planning

One of the oldest and most common themes in motion planning work is the idea of representing the problem in such a way that the object to be moved is a point. Moravec '83 for example describes path planning for the Stanford Cart. The Cart and the obstacles in its environment are approximated by enclosing them in circles. The circles bounding objects are then increased in radius by the radius of the circle enclosing the robot, and the robot's position is represented by the center of its enclosing circle. A path for this point is sought that misses the enlarged obstacles.

Another common theme that goes hand in hand with the point representation of the moving object is the idea of having the point move through a *configuration space* not necessarily equal to the 2- or 3- dimensional space of the physical problem, which is often called the *Cartesian space* (regardless of the chosen coordinate system). For example, the position and orientation of a rigid object moving in the plane can be represented by a point (x,y,θ) in a 3-dimensional parameter space: x and y represent the position of a specified reference point of the object relative to a fixed Cartesian coordinate system, and θ represents the angle made by a reference line on the object with, say, the positive x axis. Similarly, a rigid object translating and rotating in 3-dimensional space can be represented as a point moving in a 6-dimensional configuration space.

An early example of a configuration space approach for motion planning is Howden '68. Adamowicz '70, Freeman '75, Adamowicz and Albano '76, and Albano and Sapuppo '80 use the technique for template layout problems. Udupa '77a,b solves obstacle avoidance problems for a manipulator by using a configuration space in which the moving object is represented by a moving point.

Of course there are several natural ways to represent the configurations of an object. We will use the term *Cspace* to refer to a given parameter space for the configurations of a given object. In general, some points in Cspace will correspond to placements of the object where it collides with obstacles in the physical space of the problem. Such Cspace points will be called *illegal* or *forbidden*; the Cspace points that are not illegal will be called *legal*.

To find a path for an object, one common approach is to "grow" physical obstacles into corresponding obstacles in the Cspace of the object. For example, if A is an object to be moved and B is a "physical" obstacle, then the obstacle for the point representing A in the Cspace of A is just the set of all tuples of parameters that correspond to a collision between A and B in physical, Cartesian space. This set of parameter tuples is called the *Cspace obstacle for A due to B.* (Note that since the orientation of a circle moving in 2-dimensional space can be ignored, the configuration space of the circle is also 2-dimensional. Hence the technique described by Moravec of enlarging circles in the plane can be viewed as a construction of Cspace obstacles for the center of the circle.) There are several techniques for computing or approximating the Cspace obstacles; these will be discussed in following subsections.

Once configuration obstacles or their approximations have been obtained, there are several strategies for finding a path for the moving point that represents object A. One method is to construct a *visibility graph* whose vertices are the Cspace representations for the initial and final configurations of A together with points on the Cspace obstacle boundaries; its edges represent the existence of straight lines of legal positions in Cspace connecting those positions. Path planning is thereby converted to searching for a path between two vertices of a graph.

Another approach is to find a decomposition into cells of the set of points in Cspace that represent the legal, collision-free configurations of the moving object. Then a graph is constructed whose vertices and edges represent these cells and their adjacencies. Again, the motion planning problem becomes a search for a path joining two vertices of a graph. A computational difficulty here is that the Cspace obstacles have in general very complicated shapes. Nevertheless, this approach has yielded a good exact algorithm (Lozano-Pérez '83a) that finds shortest paths for the special case of a polygon translating with fixed orientation in a planar space cluttered with polygonal obstacles. A practical difficulty with this approach is that it finds paths that hug obstacle boundaries.

A third approach is to have the point representing the object move along edges or curves in a special set such as a *generalized Voronoi diagram.*

As will become apparent in later discussion, several computational geometry problems arise in making use of these ideas. For example, motion planning for convex polygonal or polyhedral objects and obstacles gives rise to manageable Cspace obstacles. Consequently, it is desirable to be able to decompose polygonal objects into a small number of possibly overlapping convex polygonal pieces in order to represent Cspace obstacles as unions of the Cspace obstacles arising from the convex pieces.

We now give a preview of some of the papers we are about to discuss in subsections 2.2-2.5.

While work on motion planning for particular robots or robot manipulators such as Nilsson '69 and Udupa '77a,b contained the basic idea of representing a moving object by a point moving past "grown obstacles" in physical or configuration space as well as the idea of a visibility graph computation, it was Lozano-Pérez and Wesley '79 who first made a smooth transition from the problems of controlling particular machines to the problem of finding general algorithmic methods for solving the Mover's Problem. (See Udupa '77b for a survey of heuristics and previous work on collision avoidance for computer-controlled manipulators.) This work was motivated in part by the failure of early motion planning strategies for robots to work consistently well in a global way. For example, in a "generate and test" paradigm for path finding, a straight line from start to goal would be refined by introducing subgoals if the line intersected obstacles. Unfortunately, this method often leads to dead ends, long paths, and expensive computations. Consequently, it seemed desirable to develop good path finding algorithms for the situation in which the descriptions of the obstacles in the environment of the automaton are known.

The paper of Reif '79, which gave the first complexity results in motion planning, helped to bring the Mover's Problem to the attention of the theoretical computer science community by sketching an algorithmic solution.

The papers Schwartz and Sharir '83a,b,c showed the possibility of using analytical and topological, rather than purely geometric, methods in motion planning. This work stimulated approaches with a more topological flavor in papers such as O'Dúnlaing, Sharir and Yap '83, O'Dúnlaing and Yap '83, and Yap '84.

Because the papers of Lozano-Pérez and Wesley '79 and Schwartz and Sharir 83a,b,c have been especially influential, these papers and their approaches are the basis for the organization of the rest of this section. First we discuss papers that are related to Lozano-Pérez and Wesley '79 and then papers that stem from Schwartz and Sharir '83a,b,c. While approaches in both categories have elements of similarity (often problems are reduced to a graph search, for example), the papers of subsection 2.2 have a geometric flavor, a motivation to find solutions that are workable in practice, and an interest in implementation while the papers in subsections 2.3-2.5 have a more topological flavor and an interest in developing general ways to find non-heuristic solutions, although specific algorithms may not be of direct practical value. (Using a synthesis of many of these ideas as a starting point, Donald '84 designs and implements an algorithm for a 6-degree-of-freedom Mover's Problem for polyhedral objects.)

2.2 Geometric motion planning

The purpose of this subsection is to summarize the contents of Lozano-Pérez and Wesley '79 and some subsequent papers that have further developed the ideas in it or that have been inspired by it in a significant way.

Lozano-Pérez and Wesley '79: This paper develops methods for motion planning in the plane or 3-space for the case of objects that can be modelled as the union of convex 2-dimensional polyhedra (polygons) or 3-dimensional polyhedra, respectively. The observation is made that if the moving object A is a convex polyhedron whose orientation is fixed, that is, whose motion is purely translational, then the configuration space obstacle for A due to a convex polyhedral object B is again a convex polyhedron. Furthermore, this Cspace obstacle is easy to describe in terms of the original objects: its boundary can be traced out by a reference point of A while A is moved with fixed orientation around B, touching B but not intersecting its interior. If A and B are unions of convex polyhedra, A moving with fixed orientation, then the Cspace obstacle is the union of the Cspace obstacles obtained from the various combinations of constituent pieces of A and B.

For the 2-dimensional problem, Cspace is also 2-dimensional, and the Mover's Problem can be reduced to a path search in a *visibility graph* once all the Cspace obstacles have been obtained. The vertices of the graph are the vertices of the Cspace obstacles together with representations of the start and goal positions of the moving object. Two vertices are joined whenever the line segment they determine does not pass through the interior of any Cspace obstacle. Object A in its fixed

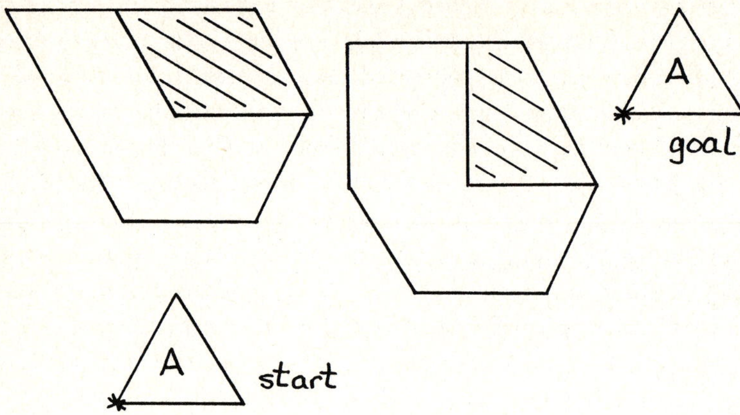

Fig. 1a. If the orientation of A is held fixed, then moving A from start to goal without collision with the shaded obstacles is equivalent to moving the starred vertex from start to goal without collision with the "grown" obstacles.

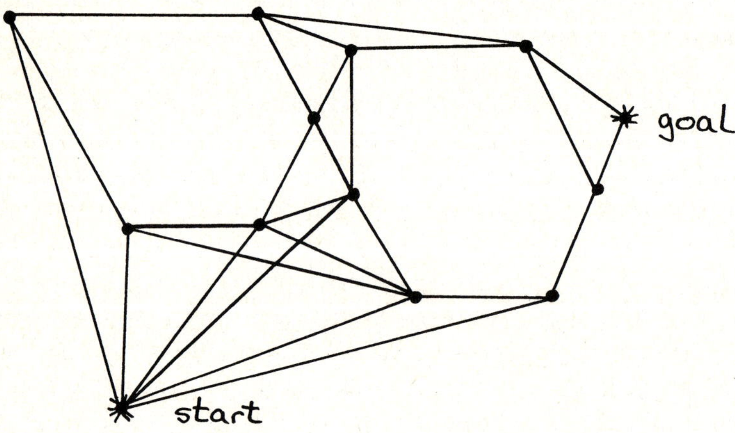

Fig. 1b. The visibility graph based on the Cspace obstacles for A.

orientation can be translated from its start to its goal position if, and only if, the vertices of the visibility graph corresponding to these positions are connected by a path. (For visibility graph computation in the plane, see Lee and Chen '85.) Consequently, motion planning for the class of objects under consideration, moving with fixed orientation in the plane, can be solved in an algorithmic way. Furthermore, any path found by this method has the minimum length possible. See Fig. 1.

For an object with fixed orientation moving in 3-dimensional space, Cspace is 3-dimensional, but the strategy described above is no longer optimal. In fact, it may lead incorrectly to the conclusion that no path from start to goal exists. This is because shortest paths do not necessarily go from obstacle vertex to obstacle vertex; instead, they may go from a point on one obstacle edge by a straight line to point on another edge. Paths of this sort may exist while legal paths moving from vertex to vertex may fail to exist. The paper suggests considering points spaced at regular intervals along obstacle edges. In this way an approximate algorithm can be obtained using the visibility graph. (Sharir and Schorr '84 give a doubly exponential algorithm for determining the sequence of boundary edges through which the shortest path goes in 3-dimensional space.)

For a convex polygon moving in the plane and allowed to rotate and translate simultaneously, the 3-dimensional Cspace obstacles have convex polygonal cross-sections but are not in general convex. Hence one might approximate these obstacles by enclosing them in convex polygons in order to continue to use the visibility graph method. Another alternative suggested by the paper is to approximate by an enclosing convex polygon the area swept over by the object A as it rotates from one orientation θ_1 to another θ_2 with the position of its reference point fixed in space, and then to compute Cspace obstacles relative to the enclosing convex polygon moving with fixed orientation. This idea provides a way to look for paths for A that are safe as long as the orientation of A belongs to the interval $[\theta_1,\theta_2]$. To refine the strategy, one can construct Cspace obstacles based on each of several orientation intervals and then make a visibility graph whose vertices include all the vertices of all the graphs obtained for the individual intervals, adding some "transition" vertices that represent positions where A can be safely rotated through a union of orientation ranges. The edges of this type of graph include edges representing safe paths to positions corresponding to transition vertices as well as the edges of graphs based on fixed orientation intervals.

For problems involving configuration spaces of dimension higher than three, the paper suggests working in some convenient subspace by fixing some of the parameter values, and then working with convex polyhedral approximations to the

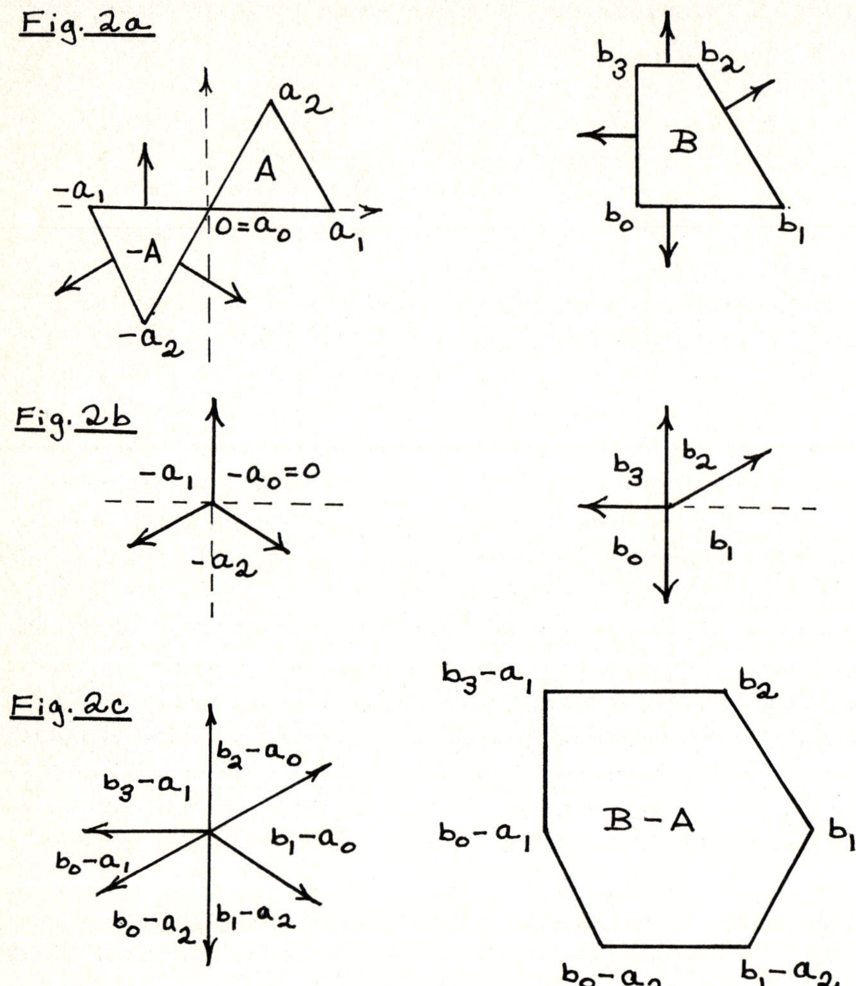

Fig. 2a shows -A, B and their outward normals. In Fig. 2b, the wedge labels indicate the contact point for all support lines to -A or B whose normals lie in the wedge. The wedge labels in Fig. 2c are the contact points of support lines to Cspace obstacle B-A.

volume swept by A as some vector of parameters ranges over an interval between fixed vectors.

Lozano-Pérez '83: This paper develops some ideas from the preceding paper in more detail. For example, it points out that the configuration space obstacle for a convex polygon (polyhedron) A moving in fixed orientation with respect to a convex polygonal or polyhedral obstacle B can be obtained by taking the convex hull (Graham '72 and Preparata and Hong '77 are cited) of the following set of points:

$\{v_b - v_a \mid v_b$ ranges over the position vectors of vertices of B and v_a ranges over the position vectors of vertices of A when the reference point of A is at the origin$\}$.

This gives an $O(v \log v)$ algorithm, where v is equal to the product of the numbers of vertices of A and B. In the 2-dimensional case, an improvement is obtained by merging the list of edges of B, sorted by orientation, with the list of sorted edges of -A. (Here, -A is the polygon whose vertex vectors are the negations of the vertex vectors of A when the reference vertex of A is at the origin.) See Fig. 2. Each pair of adjacent edges on the sorted list gives rise to a Cspace obstacle vertex obtained as follows. Let θ_1 and θ_2 be the orientations of the outward normals to two adjacent edges on the list. For all θ in the range (θ_1, θ_2), the support lines to -A and B with outward normal orientation θ contact -A and B at some fixed pair of vertices v_{-a} and v_b respectively. The vector sum of these two vertices is a vertex of the Cspace obstacle. This gives a way to find the Cspace obstacle in time $O(n+m)$, where n and m are the numbers of vertices of A and B. This idea also appears in Schwartz '81.

Brooks and Lozano-Pérez '83: This paper gives a procedure for solving the Mover's Problem in the case of an object A moving in the plane with variable orientation, where the object and obstacles are unions of convex polygons. (For simplicity, we assume in the following discussion that the object and obstacles are convex polygons; the extension to the more general case of unions of convex polygons is straightforward.) The procedure is to construct a collection of rectanguloids, with faces perpendicular to the x, y or z axes, that cover the Cspace obstacles. These rectanguloids, whose interiors do not intersect, are refined and extended to a decomposition of the whole configuration space into rectanguloid "cells". The cells are labeled as "empty", "full", or "mixed" depending on whether they lie entirely inside or outside the union of the obstacles. Cells are also labeled by lists of those constraints that define the Cspace obstacles and that are satisfied by some, but not all, of the points in the cells. (These lists are updated as rectanguloids are refined.) Then a graph is constructed whose vertices represent "empty" and "mixed" cells and whose edges represent cell adjacencies. If a path cannot be found from the vertex representing the cell that contains the starting configuration to the vertex representing the cell that contains the goal configuration so that the path passes only through vertices that represent "empty" cells, then the "mixed" cells are refined, the resulting new cells are labeled, and the graph is updated for a new attempt to find a path. Cell refinement stops when a safe path is found or when cells become smaller than some user-defined limit.

While the paper discusses heuristics for cell refinement and path searching in the graph, what is particularly interesting from the geometry point of view is the way in which the initial rectanguloids are created and the way in which cells are labeled. This is done in a way that avoids representing the Cspace obstacles explicitly and is accomplished by taking advantage of special properties that the Cspace obstacles have in this situation. As Lozano-Pérez '83 points out, the cross-section of a Cspace obstacle due to B defined by setting the orientation θ of moving object A equal to a constant value is a convex polygon whose edges, except at certain critical θ values, are of two types. Type A edges are obtained by the vector addition of an edge of -A to a vertex of B. Type B edges are obtained by the vector addition of an edge of B to a vertex of -A. See Fig. 3. Between two adjacent values θ_1 and θ_2 from a finite list of certain critical orientation values, the nature of the edges of the obstacle cross-section does not change. That is, the pairings of vertices and edges of -A and B that produce the edges of the Cspace obstacle does not change. As θ varies from θ_1 to θ_2, each type A edge looks like an edge of -A rotating around a vertex of B. Each type B edge looks like an edge of B being translated parallel to itself by a rotating vertex vector from -A . The critical values of θ occur at orientations for which an edge of A is parallel to an edge of B. See Fig. 4.

To construct the rectanguloids that cover the Cspace obstacle for A due to B, one slices the Cspace obstacle with planes defined by setting θ equal to the critical values. For values of θ between neighboring critical values, the coordinates of the vertices of the convex polygonal cross-sections can be expressed as functions of $\cos\theta$ and $\sin\theta$. These can be differentiated to determine the minimum and maximum values of x and y for points on the surface of the Cspace obstacle having their θ coordinate between the two neighboring critical values. Planes defined by setting x and y equal to their maximum and minimum values then determine, with the θ equals constant planes, six half-spaces whose intersection is a rectanguloid that covers a portion of the Cspace obstacle.

Type A edges: $[0,-a_1] + b_3$, $[-a_1,-a_2] + b_0$, $[0,-a_2] + b_1$

Type B edges: $[b_3,b_0] - a_1(\theta)$, $[b_0,b_1] - a_2(\theta)$, $[b_1,b_2]$, $[b_2,b_3]$

Note change in pairing of edges and vertices below.

Type A edges: $[0,-a_1] + b_3$, $[-a_1,a_2] + b_1$, $[-a_2,0] + b_2$

Type B edges: $[b_3,b_0] - a_1(\theta)$, $[b_0,b_1] - a_1(\theta)$, $[b_1,b_2] - a_2(\theta)$, $[b_2,b_3]$

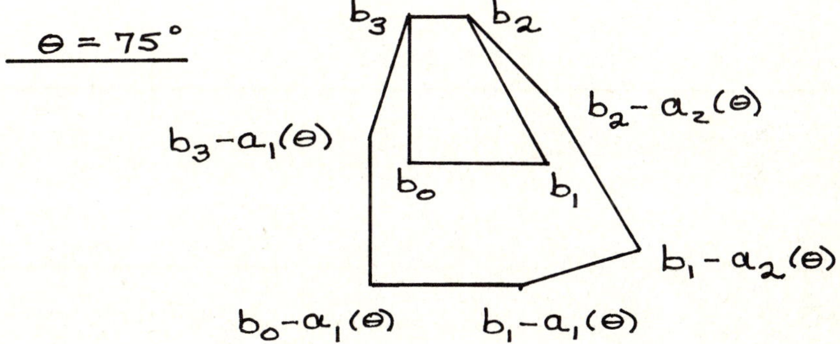

Fig. 3. Cross sections B-A(θ) for θ=45° and θ=75°, where A and -A have been rotated counter-clockwise by θ degrees from the initial θ=0° orientation shown in Fig. 2.

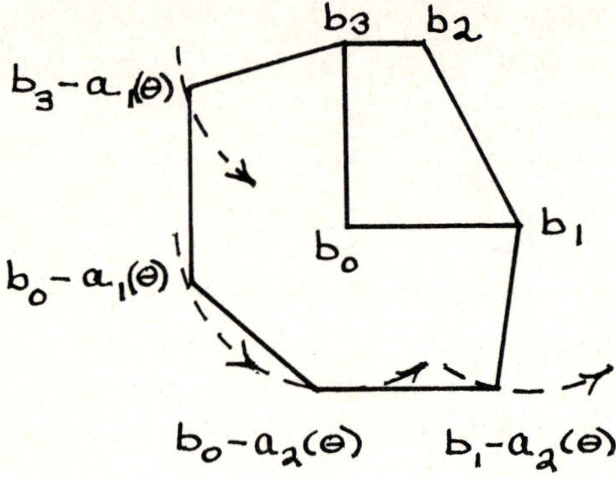

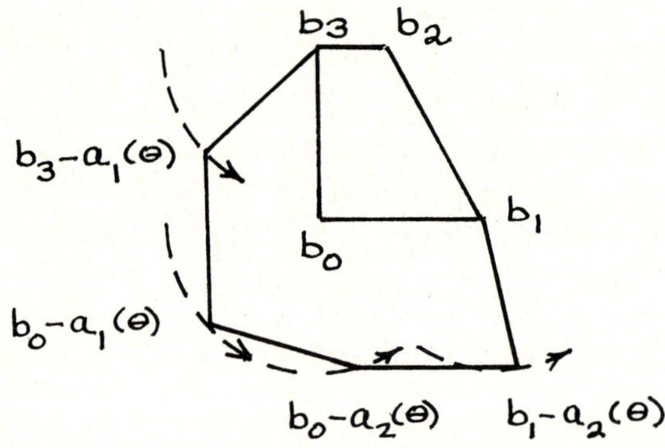

Fig. 4. Type A edges such as $(0,-a_1(\theta)) + b_3$ are edges of -A rotating around vertices of B; type B edges such as $(b_3,b_0) - a_1$, are edges of B translated by a rotating vertex vector of -A.

Of course rectanguloids from one Cspace obstacle may intersect those from another. In order to get an initial decomposition of configuration space into rectanguloids whose interiors have empty intersection, Cspace is sliced by planes perpendicular to the x axis at the maximum and minimum x values of the various rectanguloids. (This refines those rectanguloids.) In each sandwich formed by adjacent parallel slices, planes perpendicular to the y axis are added at the extreme y values of the rectanguloids in the sandwich (these new planes are not extended beyond the sandwich boundary into other sandwiches). This operation further refines rectanguloids. However, if two of them have a 3-dimensional intersection, then the intersection is a rectanguloid that can be used in a refinement because its subtraction from the original rectanguloids leaves two rectanguloids. Consequently, the desired initial decomposition of configuration space can be achieved.

This cell decomposition, and later refinements of it, has the property that testing whether a cell is full, mixed, or empty, and testing which constraints are satisfied by some but not all of the cell points, can be done by testing the four rectanguloid corners at the lowest θ value. The corners can be tested using the following observations. As θ varies from the minimum value θ_1 for the box to the maximum value θ_2, the projection of the type B constraint lines onto the plane at $\theta=\theta_1$ is a strip bounded by the parallel lines corresponding to θ_1 and θ_2. If a box corner (x, y, θ_1) lies in this strip, this means that there is a point (x, y, θ) with $\theta_1<\theta<\theta_2$ on the edge of the box that simultaneously lies on the surface of the Cspace obstacle. See Fig. 5. For a type A constraint line, the points swept over by the projection of the constraint as θ varies from one critical value to the next form a pair of cones with an extra wedge, as shown in Fig. 5. Again, if corner (x, y, θ_1) lies in the area swept by the projections, then there must be a point on the surface of the Cspace obstacle that lies on an edge of the box. *All* the points in a rectanguloid cell satisfy a given type A or type B constraint for the portion of the Cspace obstacle between θ_1 and θ_2 if, and only if, the (x,y) coordinates of the four corners of the rectanguloid at the θ_1 level lie in the region, which is convex, consisting of the points (x, y) that are always on the *obstacle* side of the constraint line as it varies through all its positions between θ_1 and θ_2. Consequently, a rectanguloid lies entirely inside a particular obstacle if, and only if, the four bottom corners all lie in the appropriate convex regions determined by sweeping the constraint line projections. In similar fashion, it can be determined whether a rectanguloid lies completely outside an obstacle.

Brooks '83: This paper gives a method for finding a path for a convex polygon moving in the plane past objects that can be modeled as unions of convex

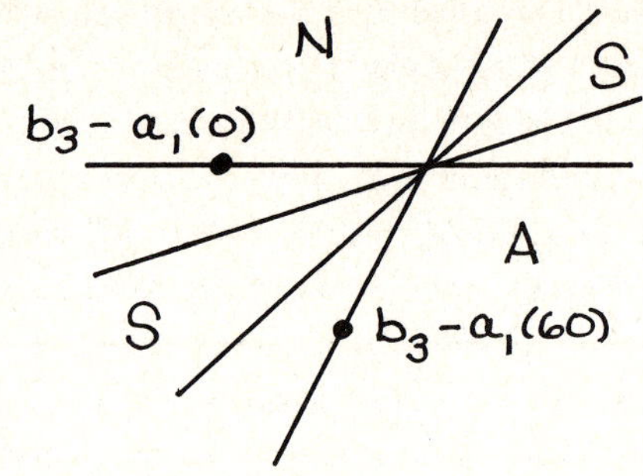

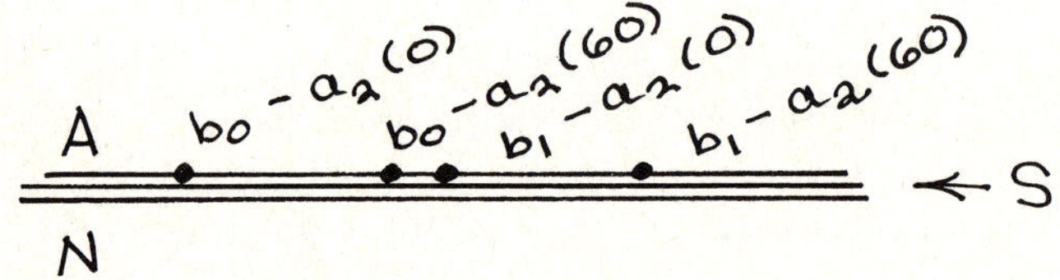

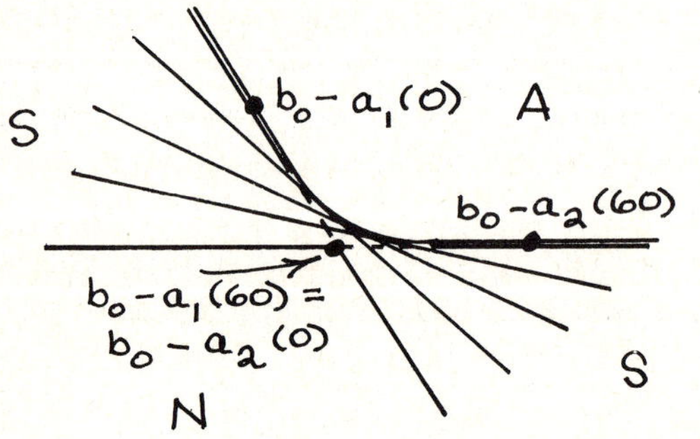

Fig. 5. Three of the 7 constraint lines through the edges of B-A(θ) shown as θ varies from 0° to 60°. Regions A and N are always or never on the obstacle side of the constraint. Points in S regions are sometimes on the constraint line. A box bounded by θ=0° and θ=60° is strictly inside the Cspace obstacle iff each corner at level θ=0° is in the (convex) A region of each constraint.

polygons. It takes a different approach than that of the previously described papers in that it does not work with obstacles in configuration space. The idea is to keep the reference point of the object far away from the obstacles. This is done by first finding a collection of overlapping cone-like shapes in free space (that is, the set of points not belonging to any obstacle) whose sides, roughly speaking, come from obstacle edges that face each other. See Fig. 6. (The union of the cones does not in general cover all of the free space.) Then the "spines" of the cones are intersected in pairs, the spine of a cone being a portion of the bisector of the angle formed by the cone sides. At intersection points of spines, intervals of safe orientations for a rectangular approximation of the moving object are computed. (The reference point of the rectangle is located at the spine intersection point.) Then a graph is constructed whose vertices correspond to triples (p, C, I), where p is an intersection point of spines, C is a cone containing one of the spines, and I is a legal orientation range for the approximating rectangle at p relative to C. Two distinct vertices (p_1, C_1, I_1) and (p_2, C_2, I_2) of the graph are joined if either $p_1 = p_2$ and I_1 intersects I_2 (meaning that the moving object can be transferred from one cone to another by undergoing an appropriate rotation at point p in the intersection of the cones) or $C_1 = C_2$ and I_1 intersects I_2 (meaning that the object can be translated from p_1 to p_2 along the cone spine). Path finding is thereby reduced to a search for a path in this graph.

This method may fail to find a path even though one exists: the cones may not cover the entire space, and in any case, the motions being considered consist of a finite sequence of pure translational and pure rotational movements. Implementations of the method show that it works very fast in relatively uncluttered environments where cell decomposition methods can be costly. The method also has the attractive property that the moving object stays away from obstacles. The method is further pursued by Nguyen '84.

2.3 The "projection" approach

The projection approach to the Mover's Problem involves reducing the problem to searching for a path in a graph that represents connectivity properties of the space FP of free "legal" configurations for the moving object. The basic strategy is as follows. First, FP is projected onto a subspace H of smaller dimension, chosen, say, for conceptual or computational convenience. Next, the projection is partitioned into connected regions that have the following property. For each region R in the partition, as a point X varies over R, the pre-image of X does not change

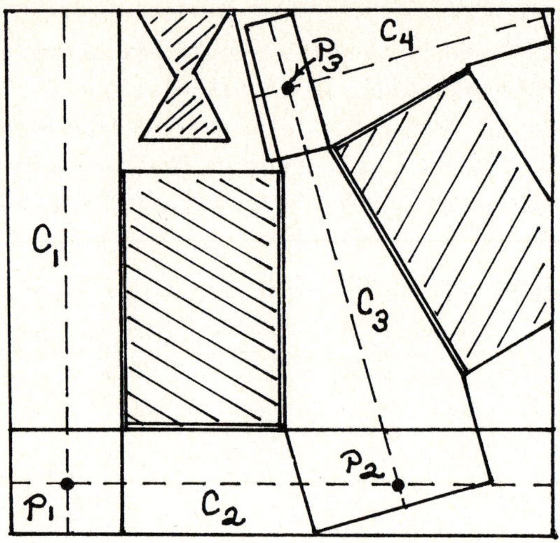

Fig. 6a (from Brooks '83): Generalized cones C_1, C_2, C_3, C_4 are generated by obstacles and work boundary (not all cones are shown).

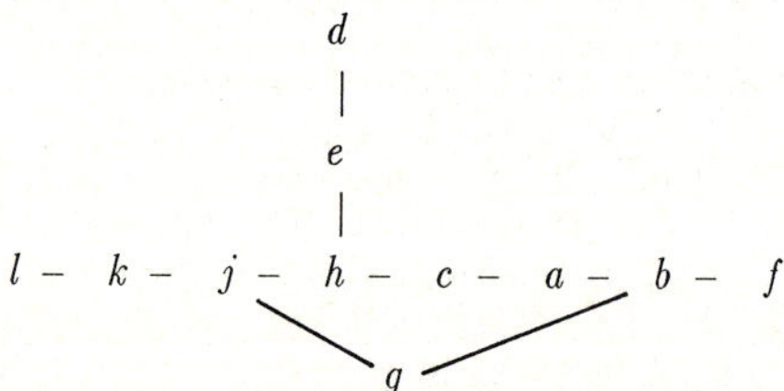

Fig. 6b. Connectivity graph for cones shown in 6a, where:

$a = (p_1, c_1, [30,150])$

$b = (p_1, c_2, [300,60])$

$c = (p_1, c_2, [120,150])$

$d = (p_1, c_1, [210,240])$

$e = (p_1, c_2, [210,240])$

$f = (p_1, c_1, [300,330])$

$g = (p_2, c_2, [300,60])$

$h = (p_2, c_2, [120,240])$

$i = (p_2, c_3, [355,220])$

$j = (p_3, c_3, [80,135])$

$k = (p_3, c_4, [240,135])$

$l = (p_3, c_3, [255,315])$

"qualitatively". For example, the number of connected components of the set of points of FP that project to X does not change, and the components vary continuously with X. (Note that in general, a region R is not equal to an entire connected component of the projection of FP). Finally, a "connectivity" graph is constructed whose vertices represent connected components of pre-images of the various regions in the partition and whose edges represent an adjacency relationship among all these components.

Schwartz and Sharir '83a describe this approach and analyze the case of a polygonal object moving in the plane amidst polygonal obstacles. They give an $O(n^5)$ algorithm for planning the motion of a "ladder" consisting of a line segment moving past obstacles whose boundaries are composed of n line segments. Schwartz and Sharir '83d analyze a rod moving in three dimensions. (See also Chien, Zhang, and Zhang '84 for the case of a rod moving amongst polygons in the plane.) Schwartz and Sharir '83c apply the projection approach to a motion coordination problem, as a subsequent section describes.

Sharir and Ariel-Sheffi '84 employ a simple generalization of the projection approach of Schwartz and Sharir '83a to give an $O(k^2 n^{k+4})$ algorithm for the Mover's Prroblem for a "spider" moving past polygonal walls in the plane. Here, n is the number of straight lines in the walls, and k is the number of legs of the spider. A spider consists of a set of straight line segments joined at a common endpoint about which they are free to rotate. Legs are free to rotate past one another.

For the specific case of a ladder moving past planar, polygonal obstacles, the $O(n^5)$ result of Schwartz and Sharir '83a is improved by O'Dúnlaing, Sharir and Yap '83, who obtain an $O(n^2 \log n)$ algorithm by applying what they describe as a "retraction" approach. After a description of this method in the next subsection, we will return to describe a kind of projection approach motion planning algorithm proposed by Schwartz and Sharir '83b. It is extremely general, though inefficient in practice, and works by constructing a connectivity graph obtained from recursive projections of legal positions in configuration space. Other projection approach results can be thought of as streamlined versions of this general approach applied to special cases.

2.4 The "retraction" approach

O'Dúnlaing, Sharir and Yap '83 show that the idea of a retraction mapping, a standard concept in topology, can provide a fruitful approach for solving the Mover's Problem. (Yap '84 also uses the approach to solve a motion coordination

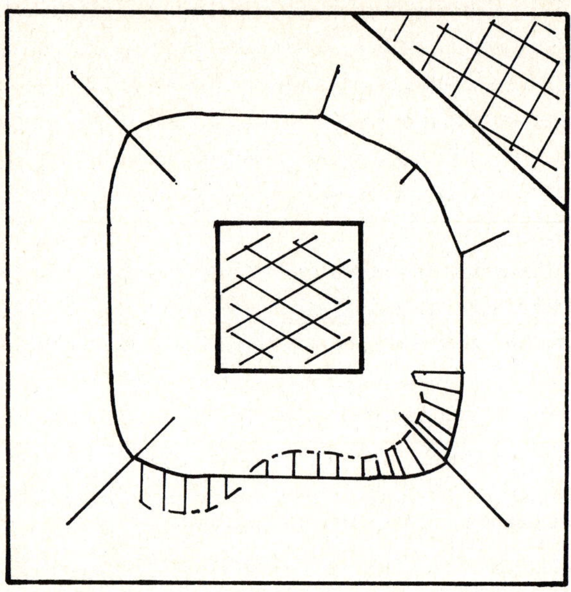

Fig. 7. The generalized Voronoi diagram for a disc (not shown) moving inside a square occupied by obstacles (shown hashed). The generalized diagram consists of those points on the standard Voronoi diagram that are legal positions for the center of the disc. A path for the center of the disc is shown retracted to the diagram.

problem, as will be discussed in a subsequent section.) A retraction mapping is defined as follows.

Definition: A *retraction* r of a space S to a subspace R of S is a continuous mapping of S to R satisfying $r(x)=x$ for all x in R.

O'Dúnlaing, Sharir and Yap '83 obtain $O(n \log n)$ and $O(n^2 \log n)$ algorithms for moving a disc and a line segment, respectively, past polygonal obstacles in the plane (here, n is the number of line segments in the obstacle boundaries). They do this by using the notion of a *Voronoi* diagram and its generalization to 3-dimensional Cspace; they do a retraction mapping to this diagram to reduce the Mover's Problem to a path search in the diagram. Their definition of the (standard) Voronoi diagram, which they generalize, is as follows.

Definition: Let S be the union of a collection of polygonal obstacles in the plane, where S is a closed set whose complement is bounded. The standard Voronoi diagram for S is the set of all points in the complement of S with more than one closest neighbor in the boundary of S.

Kirkpatrick '79 shows that the Voronoi diagram for n points and open line segments in the plane can be constructed in $O(n \log n)$ time, and O'Dúnlaing and Yap '83 use this result as a preprocessing step for their disc motion planning algorithm. (Also see Lee and Drysdale '81 for Voronoi diagram computation.) O'Dúnlaing, Sharir and Yap's notion of a "generalized" Voronoi diagram is as follows.

Definition: Let S be a closed set consisting of the union of a collection of polygonal obstacles in the plane, where S has bounded complement, and let B be a rigid two-dimensional body. The *generalized* Voronoi diagram for B and S is the subset of legal points *in configuration space* that correspond to configurations of B such that the points of S that are closest to B form a disconnected set.

For the case of a disc B, configuration space can be identified with the plane in which B and S lie, and the generalized Voronoi diagram is then the intersection of the legal positions for B with the standard Voronoi diagram for S. Each legal point x in configuration space that does not lie on the standard Voronoi diagram Vor(S) for S has a unique nearest neighbor y in the boundary of S. O'Dúnlaing Yap '83 define a continuous retraction by mapping x to the nearest intersection of Vor(S) with the straight line through x and y. Points in Vor(S) are mapped to themselves. See Fig. 7. They then show that there exists a legal path for B between two positions in the plane if, and only if, there is a path between the retractions of those positions in the generalized Voronoi diagram. In fact, after the initial $O(n \log n)$ preprocessing time used to compute Vor(S), path searching is done in O(n) time.

O'Dúnlaing, Sharir and Yap '83 go on to give an O($n^2\log n$) algorithm for moving a "ladder" consisting of a single line segment. This case is considerably complicated by the fact that the generalized Voronoi diagram for the ladder is a 2-dimensional subset of the legal points in a 3-dimensional Cspace. It is handled by defining a second retraction that maps the generalized Voronoi diagram to a network of curves lying in the generalized diagram. The network has the property that a path between two configurations of the ladder in the plane exists if, and only if, there is a path through the network that connects the images under the composition of the two retractions of the points representing the initial and desired configurations.

Rowat '79 was perhaps the first to use the Voronoi diagram for motion planning.

2.5 Cylindrical algebraic cell decomposition

Schwartz and Sharir '83b give an algorithmic solution to the Mover's Problem in a very general situation that does *not* require that the objects be expressible as unions of convex polyhedra and that allows the moving object to consist of parts that move relative to one another. They use the Tarski '51 algorithm for deciding statements in the quantified elementary theory of real numbers, together with crucial technical modifications introduced by Collins '75, which are reviewed by Arnon '81.

The running time of the Schwartz and Sharir algorithm is exponential in the number of degrees of freedom of the problem but "polynomial in the geometric constraints". Even for a fixed number of degrees of freedom, the algorithm is not useful in practice. Nevertheless, it is impressive because its methods are powerful enough to handle a situation of utmost generality, thus providing important insight into the Mover's Problem. Perhaps the methods will be useful for other geometric problems as well. In any case, results such as Schwartz and Sharir '83a can be regarded as streamlined versions of Schwartz and Sharir '83b for special cases.

Since Schwartz and Sharir '83b is long and technical, we shall motivate and describe the main ideas in the context of the following extremely simple situation, chosen to avoid the technical difficulties while still suitable for revealing many of the main ideas. Let the obstacle set consist of a single sphere of radius 1 centered at the origin in E^3, and let the object which is to be moved from a starting to a goal configuration also be a sphere of radius 1. In this case, Cspace consists of the positions for the center of the moving sphere and is hence 3-dimensional. Representing orientation is not necessary here, nor is it necessary to represent the orientation of one part of a jointed body relative to another.

Obviously the moving sphere avoids the fixed sphere as long as the center of the moving sphere stays outside a closed sphere of radius 2 centered at the origin 0. Schwartz and Sharir '83b require that the set XP of forbidden configurations include its boundary points, so here XP is the closed sphere of radius 2 centered at the origin. It is easy to devise common sense motion planning algorithms for this situation, but we want instead to see how their cell decomposition approach would work.

The first step is to describe the set XP as the set of all points that satisfy conditions that can be given as conjunctions and disjunctions of multivariable polynomial expressions over the rationals of the form $P>0$ or $P=0$:

$$XP=\{(x,y,z) \mid (-x^2-y^2-z^2+4>0) \; or \; (-x^2-y^2-z^2+4=0)\}$$

Next, consider the following decomposition of E^3 into three cells (this is not the decomposition given by the algorithm; we consider it only to motivate the procedure): $E^3=C_1 \cup C_2 \cup C_3$, where C_1 is the open sphere of radius 2 centered at the origin, C_2 is its boundary, and C_3 is the complement of the closure of C_1. This decomposition has the property that for each choice of one of the two formulas for the disjuncts in the logical expression P_1 *or* P_2 used to define XP, where

$$P_1=(-x^2-y^2-z^2+4>0) \text{ and } P_2=(-x^2-y^2-z^2+4=0),$$

and for each choice of one of the cells C_1, C_2, C_3, the formula is true everywhere in the cell or false everywhere in the cell. Hence XP must be a union of cells. P_1 (and hence P_1 *or* P_2) is satisfied throughout C_1; P_2 (and hence P_1 *or* P_2) is satisfied throughout C_2; both fail throughout C_3. So $XP=C_1 \cup C_2$.

The property just described illustrates a fundamental strategy. Decompose configuration space into cells having the property that in each cell, each formula in the description of XP is true throughout or false throughout the cell. Then determine whether the cell belongs to XP by evaluating at some representative point the boolean combination of truth values given by the defining expression for XP. Of course the complement of XP is the set of free, legal configurations. It is composed of the remaining cells.

The next step is to construct a graph whose vertices represent cells and whose edges represent cell adjacencies, and then to look for a path in this graph connecting the vertex representing the cell containing the initial configuration to the vertex

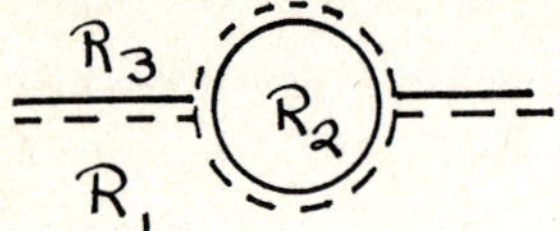

Fig. 8. Moving from R_1 to R_3 requires knowing what portions of the boundary can be used.

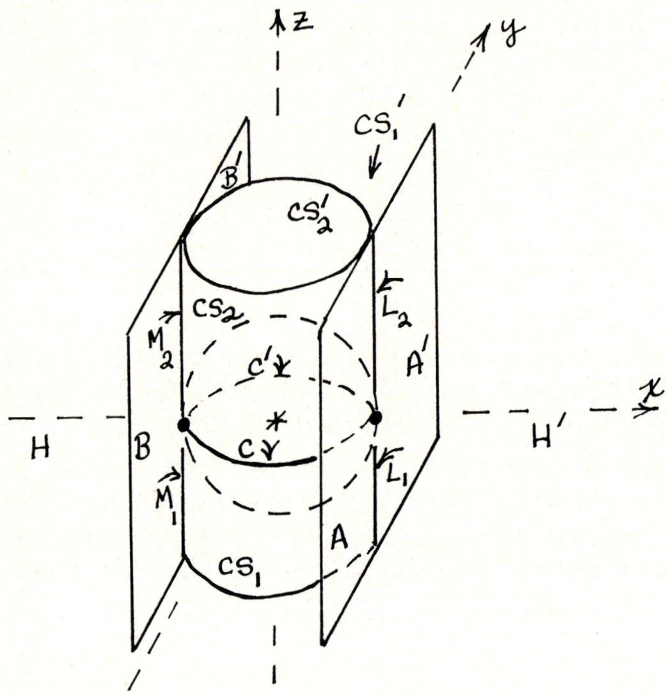

Fig. 9. Refinement of initial decomposition of E^3.

Fig. 10. Decomposition for E^2 obtained by lifting cells of E^1 and slicing the cylinders obtained by the real roots of $Q_x(y)$. There are 13 cells.

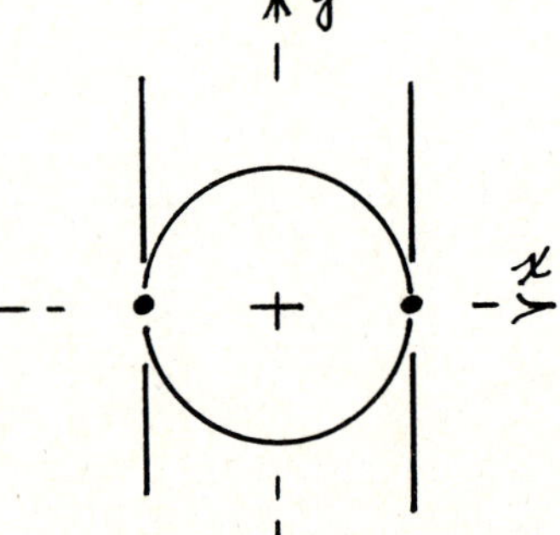

representing the cell of the desired configuration.

Of course carrying out this plan in general requires careful attention to many details. For example, how are cell adjacencies to be determined? How can cells be constructed so that it is easy to specify motions inside the cells as well as from one cell to a neighboring cell? How can the nature of cell adjacencies be controlled? For example, in the case of the regions of the partition of E^2 shown in Fig. 8, moving directly from R_1 to R_3 requires knowing what portions of the boundary of R_1 can be used. Also, specifying how to move from one point in R_1 to another involves knowing how to avoid the hole R_2.

The Schwartz and Sharir algorithm handles these problems in the general situation by using Collins' '75 method of constructing a cell decomposition recursively. This means that for a particular instance of the Mover's Problem such as the simple one we are presently considering, the algorithm produces a much more complicated decomposition than ones that a common sense, ad hoc approach would suggest. In order to convince the reader that this seemingly complex decomposition is in fact very natural, we are about to suggest a second cell decomposition for the present problem, point out that it has some features that are useful for decompositions in general, and then point out some of its short comings. This will motivate an approximate description of the actual Collins decomposition which is very complicated even for this simple case. Then we will conclude the section by sketching how the cell decompositions are obtained in general. The reader is urged to consider the problem of constructing desirable cell decompositions in general while reading the following discussion.

Returning now to the cell decomposition of E^3 that was initially proposed, notice that the decomposition has a nice property: C_2, the surface of the sphere, belongs entirely to the boundary of C_1 and C_3 (in fact, it is equal to the boundaries). Hence we need not worry about legal points of entry from one cell to another. This property also holds in general for cell decompositions used by Schwartz and Sharir '83b: each cell is an open set whose boundary is a union of entire cells.

Unfortunately, moving from one point of C_3 (the complement of the closed, solid sphere) to another point in C_3 requires knowing how to avoid the hole (this is essentially the original problem, after all). To facilitate this, we might refine the original decomposition as in Fig. 9. This decomposition is still not the decomposition produced by the Schwartz and Sharir algorithm, but it shares some important properties with the true decomposition and will serve to illustrate some fundamental ideas. In the new decomposition, E^3 has been partitioned into five 3-

dimensional cells, ten 2-dimensional cells, six 1-dimensional cells and two points. The 3-dimensional cells are the half-spaces H and H', the inside of the cylinder below the sphere, the inside of the sphere, and the inside of the cylinder above the sphere. The 2-dimensional cells are the open half-planes A, A', B and B', and the open half-cylinder surfaces CS_1, CS_1', CS_2, CS_2' and the upper and lower half-sphere surfaces. The 1-dimensional cells are the half-lines L_1 and L_2 in the boundary of A and A', the half-lines M_1 and M_2 in the boundary of B and B', and the open half-circles C and C' lying in the intersection of the xy-plane with the cylinder surface. The 0-dimensional cells are the points separating L_1 and L_2 and separating M_1 and M_2. Note that each cell boundary consists of a union of entire cells, as before. The boundary of the half-plane A, for example, is L_1, L_2, and a point. The boundary of the interior of the sphere is the union of the upper and lower sphere surfaces, the front and rear open half-circles, and two points. This is a *refinement* of the original decomposition, so each defining formula is true throughout each cell if it is true anywhere in the cell. Of course we have generated many new cells, but we have gained an important general advantage: no cells have holes in them. Also, each cell is either unbounded in the +z (-z) direction or is bounded in that direction by a single cell (possibly itself). Hence to specify how to move within a cell, we can specify the projection of a path on the xy-plane and then specify the z coordinates for the path by linearly interpolating between starting and final values, using the proportion of distance between the boundaries in the +z and -z directions or, in the case of a semi-infinite cell, adding (subtracting) an appropriate amount to the z values of points on the boundary below (above) the path.

Of course there remains the problem of specifying the path projection in the xy-plane. This raises the recursive nature of the Collins cell decomposition strategy. Instead of refining a decomposition of E^3 as we have been doing, it first decomposes E^1, extends this decomposition by "lifting" it into a decomposition of E^2 that has nice properties, and finally it lifts the decomposition of E^2 to a decomposition of E^3. To obtain the requirements that the E^1 decomposition should have, however, calls for first finding the requirements for an E^2 decomposition that will lift to an E^3 decomposition, because the E^2 decomposition requirements impose requirements on the E^1 decomposition. In other words, the procedure begins with requirements for the cell decomposition of the full-dimensional space (E^3 in our case) and uses these eventually to compute requirements for the decomposition of E^1. Then it constructs a decomposition for E^1, using this as a foundation for the E^2 cell decomposition (by "lifting" E^1 cells and dividing the resulting "towers", called "cylinders", into pieces), and so forth, eventually recursively constructing the desired cell decomposition of the full dimensional space at the last step.

To see how this idea works (while still avoiding the complexities of the true Collins decomposition) suppose that we ask for a decomposition of E^2 that we can use to generate a decomposition of E^3. To see how we might obtain the decomposition of E^3 shown in Fig. 9 from the decomposition of E^2 shown in Fig. 10, we observe that this latter decomposition has the following property: as (x, y) ranges over any cell in the E^2 decomposition, the number of real roots of $P_{xy}(z) = -z^2+(4-x^2-y^2)$ remains constant. (Here $P_{xy}(z)$ is a polynomial in z whose coefficients depend on its index x, y. Hence every point (x, y) in E^2 gives rise to a polynomial $P_{xy}(z)$ of degree 2 in z.) For example, for any choice of (x, y) lying inside the circle, $P_{xy}(z)$ has two distinct real roots. For each choice of (x, y) in the open half-planes H and H' and on the open half-lines, $P_{xy}(z)$ has no real roots. On the half-circles and on the points, $P_{xy}(z)$ has exactly 1 distinct real root. Given this decomposition, we can increase the dimension of any cell C by "lifting" it. The point cells become lines parallel to the z axis, the half-lines become half-planes perpendicular to the xy-plane, the open half-circles become open half-cylinder surfaces, the open disc becomes an open cylinder and so forth. These lifted cells are called "cylinders" and are said to have the cells of the decomposition of the plane as a "base". However, they are not quite the cells we seek. To construct the decomposition for E^3, we slice each "cylinder" by the two roots of $P_{xy}(z)$ as (x, y) varies over the "base" of the cylinder. For example, the "cylinder" $\{(2,0,z) \mid z \text{ real}\}$ formed by "lifting" the point cell (2,0) is divided into three pieces by the single real root $z=0$ of $P_{2,0}(z)$: two open half-lines $\{(2,0,z) \mid z>0\}$ and $\{(2,0,z) \mid z<0\}$ and one point (2,0,0). The cylindrical "cylinder" based on the open disc is divided into five cells by the two real roots of $P_{xy}(z)$ as (x, y) ranges over the disc, namely:

$$\{(x, y, z) \mid x^2+y^2<4,\ x^2+y^2+z^2>4,\ z>0\}$$

$$\{(x, y, z) \mid x^2+y^2<4,\ x^2+y^2+z^2=4,\ z>0\}$$

$$\{(x, y, z) \mid x^2+y^2<4,\ x^2+y^2+z^2<4\}$$

$$\{(x, y, z) \mid x^2+y^2<4,\ x^2+y^2+z^2=4,\ z<0\}$$

$$\text{and } \{(x, y, z) \mid x^2+y^2<4,\ x^2+y^2+z^2>4,\ z<0\}.$$

The advantage of starting with a decomposition of E^2 within whose cells the number of distinct real roots of $P_{xy}(z)$ is constant is that the roots of $P_{xy}(z)$ slice the cylinders obtained by the lifting operation into cells in a simple way. The cells have no holes in them if the base cells do not. We can move about in a cell by interpolating z values and planning a path projection on the base cell. Furthermore, we are guaranteed that in each cell of the E^3 decomposition, $P(x,y,z)$ is either greater than 0, equal to 0, or less than 0 throughout the cell. Also, note that if we already know which cells are adjacent to one another in the decomposition of E^2, it is easy to use this information to find what the cell adjacencies are for the E^3 decomposition. Another important feature is that any representative point (x,y) in a base cell in the E^2 decomposition can have appropriate z values adjoined to it to form representative points of the form (x,y,z) for the E^3 decomposition cells having that base. For example, if (0,0) represents the disc, then (0,0,0) can represent the sphere ($z=0$ is chosen because it is the average of the two roots of $P_{0,0}(z)$); (0,0,2) can represent the open half-sphere above the xy-plane.

To get the decomposition of E^2, we might have started with the decomposition $(-\infty,-2)\cup[-2]\cup(-2,2)\cup[2]\cup(2,\infty)$ of E^1, using the roots of the polynomial $P_x(y)=-y^2+4-x^2$ to slice the cylinders: $P_x(y)$ has no real roots for x in $(-\infty,-2)$, one real root for $x=-2$ etc.

At this point we can see how having a polynomial $P(x_1,...,x_n,x_{n+1})$ that we regard as an indexed family of polynomials $P_{x1,...,xn}(x_{n+1})$ in x_{n+1}, and a base decomposition of E^n that behaves nicely with respect to P allows us to construct a decomposition of E^{n+1} such that in each cell, P is >0, $=0$, or <0 throughout. So far, we have guessed decompositions by using hindsight, by guessing a decomposition of E^3 and noticing that it could be regarded as based in E^2 and so back to E^1; we would like a recursive algorithm computing cell decomposition in general. At this point we show roughly what the algebraic cell decomposition algorithm does in this example, where $P(x,y,z)=-x^2-y^2-z^2+4$. Although some details will be omitted even for the example, the discussion will give a good idea of how the general algorithm works.

First, recall that the number of distinct roots (real or complex) of a polynomial P in a single variable can be found by the formula $k-m$, where k is the degree of P and m is the degree of the greatest common divisor of P and its derivative P'. If the number of distinct real and complex roots of $P_{x1,...,xn}(x_{n+1})$ is constant as $(x_1,...,x_n)$ varies throughout a region in E^n, then the number of distinct real roots is also constant there. Hence one way to devise a cell decomposition for E^2 that can serve as a base for an E^3 cell decomposition is to find a way to insure that the degree of $P_{x1,...,xn}(x_{n+1})$ stays constant on each E^n cell, and that the degree of the GCD of P and P' is also constant on each cell.

In the example, $P_{xy}(z)$ has degree $k=2$ for all values of (x,y). In general, we can keep the degree of a polynomial constant by insisting that the coefficients of the powers of x_{n+1} are >0, $=0$, or <0 on each cell, where the coefficients are polynomials in $x_1,\ldots,x_n$. (Schwartz and Sharir 83b describe how to handle the situation where a polynomial becomes identically 0.)

The derivative $P_{xy}'(z)$ of $P_{xy}(z)$ with respect to z is $-2z$, and the greatest common divisor $GCD(P,P')$ is 1 (so $m=0$) except when $x^2+y^2=4$: then it is z (and $m=1$). Consequently the cell decomposition for E^2 needs to have the property that throughout each cell, the polynomial $Q(x,y)=-x^2-y^2+4$ is >0, $=0$, or <0. Thus we have moved from E^3 to E^2, replacing $P(x,y,z)$ by $Q(x,y)$. To continue, note that $Q_x(y)$ has degree 2 for all values of x, so now $k=2$. $Q_x'(y)=-2y$ has degree 1, and the $GCD(Q,Q')$ is 1 $(m=0)$ except when $4-x=0$, when it is y $(m=1)$. Consequently, we need a decomposition for E^1 that satisfies $R(x)=4-x>0$, $=0$, or <0 throughout each cell. Choosing cells $(-\infty,-2)$,$[-2]$,$(-2,2)$,$[2]$,$(2,\infty)$ gives this decomposition. To obtain the decomposition we need for E^2, we "lift" these base cells by using the roots of $Q_x(y)$ (we know that on each cell of E^1, the roots of $Q_x(y)$ will slice the "cylinder" having that base into pieces because the roots cannot cross as x varies inside a cell of E^1). This generates the cells shown in Fig. 10. By construction we know already that on these cells, $P_{xy}(z)$ has a constant number of distinct real roots. Therefore, we can use these roots to slice the cylinders over these base cells, obtaining the decomposition of E^3 shown in Fig. 9.

In general, we must be able to handle the degree variations of the polynomials $P_{x1,\ldots,xn}(x_{n+1})$ and $GCD(P,P')$ as $x_1,\ldots,x_n$ varies over E^n (here n ranges from 1 less than the full dimension of Cspace down to 1). We have already pointed out conditions on a cell decomposition of E^n that guarantee that the degree of P stays constant in each cell. To see how to insure that the degree of the GCD does not change, we consider the following polynomial equation:

$$P(z)U_j(z) - P'(z)V_j(z) = C_j(z).$$

In this equation,

i) P(z) is a given polynomial with z-derivative $P'(z)$. The index indicating the variables that may appear in the coefficients of P and P' is not shown, but it is assumed to have a fixed but arbitrary value;

ii) $P(z)$ and $P'(z)$ are known to have at least j common factors, so $m \geq j$, where $m=\deg(GCD)$;

iii) $U_j(z)$, $V_j(z)$ and $C_j(z)$ are unknown polynomials such that

$\deg(U_j) \le \deg(P') - j - 1$, $\deg(V_j) \le \deg(P) - j - 1$, and $\deg(C_j) \le j - 1$. ($C_j = 0$ if $j = 0$)

Since $\deg(C_j) \le j-1$ if $j \ne 0$, solutions to the polynomial equation correspond to solutions to the homogeneous system of equations that equate to 0 the coefficients of the terms of degree j or greater on the left side of the polynomial equation. This system has unknowns (the coefficients of $U_j(z)$ and $V_j(z)$) equal in number to $\deg(P)+\deg(P')-2j$, and it has the same number of equations. The determinant of the matrix M_j of its coefficients is non-zero if, and only if, the trivial solution $U_j(z) = V_j(z) = C_j(z) = 0$ is unique.

If $P(z)$ and $P'(z)$ have a $(j+1)$st common factor, then setting $V_j(z)$ equal to the result of dividing $P(z)$ by $j+1$ common factors and setting $U_j(z)$ equal to the result of dividing P' by the same common factors gives a non-zero solution. On the other hand, if P and P' have *exactly* j common factors, then dividing the polynomial equation by the corresponding common factors shows that the zero solution must be unique. Therefore, given that P and P' have at least j common factors, they in fact have at least $j+1$ common factors if, and only if, the determinant of matrix M_j is 0. Thus m, the degree of the GCD, is equal to the smallest value of j for which the determinant of M_j is non-zero.

Now we can specify a set S_n of polynomials in the index variables $x_1, \ldots, x_n$ of P such that $P_{x1,\ldots,xn}(x_{n+1})$ has a constant number of distinct real roots as the index varies over any connected cell of E^n in which the polynomials in S_n are each >0, $=0$, or <0. This will reduce the original problem to one of the same form for a new set of polynomials in fewer variables, making a recursive solution possible (E^1 will be decomposed into cells by the roots of the polynomials in its associated set S_1). S_n should contain each coefficient polynomial of $P(x_{n+1})$ to insure that the degree of $P(x_{n+1})$ does not change as its index varies over an E^n cell. To make sure we are computing the correct polynomials $D_j(P, P')$, where P may have less than its full degree k for some index values, we conservatively put into S_n the D_j's based not only on P but also on all polynomials obtained from P by deleting one or more leading high order terms. In other words, if

$$P_{x1,\ldots,xn}(x_{n+1}) = \sum_{i=0}^{k} a_i(x_1,\ldots,x_n)x_{n+1}^i$$

then we define

$$T_r(x_{n+1}) = \sum_{i=0}^{r} a_i(x_1,\ldots,x_n)x_{n+1}^i \quad \text{for } r=2,\ldots,k$$

and put the polynomials $D_j(T_r, T_r')$ for $r=2,\ldots,k$ and $j=0,\ldots,r-2$ together with the polynomials $a_i(x_1,\ldots,x_n)$ for $i=1,\ldots,k$ into S_n. (We omit $a_0(x_1,\ldots,x_n)$ from S_n as it can be assumed that P is never identically 0 for any choice of index.) Note that in general, the starting set S_{n+1} consists of several polynomials P.

In our example, the procedure goes as follows. The initial set S_3 of polynomials that must be >0, $=0$, or <0 on cells of E^3 is $\{P(x,y,z)=-x^2-y^2-z^2+4\}$. Since the coefficient polynomial $a_2(x,y)$ is constant, the degree of $P_{xy}(z)$ is always 2, so the coefficient polynomials do not need to be included in S_2. Hence S_2 consists entirely of $D_0(T_2, T_2')$ where $T_2(z)=P_{xy}(z)$. To determine whether $P_{xy}(z)$ and $P_{xy}'(z)$ have a common factor, we write

$$[-z^2+(4-x^2-y^2)]U_0(z) - [-2z]V_0(z)=0$$

where $U_0(z)$ is a constant u_0 and $V_0(z)=v_1z+v_0$. This equation has a non-zero solution if, and only if, P and P' have a common factor. The system of linear equations that match the coefficients is

$$\begin{bmatrix} -1 & 2 & 0 \\ 0 & 0 & 2 \\ 4-x^2-y^2 & 0 & 0 \end{bmatrix} \begin{bmatrix} u_0 \\ v_1 \\ v_0 \end{bmatrix} = \begin{bmatrix} 0 \\ 0 \\ 0 \end{bmatrix}$$

The determinant of the square matrix is the polynomial $D_0(T_2, T_2')=4(4-x^2-y^2)$. Hence P and P' have a common root if, and only if, $(4-x^2-y^2)=0$, so we let $S_2=\{-y^2+(4-x^2)\}$. To find S_1, we note that $-y^2+(4-x^2)$ has degree 2 in y no matter what value x has, so S_1 does not need coefficient polynomials. Hence S_1 consists of a $D_0(T_2, T_2')$ based on the polynomial in S_1. Now we consider $[-y^2+(4-x^2)]u_0 - [-2y][v_1y+v_0]=0$ and note that $-y^2+(4-x^2)$ has a factor in common with its y-derivative if, and only if, the determinant

$$\begin{vmatrix} -1 & 2 & 0 \\ 0 & 0 & 2 \\ (4-x^2) & 0 & 0 \end{vmatrix} = 4(4-x^2) \text{ is zero.}$$

Hence we set $S_1=\{4-x^2\}$. The real roots of the polynomial in S_1 give the cell decomposition for E^1 that we "guessed" previously, and this cell decomposition lifts to E^2 and then to E^3 as described earlier and as Fig.s 10 and 9, respectively, show.

There are many issues we have not discussed, such as: how cell adjacencies are determined; how representative points in cells can be extended, starting with representatives for the E^1 decomposition, to representatives for the E^{n+1} decomposition; what to do if a polynomial becomes identically 0 for some values of its index variables; how to represent orientations in configuration space so that these techniques are valid, as the identification of 0 degrees with 360 degrees introduces some technical problems; how coefficients of the polynomials that arise can be represented in the computer. The reader is referred to Schwartz and Sharir '83b for details.

At least we can now understand the expression that appears in the running time of the algorithm, which the authors say is "polynomial in the geometric complexity of the problem for each fixed number of degrees of freedom, but exponential in the number of degrees of freedom." In particular, we need to understand what "geometric complexity" means. The data (which may be approximations to real objects or which may exactly describe ideal objects, depending on the algorithm user's point of view) are to be given as a set of multivariable polynomials with rational coefficients. The user must also specify how to combine these expressions with logical connectives to form a description of the set of forbidden configurations XP. Collins showed that the number of cells in a decomposition is $O((2n)^{3^{r+1}} m^{2^r})$ where m is the number of polynomials defining XP, n is the maximum degree of any one such polynomial, and r is related to the number of degrees of freedom of the system (the number of variables needed to describe the position and orientation of all the moving objects).

An important property of the algorithm is that it is complete. That is, once given the data, it always finds a path if one exists; it does not make use of a user-defined resolution. It is able to do this by performing exact calculations with algebraic numbers. Schwartz and Sharir discuss the problem of growth in the size of the integers involved in doing the computations, and they state that the extra cost incurred by taking into account the size of the integers needed (as opposed to assigning every elementary operation on integers a nominal, unit cost) still leads to an algorithm that is of order polynomial in the degree of the polynomials and the size of their coefficients when the number of degrees of freedom is fixed.

3. Motion Coordination Problems

A special type of Mover's Problem that has started to receive considerable attention is the problem of choreographing the motion of several disjoint bodies so that they move from an initial configuration to a desired final one without collision with each other or with obstacles. This can be viewed as a special case of the

Mover's Problem in which the body being moved consists of several disjoint pieces whose relative motions are not restricted except to require that they do not collide. Many authors refer to this as a *motion coordination* problem.

Motion coordination problems are related to *separability problems*, which are surveyed in Toussaint '85. In separability problems, however, the goal is to spread objects far apart, and a specific final configuration is not specified. Also, separability problems do not involve fixed obstacles, and pieces are not typically required to lie in some bounded, closed restricting region. Often, requirements are put on the type of motion (e.g., translation) that is to be made. Obviously motion coordination problems in which there are no obstacles or confining regions can be reduced to a pair of separability problems in which the pieces of the initial and final configurations of the coordination problem are to be spread apart.

We now describe several motion coordination results.

Schwartz and Sharir '83c: This is an early paper in motion coordination. It concerns the problem of coordinating the motion of several discs $B_1,...,B_k$ moving in the plane. Fig. 11, which is taken from the paper, shows an instance of this problem for two discs. Although this problem can be solved in polynomial time for a fixed number of discs by the techniques of Schwartz and Sharir '83b, solutions obtained in that manner are slow in practice. Consequently, an improved method for the special case of discs moving in the plane was sought. The authors point out that Moravec, for example, suggests as a heuristic for planning the motion of objects, enclosing the objects in circles and coordinating the movements of the circles. (See Moravec '83.) Only in case this method cannot find a path does one use better object approximations and finer analysis. Hence disc coordination is an interesting problem in its own right.

The authors discuss the motion coordination problem for two discs that must avoid colliding with "walls" made up of line segments that can be partitioned into a disjoint collection of simple polygonal closed curves. It is assumed that the free space (the set of points not belonging to obstacles) available to the discs is open and has bounded closure. Additional assumptions are also made to ensure that the free space has no isolated points or "slits". For boundaries composed of n straight line segments, the authors give an $O(n^3)$ algorithm for two discs and an $O(n^{13})$ algorithm for three discs. Since their techniques are recursive, they point out that polynomial algorithms for an arbitrary but fixed number of discs can be obtained in a similar fashion.

For two discs in the plane, Cspace is 4-dimensional. The set of legal configurations $\{x_1,y_1,x_2,y_2 \mid B_1$ has center (x_1,y_1), B_2 has center (x_2,y_2), B_1 and

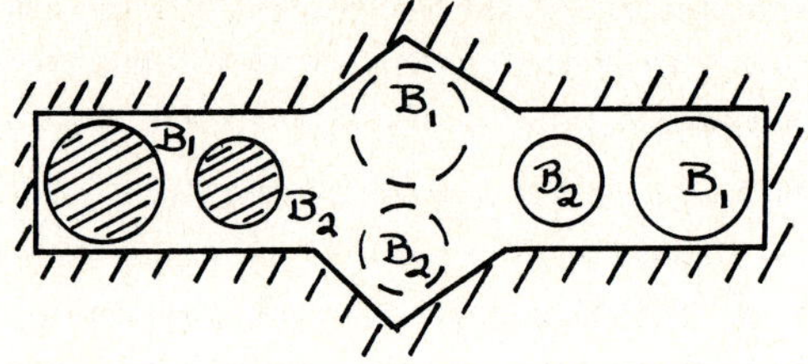

Fig. 11. Motion coordination for two discs (from Schwartz and Sharir '83c).

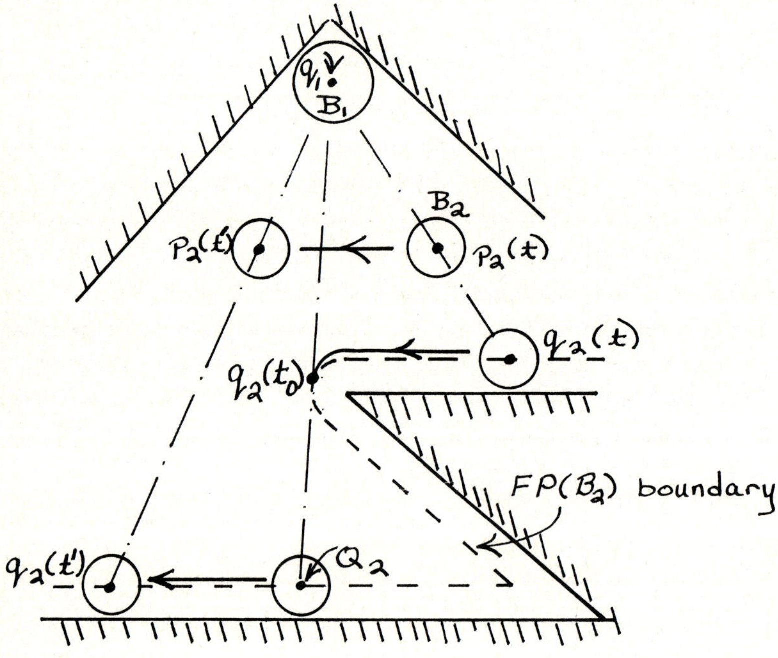

Fig. 12. The motion $(p_1(t), p_2(t))$ is retracted to a motion $(q_1(t), q_2(t))$ that has a discontinuity at $t = t_0$. Here B_1 stays fixed, $p_1(t) = q_1(t) = constant$. (Based on Yap '84).

B_2 do not intersect } is projected into the 2-dimensional Cspace for B_1 alone. For each fixed point (x_1,y_1) in this 2-dimensional space, the legal positions for the center of B_2 form a collection of connected components. For certain critical positions of B_1, the nature of these associated connected components changes. In fact, critical positions for B_1 determine critical curves that divide the legal positions for B_1 into noncritical regions. A graph is constructed that represents the noncritical regions, the associated connected components for B_2, and the possible ways of crossing from one region into another. Then disc coordination is reduced to a search for a path in this connectivity graph.

The case of three discs is handled by projecting the legal configurations in 6-dimensional space to the 2-dimensional space of configurations for one disc. Then for each legal (x_1,y_1) value, the components of the set of legal positions (x_2,y_2,x_3,y_3) for the other two discs are considered.

Yap '84: This is a technical report that proves that by using a "retraction" approach, the results cited above for two and three discs can be improved to $O(n^2)$ and $O(n^3)$. The conjecture is made that $O(n^k)$ is sufficient for planning the motion of k open discs in the plane. (Since there are $2k$ degrees of freedom, one might expect a lower bound of at least $\Omega(n^{2k})$.) Here it is assumed that the set of legal or "free" positions $FP(B_i)$ for the center of each open disc B_i is closed and bounded and that each connected component of the boundary of $FP(B_i)$ is a simple closed path. Such a path consists of straight line segments, which arise from displacing edges of polygons by the radius of the disc B_i, and circular arcs, which are centered at the vertices of the polygons.

The general idea of the "retraction" method in this situation is as follows. For each legal placement of two discs with centers at points p_1 and p_2, the line segment p_1p_2 through their centers can be extended in each direction until it meets the boundaries of the $FP(B_i)$ (intersections of the original line segment with obstacles are ignored). It is as though a mutually repulsive force pushed the discs away from each other until they collided with the walls. Consequently a legal motion $(p_1(t),p_2(t))$ of the discs gives rise to a moving pair of image points $(q_1(t),q_2(t))$, where $q_i(t)$ moves along the boundary of $FP(B_i)$ (i.e., it corresponds to the disc B_i sliding along an obstacle edge or rotating around a corner). Such a pair of points moving along the boundaries of the $FP(B_i)$ is called "stable". Of course the "motion" $(q_1(t),q_2(t))$ may be discontinuous, as Fig. 12 shows. As long as $q_1(t)$ and $q_2(t)$ are moving along some fixed pair of line or arc segments from the boundaries of $FP(B_1)$ and $FP(B_2)$, their motion is said to be "elementary". The idea is to push the discs to the walls and then to piece together elementary motions to reach the final configuration. The paper shows that this particular strategy works whenever the coordination problem has a solution.

The strategy for piecing together elementary motions involves finding the *critical values* for stable positions. One example of a critical value is a pair (q_1,q_2), where q_1 is at an end point of a line or arc segment constituent of the boundary of $FP(B_1)$ and the line through q_1 and q_2 is tangent to an arc segment in the boundary of $FP(B_2)$ at q_2. Such a critical placement has associated with it a complementary placement. For example, in Fig. 12, suppose B_1 stays fixed $(p_1(t)=q_1(t)=q_1)$ and that the center of B_2 follows the indicated path, so that $(q_1,q_2(t))$ arrives at the middle, critical position $(q_1,q_2(t_0))$, at time t_0. Note that the limit as t approaches t_0 from *above* is (q_1,Q_2). This is the associated *complementary* position of $(q_1,q_2(t_0))$. A graph is constructed whose vertices are critical positions and their complements. (Fig. 12, taken from the report, shows just one type of critical position and stable motion discontinuity.) Each edge in the graph indicates that the two positions corresponding to its endpoints can be reached one from the other by elementary motions or by "jumps" such as the one shown in the Fig. 12.

The report uses the Voronoi diagram, citing the $O(n \log n)$ algorithm of Kirkpatrick '79 to compute the boundaries of each $FP(B_i)$. The *generalized* Voronoi diagram in Cspace used by O'Dúnlaing, Sharir and Yap '83 for planning the motion of a line segment is not needed to construct the graph of critical placements and their complements.

4. Complexity Results

The Schwartz and Sharir '83b general algorithm for the Mover's Problem is polynomial for a fixed number of degrees of freedom and hence puts many motion problems in the family P of problems solvable in polynomial time. (See Garey and Johnson '79 for complexity terminology.) Since the algorithm is exponential in the number of degrees of freedom, however, it is interesting to study motion problems in which the number of degrees of freedom is an input variable in order to understand how problem difficulty is related to the nature of the objects, obstacles, and type of motion allowed.

The first motion complexity result was obtained by Reif '79, who showed the PSPACE-hardness of determining whether a certain bug-like object with an arbitrary number of legs hinged to a spine could crawl through holes in a complicated swiss-cheese-like obstacle.

Hopcroft, Joseph and Whitesides '84 show the PSPACE-hardness of determining whether a given joint of a planar linkage can move to a specified position. A planar linkage consists of rigid rods that are free to rotate about joints at their endpoints. Each joint connects two or more rods, and some joints are fastened to the plane. (The environment is obstacle-free.)

Spirakis and Yap '84, who refer to the motion coordination problem as the *Many (Moving) Bodies Problem*, show by a reduction from 3-PARTITION that the problem of coordinating the motion of discs in the plane is NP-hard in the "strong sense" They assume that the obstacles are bounded by polygons and that a problem instance consists of data giving the radii for an arbitrary number of discs, a description of the polygons bounding the obstacles, and sets of initial and final placements for the discs.

Hopcroft, Schwartz and Sharir '84 assert that the motion coordination problem for 2-dimensional rectangles confined inside a bounding rectangular box is PSPACE-hard by showing that arbitrary linear bounded automata can be simulated by the motions of such rectangles. Then Hopcroft and Wilfong '84b put this problem in PSPACE, making the problem PSPACE-complete, by applying the strategy of following the features of Cspace obstacle boundaries to find legal paths. Chazelle, Ottoman, Soisalon-Soininen and Wood '83 study a family of sliding block puzzles in which the pieces are planar, rectilinear (but not necessarily rectangular) "blocks". There are no obstacles. The problem is to spread the pieces apart by a finite sequence of *translations* to the east, west, north and south. It is shown that the problem is decidable but NP-hard, and that some puzzles require exponentially many moves.

Hopcroft, Joseph and Whitesides '82a, '85 show that the problem of folding a carpenter's rule is NP-complete although solvable in pseudo-polynomial time by dynamic programming. The data consists of integers $l_1, \ldots, l_n$, giving the lengths of rule links that are allowed to rotate about their endpoints, and an integer k, giving the maximum desired folded length of the rule. Hopcroft, Joseph and Whitesides '82a,'85 also give an example of a Mover's Problem that is solvable in polynomial time in spite of the fact that the number of degrees of freedom is arbitrary. The problem is to decide whether an n-link "arm", fastened at one end to the plane and confined inside a circle, can be moved from one given configuration to another. An arm consists of a set of links that are hinged together by rotary joints at their endpoints to form a linear sequence. Links are free to pass over one another. There are no obstacles in the problem considered other than the confining circle.

Hopcroft, Joseph and Whitesides '82b,'85 show that the boundary of the set of points that can be reached by a given joint of an arm confined inside a circle consists of arcs of circles, and that the number of circles needed to cover the boundary is at most 148, *no matter how many links are in the arm.*

5. Motion Related Problem Areas

The purpose of this section is to point out that the Mover's Problem, or FINDPATH, belongs to a hierarchy of interrelated motion problems in robotics, and also that there are several other problem areas, inside and outside of robotics, that are relevant to, or concerned with, motion in some way.

Several references that discuss motion planning in the larger context of robotics are: Brady '84, who discusses the relationship between artificial intelligence and robotics; Birk and Kelley '81, who summarize the conclusions of a workshop held to identify the scientific knowledge needed to provide a foundation for robotics; the Proceedings of the IEEE Special Issue on Robotics edited by Reddy '83; Brady, Hollerbach, Johnson, Lozano-Pérez and Mason '82, which is a collected volume of several important papers on robotics introduced by supplementary chapters; and Schwartz and Sharir '83e, who identify mathematical problems in robotics.

Within the context of the planning and control of the motion of robot manipulators, the FINDPATH problem is a major aspect of what has been called *task planning.* The goal of task planning is to enable a robot user to specify the desired effects of a robot on the objects it is manipulating without having to specify the details of the robot's motions. Achieving this goal is an essential prerequisite for the implementation of truly high-level robot programming languages. Task planning involves not only finding solutions to FINDPATH but also finding solutions to a variety of additional problems. One of these is to plan compliant motions of objects in contact, using data from sensing devices. Another problem is to find appropriate grasp positions for the end effector of the robot on the object it is manipulating, taking into consideration, say, that the object might be a pin that is to be inserted into a hole.

How objects, including the manipulator itself, are modeled geometrically (by edge descriptions, surface descriptions, or as combinations of primitive solids, for example) is another important issue relevant to task planning. Of course geometric information alone is insufficient for task planning needs. It is necessary to have, for example, information about the characteristics of joint actuators and about the masses and moments of inertia of objects, including the manipulator.

Boyse '79 emphasizes the importance of having object representation schemes that support motion planning. Eastman and Preiss '84 and Requicha '80 survey object representation schemes. Lee and Requicha '82a,b discuss algorithms for computing properties of solids based on their representation. Tilove '84 presents algorithms that operate on Constructive Solid Geometry representations of objects to determine equality and to check for intersection.

An overview of the subject of task planning is given by Lozano-Pérez '82. This is a chapter in Brady et al. '82 that introduces two additional papers on that subject. One of these, reprinted from Lozano-Pérez '81, includes a review of Cspace techniques and applications as well as references to model-based manipulator systems. Lozano-Pérez '83b places task planning in the context of robot programming languages.

Task planning itself can be placed in the larger context of robot motion planning and control. Brady et al. '82 contains, in addition to the section on task planning, sections on dynamics, feedback control, trajectory planning, and a separate section on compliance. (Trajectory planning involves specifying the desired positions, velocities and accelerations of a robot at a sequence of moments in time. Dynamics, fundamentally concerned with torques and forces, includes the "inverse dynamics" problem, which is to specify what torques to apply at joints in order to achieve a specified trajectory for the end effector. Feedback control involves using the measured difference between actual motion and planned motion to reduce the error.) The book, which brings together many important papers, provides useful introductory material for each section together with an extensive bibliography. It also provides in the first chapter a discussion of motion planning and control for a planar, 2-link manipulator with rotary joints. This simple example illustrates many of the main concepts and ideas that appear in later sections and shows how the various problem areas of robotics interact.

An important book devoted to providing an introduction to the theory behind robot programming and control systems is Paul '81.

An underlying assumption behind viewing motion planning as a Mover's Problem is that an accurate model of the environment is available as input data for path finding algorithms; a further assumption is that a motion, once specified, can be carried out perfectly. In practice, manipulator uncertainty is a serious problem. Also, errors may arise from unforeseen events as well as from the positioning limitations of the robot and from the inaccuracies in the environment model. Consequently sensing the environment is essential, even when the environment is in principle well known. In addition to the extensive research done in computer vision (see Ballard and Brown '82, Kanade '83) there is emerging a body of work on tactile sensing (see Harmon '82 for a survey of a large number of tactile sensing applications). Typically, vision strategies rely on large amounts of data that can support some kind of statistical pattern recognition approach. This holds true for much work on tactile recognition strategies also.

Gaston and Lozano-Pérez '84 and Grimson and Lozano-Pérez '83 study the problem of identifying and locating objects using sparse range or tactile data that provide approximate surface points and their normals. They assume that the objects belong to a set of known objects whose models are available; the objects in these two papers are (not necessarily convex) polyhedra with 3 and 6 degrees of positional freedom, respectively. These papers focus on the problem of generating feasible interpretations of the data and testing these interpretations for consistency with the object models.

Schwartz and Sharir '84 and Bernstein '84 are technical reports that suggest tactile probing algorithms that involve choosing a next probe direction based on previous probe contact locations. (Surface normals are not used.) Schwartz and Sharir '84 show that the shape and position of a convex polygon known to contain the origin can be obtained by at most 3n tactile probes, where n is the number of edges of the polygon. Bernstein '84 shows that if the object belongs to a finite set of polygons, then $2n+k$ (k constant) probes suffice.

Locating and recognizing objects is but one of several uses of sensing in motion planning. Another example is "fine motion" strategy: sensory information such as force data is essential for directing the motion of an object such as a pin to be fitted into a hole when it is very close to its goal position. See Lozano-Pérez, Mason and Taylor '84 for work on fine motion strategy.

Computing the set of points that a robot manipulator can sweep out during its operation is important both for safety reasons and for planning the set-up of jobs to be performed by the robot. Hansen, Gupta and Kazerounian '83 study this problem. There are a variety of related problems, such as determining the set of points the end effector can reach, perhaps in specified ways (approach angles might be given, for example).

Another problem area related to task planning is that of finding a place to put down a held object. Lozano-Pérez '83a calls this the FINDSPACE problem. It is related to bin packing and pattern layout problems. It is interesting to note that in this latter context, the method of Cspace has been used (Lozano-Pérez '81 cites several references.)

Additional motion problem areas include the following.

Dynamic scene analysis: The problem is to determine the motion of a body moving through a stationary environment by using images seen by a camera attached to the body, or to detect and interpret movement in the environment of a camera from the images it sees. (See for example Ballard and Brown '82, Bruss and Horn '83, Jerian and Jain '84.)

Plan generation: Early work in artificial intelligence considered the high level planning of motion in "blocks world" or similarly constrained environments. Problems considered were of the following type. Given a description of the environment in terms of facts (such as "the green block is on top of the blue block", "the robot hand is empty") and rules (such as "if the hand is empty and block X is clear then the robot can pick up block X"), the problem is to determine a sequence of moves that arrives at some specified goal state (such as "the red block is on top of the green block"). See Nilsson '80.

The position-location problem: The problem is to compute the coordinates of a set of points in space from a set of distances measured between the points. The problem has a finite set of solutions modulo translations and rotations only when it corresponds to a rigid structure, and is analogous to the problem of constructing pin-jointed structures from rigid bars. Hence as Yemini '79 points out, position location problems are closely related to *structural rigidity* (see Whitely '84).

Representation of moving objects: Klinger and Dyer '75 is an early paper calling for translation and rotation algorithms for decomposition representations for moving image regions. Boyse '79 outlines the need for object representation schemes that support motion planning. Ahuja and Nash '84 give an algorithm for updating the octree representation of an object translating through space.

Dynamic or dynamical point problems: Given a set of points in space whose trajectories (path and velocity) are known, the problem is to find various properties of the point set. For example, one might want to predict collisions, or to find closest or farthest pairs of points. There are many variations. See Ottmann and Wood '84 and Atallah '83.

6. Conclusion

This paper has examined several algorithms and complexity results for the Mover's Problem and then, to put these results in a larger context, has enumerated other types of motion planning problems, particularly in robotics. We conclude with a few remarks on the interaction between computational geometry and motion planning, and in particular, on their common geometric themes.

1) The first remark is that existing strategies and algorithms for the Mover's Problem have used solutions to several familiar computational geometry problems: the computation of object intersections, object decompositions, convex hulls, visibility graphs, and Voronoi diagrams. (This paper has not attempted to survey results on computational geometry problems. Classic papers that are frequently cited by papers on motion planning include Shamos '75,'77 and Shamos and Hoey '75,'76. Graham '72 and Preparata and Hong '77 are cited for finding the convex hull of a set of points in 2 and 3 dimensions, respectively. Mueller and Preparata '78 are cited for finding the intersection of two convex polyhedra.)

Intersection detection was used early in robot navigation when objects were approximated by circles. Motivated by this application Hopcroft, Schwartz and Sharir '83 give an algorithm for testing for intersection of spheres. Intersection detection (sometimes called the Null Object Detection problem, where the null object is the intersection of two objects whose collision is to be avoided) has long been important in the planning and verifying of tool paths for numerically controlled machine tools and will certainly continue to be important for the verification of robot plans. (See Boyse '79, Ahuja, Chien, Yen and Bridwell '80, Hunt and Voelcker '82 and Tilove '84.)

Decomposition of objects into pieces allows the generalization of several FINDPATH algorithms that initially assume that the object being moved is a convex polygon or polyhedron. The object is represented as a union of convex polygons, for example, and a path that is safe for each element of the union is planned. There are numerous object decomposition algorithms in the literature. Typically, they find a *partition* of the original set, although this is not essential for the purpose just cited.

The computation of convex hulls has been used for the purpose of approximating objects for motion algorithms that assume convexity. It has also been used to compute Cspace obstacles: Lozano-Pérez '83 takes the convex hull of the set of vector differences of obstacle vertices and object vertices, where the assumption is that the object and obstacles are convex polyhedra and the object does not rotate when it moves.

Visibility graphs have been used to find shortest paths in 2-dimensional Cspace for a point moving past convex, polygonal Cspace obstacles. In 3-dimensional Cspace, this approach may fail to find legal paths, as shortest paths may touch Cspace obstacles at edges as well as vertices. Sharir and Schorr '84 discusses the problem of sequencing the edges a shortest path touches and give a doubly exponential algorithm. Larson and Li '81 look for minimum rectilinear distance paths among obstacles.

The Voronoi-diagram and variants have been used as sets to which one retracts an imagined legal path in Cspace or Cartesian space, thereby obtaining a new legal path. The generalized cone approach of Brooks '83 is similar in spirit: a path is sought that lies strictly along a network of spines that, roughly speaking, stay midway between pairs of obstacles.

2) The second remark is that work in computational geometry on separability problems fits together with results from motion coordination problems to form a hierarchy of complexity results for motion problems. (See Toussaint '85.) Viewing these results together should suggest many new problems for computational geometry and complexity theory. (See Toussaint '81 for computational geometry problems in pattern recognition.)

3) The third remark is that finding representations of objects in order to support solutions to motion related problems such as intersection detection and position updating is an important theme that is common to graphics, computer vision, CAD/CAM, and task planning in robotics. In fact, progress in robotics will require solutions that are satisfactory from all these points of view. Objects must be designed and inspected, their proposed motions must be planned and checked, and their actual motions must be detected. Computational geometry will play an important role in this area.

4) Collision avoidance problems may require the solution of standard computational geometry problems in high dimensional Cspaces whose vectors represent position and orientation information and perhaps other types of information about objects as well.

5) Finally, it is important to note that the Cspace approach and the analytical and topological methods discussed in this paper are likely to provide useful tools and new problems for computational geometry. Hopcroft and Wilfong '84a use a topological approach to study the problem of objects that are required to remain in (perhaps variable) contact during motion. Guibas, Ramshaw and Stolfi '83 suggest the utility of a "topological" viewpoint in which polygons are regarded as paths in the state space of a moving object, and they show how several computational geometry results can be obtained in new ways with this approach.

References

[1] Adamowicz, M., "The optimum two-dimensional allocation of irregular, multiply-connected shapes with linear, logical and geometric constraints", Ph.D. dissertation, Dept. Elec. Eng., New York (1970).

[2] Adamowicz, M. and Albano, A., "Nesting two-dimensional shapes in rectangular modules", *Computer Aided Design*, vol. 8 (Jan. 1976), 27-32.

[3] Ahuja, N., Chien, R.T., Yen, R. and Bridwell, N., "Interference detection and collision avoidance among three dimensional objects", *Proc. 1st National Conf. on Artificial Intelligence*, Stanford University (August 19-21, 1980) 44-48.

[4] Ahuja, N. and Nash, C., "Octree representations of moving objects", *Computer Vision, Graphics, and Image Processing* 26 (1984) 207-216.

[5] Albano, A. and Sapuppo, G., "Optimal allocation of two-dimensional irregular shapes using heuristic search methods", *IEEE Trans. Syst. Man., Cybern.*, vol. SMC-10 (May 1980), 242-248.

[6] Arnon, D.S., "Algorithms for the geometry of semi-algebraic sets", TR436, Computer Science Dept., University of Wisconsin, Madison (1981).

[7] Atallah, M.J., "Dynamic computational geometry", *Proc. 24th IEEE Symp. on the Foundations of Comp. Sci.* (1983) 92-99.

[8] Ballard, D. and Brown, C., *Computer Vision* (Prentice-Hall, 1982).

[9] Bernstein, H.J., "Determining the shape of a convex n-sided polygon by using 2n + k tactile probes", New York University Courant Institute of Math. Sciences, TR 125 (June 1984).

[10] Birk, J.R. and Kelley, R.B., "An overview of the basic research needed to advance the state of knowledge in robotics", *IEEE Trans. Systems, Man, and Cybernetics*, V. SMC-11 8 (August 1981) 574-579.

[11] Boyse, J.W., "Interference detection among solids and surfaces", *Commun. Ass. Comput. Mach.*, Vol. ACM 22 (Jan. 1979), 3-9.

[12] Brady, M., Hollerbach, J.M., Johnson, T.L., Lozano-Pérez, T. and Mason M.T., (eds.), *Robot motion: Planning and control* (MIT Press, Cambridge, 1982).

[13] Brady, M., "Artificial Intelligence and Robotics", AI Memo 756, MIT AI Lab., Cambridge (February 1984).

[14] Brooks, R.A., "Solving the find-path problem by good represention of free space", *IEEE Trans. on Systems, Man, and Cybernetics*, V. SMC-13 (March/April 1983) 190-197.

[15] Brooks, R.A. and Lozano-Pérez, T., "A subdivision algorithm in configuration space for findpath with rotation", *Proc. Eighth Int. Jt. Conf. on AI, Karlsruhe* (August 1983) 799-806.

[16] Bruss, A.R. and Horn, B.K.P., "Passive Navigation", *Computer Vision, Graphics, and Image Processing* V21 (1983) 3-20.

[17] Campbell, C.E. and Luh, J.Y.S., "A preliminary study on path planning of collision avoidance for mechanical manipulators", Sch. Elec. Eng. Purdue Univ., Rep. TR-EE 8C-48 (Dec. 1980).

[18] Chazelle, B., Ottmann, T., Soisalon-Soininen, E. and Wood D., "The complexity of decidability of SEPARATION™ ", Univ. of Waterloo Comp. Sci. Data Structuring Group report CS-83-34 (1983).

[19] Chien, R.T., Zhang, L. and Zhang, B., "Planning collision-free paths for robotic arm among obstacles", *IEEE Trans. on Pattern Anal. and Machine Intelligence* PAMI-6 (1984) 91-96.

[20] Collins, G.E., "Quantifier elimination for real closed fields by cylindrical algebraic decomposition", *Second GI Conference on Automata Theory and Formal Languages*, Kaiserslautern, 1975, Springer-Verlag Lecture Notes in Computer Science 33 (1975) 134-183.

[21] Donald, B.R. "Motion planning with six degrees of freedom", MIT AI Lab., Technical Report 791 (May 1984).

[22] Eastman, C.M. and Preiss, K., "A review of solid shape modelling based on integrity verification", *Computer-Aided Design*, v. 16 (March 1984) 66-80.

[23] Freeman, H., "On the packing of arbitrary-shaped templates", in *Proc. 2nd USA-Japan Comput. Conf.* (1975), 102-107.

[24] Garey, M. and Johnson, D., *Computers and Intractability*, (Freeman, San Francisco, 1979).

[25] Gaston, P.C. and Lozano-Pérez, T., "Tactile recognition and localization using object models: the case of polyhedra on a plane", *IEEE Trans. on*

Pat. Anal. and Machine Int., V. PAMI-6 3 (May 1984) 257-266.

[26] Giralt, G., Sobek, R. and Chatila, R., "A multilevel planning and navigation system for a mobile robot", in *Proc. 6th Int. Joint Conf. Artificial Intell.*, Tokyo, Japan (Aug. 1979), 335-338.

[27] Goldberg, M., "A solution of problem 6-11: Moving furniture through a hallway", *SIAM Review*, 11 (1968) 75-78.

[28] Graham, R.L., "An efficient algorithm for determining the convex hull of a finite planar set", *Inform. Processing Lett.*, vol. 1 (1972) 132-133.

[29] Grimson, W.E.L. and Lozano-Pérez, T., "Model-based recognition and localization from sparse range or tactile data", MIT AI Lab., AI Memo 738 (August 1983).

[30] Guibas, L.J., Ramshaw, L. and Stolfi, J., "A kinetic framework for computational geometry", *Proc. 24th IEEE Symp. on the Foundations of Comp. Sci.* (1983) 100-111.

[31] Guibas, L.J. and Yao, F.F., "On translating a set of rectangles", in *Proc. 12th Annu. ACM Symp. Theory of Computing*, Los Angeles, CA. (April 1980), 154-160.

[32] Hansen, J.A., Gupta, K.C. and Kazerounian, S.M.K., "Generation and evaluation of the workspace of a manipulator", *Int. J. Robotics Res.*, V. 2 (Fall 1983) 22-31.

[33] Harmon, L.D., "Automated tactile sensing", *Robotics Res.*, V. 1 (Summer 1982) 3-32.

[34] Hopcroft, J., Joseph, D. and Whitesides, S., "On the movement of robot arms in two-dimensional bounded regions", Cornell Univ., Dept. Comp. Sci. TR 82-486 (1982a). To appear in *Siam J. Comp.* (1985).

[35] __________, "Determining points of a circular region reachable by joints of a robot arm", Cornell Univ., Dept. Comp. Sci. TR 82-516 (Oct. 1982b). Outline to appear in *Siam J. Comp.* (1985) paper cited above.

[36] __________, "On the movement of robot arms in 2-dimensional bounded regions", *23rd IEEE Symp. FOCS* (1982c) 280-289.

[37] Hopcroft, J., Joseph, D. and Whitesides, S., "Movement problems for 2-dimensional linkages", *SIAM J. Comput. 13* (1984) 610-629.

[38] Hopcroft, J., Schwartz, J.T. and Sharir, M., "Efficient detection of intersections among spheres", *Int. J. of Robotics Research* V. 2 (1983) 77-80.

[39] __________, "On the complexity of motion planning for multiple independent objects; PSPACE hardness of the 'Warehouseman's Problem' ", TR 103, New York Univ. Courant Inst. of Math. Sciences (February 1984).

[40] Hopcroft, J. and Wilfong, G., "On the motion of objects in contact", TR 84-602, Dept. of Comp. Sci., Cornell Univ. (May 1984a).

[41] __________, "Reducing Multiple Object Motion Planning to Graph Searching", TR-84-616, Dept. of Comp. Sci., Cornell University (July 1984b).

[42] Howden, W.E., "The sofa problem", *Comput. J.*, vol. 11 (Nov. 1968), 299-301.

[43] Hunt, W.A. and Voelcker, H.B., "An exploratory study of automatic verification of programs for numerically controlled machine tools", Tech. Memo 34, Production Automation Project, University of Rochester, Rochester, N.Y. (January 1982).

[44] Jerian, C., and Jain, R., "Determining motion parameters for scenes with translation and rotation", *IEEE Trans. on Pat. Anal. and Machine Int.* PAMI-6 (July, 1984) 523-530.

[45] Kanade, T., "Geometrical aspects of interpreting images as a three-dimensional scene", *Proc. IEEE,* V.71 (July 1983) 789-802.

[46] Kirkpatrick, D., "Efficient computation of continuous skeletons", *Proc. 20th IEEE Symp. on Foundations of Comp. Sci.* (1979) 18-27.

[47] Klinger, A., and Dyer, C., "Experiments in picture representation using regular decomposition", *Comp. Graphics and Image Processing* 5 (1975) 68-105.

[48] Larson, R.C. and Li, V.O.K., "Finding minimum rectilinear distance paths in the presence of obstacles", *Networks,* vol. 11 (1981), 285-304.

[49] Lee, D.T., and Chen, I.M., "Display of visible edges of a set of convex polygons", in Toussaint, G.T. (ed.), *Computational Geometry* (North Holland, 1985).

[50] Lee, D.T. and Drysdale, S., "Generalization of Voronoi diagrams in the plane", *SIAM J. Computing* 10 (1981) 73-87.

[51] Lee, Y.T., and Requicha, A., "Algorithms for computing the volume and other integral properties of solids. I. Known methods and open issues", *Comm. ACM* 25 (1982a) 635-641.

[52] __________, "Algorithms for computing the volume and other integral properties of solids. II. A family of algorithms based on representation conversion and cellular approximation", *Comm. ACM 25* (1982b) 642-650.

[53] Lewis, R.A., "Autonomous manipulation on a robot: Summary of manipulator software functions", Jet Propulsion Lab., California Inst. Technol., TM 33-679 (Mar. 1974).

[54] Lozano-Pérez, T., "Automatic planning of manipulator transfer movements", *IEEE Trans. Syst, Man, Cyber.*, Vol. SMC-11 (Oct. 1981) 681-698.

[55] __________, "Task Planning", in: *Brady, M. et al. (eds.), Robot Motion: planning and control* (MIT Press, Cambridge, 1982).

[56] __________, "Spatial planning: a configuration space approach", *IEEE Trans. on Computers*, V. C-32 (February 1983a) 108-120.

[57] __________, "Robot Programming", *Proc. IEEE* V. 71 (July 1983b) 821-841.

[58] Lozano-Pérez, T., Mason, M. and Taylor, R., "Automatic synthesis of fine-motion strategies for robots", *Int. J. Robotics Research* 3 (1984) 3-24.

[59] Lozano-Pérez, T. and Wesley, M.A., "An Algorithm for planning collision-free paths among polyhedral obstacles", *Commun. Ass. Comput. Mach.*, vol. ACM 22 (Oct. 1979), 560-570.

[60] Mason, M.T., "Compliance", in: *Brady, M. et al. (eds.), Robot Motion: planning and control* (MIT Press, Cambridge, 1982).

[61] Moravec, H.P., "The Stanford Cart and the CMU Rover", *Proc. IEEE*, V. 71 (July 1983) 872-884.

[62] Moser, L., "Problem 6-11: Moving furniture through a hallway", *SIAM Review* 8 (1966) 381.

[63] Mueller, and Preparata, F., "Finding the intersection of two convex polyhedra", *Theoretical Comp. Sci.*, V. 7 (1978) 217-236.

[64] Nguyen, Van-Duc, "The find-path problem in the plane", MIT AI Lab., AIL Memo 760 (February 1984).

[65] Nilsson, N., "A mobile automaton: An application of artificial intelligence techniques", in *Proc. 2nd Int. Joint Conf. Artificial Intell.* (1969), 509-520.

[66] __________, *Principles of artificial intelligence*, (Tioga Publishing Co., Palo Alto, 1980).

[67] O'Dúnlaing, C., Sharir, M. and Yap, C.K., "Retraction: A new approach to motion planning", *15th ACM STOC* (1983), 207-220.

[68] O'Dúnlaing, C. and Yap, C.K., "The Voronoi method for motion planning: I. The case of a disc", Courant Institute Technical Report 53 (Mar. 1983), to appear *J. of Algorithms.*

[69] Ottmann, T. and Wood, D., "Dynamical sets of points", *Computer Vision, Graphics, and Image Processing,* 27 (1984) 157-166.

[70] Paul, R.P., *Robot Manipulators: Mathematics, programming, and control,* (MIT Press, Cambridge, MA, 1981).

[71] Pieper, D.L., "The kinematics of manipulators under computer control", Stanford Artificial Intell. Lab., AIM-72 (Oct. 1968).

[72] Preparata, F. and Hong, S., "Convex hulls of finite sets of points in two and three dimensions", *Commun. Ass. Comput. Mach.*, vol. 20 (Feb. 1977) 87-93.

[73] Reddy, R., (ed.), Special Issue on Robotics, *Proc. IEEE* 71 (1983).

[74] Reif, J., "Complexity of the mover's problem", in *Proc. 20th Annu. IEEE Symp. Foundation of Comput. Sci.* (1979), 421-427.

[75] Requicha, A.G., "Representations for rigid solids: theory, methods, and systems", *ACM Comput. Surv.* 12, 4 (Dec. 1980) 437-464.

[76] Rowat, P.F., "Representing spatial experience and solving spatial problems in a simulated robot environment", Ph.D. Thesis, Univ. British Columbia (1979).

[77] Schwartz, J.T., "Finding the minimum distance between two convex polygons", *Inform. Processing Lett.*, vol. 13 (1981), 168-170.

[78] Schwartz, J.T., and Sharir, M., , "On the piano movers' problem: I. The special case of rigid polygonal body moving amidst polygonal barriers", *Comm. Pure Appl. Math.* vol. XXXVI (1983a), 345-398.

[79] __________, "On the piano movers' problem: II. General techniques for computing topological properties of real algebraic manifolds", *Adv. Appl. Math.* 4 (1983b), 298-351.

[80] __________, "On the piano movers' problem: III. Coordinating the motion of several independent bodies: The special case of circular bodies moving amidst polygonal barriers", *Int. J. Robotics Research* 2 (Fall 1983c), 46-75.

[81] __________, "On the piano movers' problem: V. The case of a rod moving in three-dimensional space amidst polyhedral obstacles", New York University, Courant Institute Robotics Research Report 13 (Oct. 1983d).

[82] __________, "Mathematical problems and training in robotics", *Notices of the American Math. Soc.*, V. 30 (1983e) 478-481.

[83] __________, "Some remarks on robot vision", TR 119, New York Univ. Courant Institute of Math Sciences (April 1984).

[84] Sebastian, J., "A solution to problem 6-11: moving furniture through a hallway", *SIAM Review* 12 (1970) 582-586.

[85] Shamos, M.I., "Computational Geometry", Ph.D. dissertation, Yale University (1977).

[86] __________, "Geometric complexity", in *Proc. 7th Annu. ACM Symp. Theory of Computing* (1975), 224-233.

[87] Shamos, M.I. and Hoey, D., "Closest-point problems", in *Proc. 16th Annual IEEE Symp. Foundation of Comput. Sci.* (1975), 151-161.

[88] __________, "Geometric intersection problems", in *Proc. 17th Annu. IEEE Symp. on Foundations of Comput. Sci.* (October, 1976), 268-273.

[89] Sharir, M and Ariel-Sheffi, E., "On the piano movers' problem: IV. Various decomposable two-dimensional motion-planning problems", *Comm. Pure and Applied Math.* 37 (1984) 479-493.

[90] Sharir, M. and Schorr, A., "On shortest paths in polyhedral spaces", *Proc. ACM Symp. on the Theory of Comp.* (1984) 144-153.

[91] Spirakis P. and Yap, C., "Strong NP-hardness of moving many discs", *Information Processing Letters* 19 (1984) 55-59.

[92] Strang, G., "The width of a chair", *American Math. Monthly* 89 (1982) 529-534.

[93] Tarski, A., *A Domain Method for Elementary Algebra and Geometry* (University of California Press, 1951).

[94] Tilove, R.B., "A null-object detection algorithm for constructive solid geometry", *Comm. ACM*, V. 27 (July 1984) 684-694.

[95] Thompson, A.M., "The navigation system of the JPL robot", in *Proc. 5th Int. Joint Conf. Artificial Intell.*, Massachusetts Inst. Technol. (1977) 749-757.

[96] Toussaint, G.T., "Computational geometric problems in pattern recognition", in *Pattern recognition theory and applications*, J. Kittler (ed.), (NATO Advanced Study Inst., Oxford University, April 1981) 73-91.

[97] __________, "Movable separability of sets", in: Toussaint, G.T. (ed.), *Computational Geometry* (North Holland, 1985).

[98] Udupa, S., "Collision detection and avoidance in computer controlled manipulators", in *Proc. 5th Int. Joint Conf. Artificial Intell.*, Massachusetts Inst. Technol. (1977a) 737-748.

[99] __________, "Collision detection and avoidance in computer controlled manipulators", Ph.D. dissertation, Dep. Elec. Eng., California Inst. Technol. (1977b).

[100] Whitely, W., "A correspondence between scene analysis and motions of frameworks", *Discrete Applied Math.* 9 (1984) 269-295.

[101] Widdoes, C., "A heuristic collision avoider for the Stanford robot arm", Stanford CS Memo 227 (June 1974).

[102] Yap, C.K., "Coordinating the motion of several discs", TR-105, New York University Courant Institute of Mathematical Sciences (February 1984).

[103] Yemini, Y., "Some theoretical aspects of position-location problems", *IEEE Symp. Foundations of Comp. Sci.* (1979) 1-8.

COMPUTATIONAL GEOMETRY, G.T. Toussaint (Editor)
© Elsevier Science Publishers B.V. (North-Holland), 1985

AN ISOTHETIC VIEW OF COMPUTATIONAL GEOMETRY[†]

Derick Wood

Data Structuring Group
Department of Computer Science
University of Waterloo
Waterloo, Ontario N2L 3G1
Canada

ABSTRACT

We survey a number of results, some new, for isothetic polygons in the plane. Specifically we consider intersection, convexity, combinational, clustering and visibility problems. These problems occur in image processing, VLSI design and layout, circuit testing, transaction systems, and pattern recognition.

1. INTRODUCTION

Since Michael Shamos [59] carried out his pioneering work in computational geometry the field has blossomed considerably. The purpose of this chapter is to review some of the developments in computational geometry since its inception. In order to do justice to the field I have been obliged, albeit not unwillingly, to restrict my attention to the plane and to isothetic or rectilinearly-oriented polygons. Since the first problem I studied in computational geometry was the rectangle intersection problem with Jon Bentley [6] and my most recent work [69] involves isothetic polygons, this restriction is a happy one. This survey provides an opportunity to draw together the variety of problems for isothetic objects into some coherent whole, while also allowing the formal treatment of isothetic polygons in the style of Shamos [59].

Isothetic polygons occur in VLSI design, in floor and street plans, in data bases, in CAD/CAM, in NC, in circuit routing, in circuit testing, in robotics, in locked transaction systems, in image processing, in pattern recognition, and in Saskatchewan. Often they can be viewed as discrete approximations to polygons in the real plane, for example in image processing and pattern recognition, however just as often they appear in their own right, for example in VLSI designs. From a theoretical viewpoint isothetic polygons are usually (but not always) easier to handle than arbitrary polygons. Because of this it is often possible to obtain time- and space-optimal algorithms for isothetic polygons, when this is not the case in general. Finally, isothetic polygons are the source of many beautiful problems - an important issue for a researcher.

Section 2 provides the basic definitions for isothetic objects. In Sections 3 through 7 we study

[†] Work carried out under a Natural Sciences and Engineering Research Council of Canada Grant No. A-5692.

intersection, convexity, combinational, clustering, and visibility problems, respectively. Section 8 is a closing discussion of the results, the methodology, and some of the omissions.

2. DEFINITIONS

We begin by defining our objects of study, namely isothetic lines, line segments, curves and polygons. We assume all objects are in the plane except when explicitly stated otherwise. We assume the plane has an embedded Cartesian coordinate system.

An *isothetic* (*straight*) *line* is a straight line parallel to either the x-axis or y-axis. Similarly, an *isothetic line segment* is a line segment parallel to either the x- or y-axis, and an *isothetic ray* is a ray parallel to either the x- or y-axis. See Figures 2.1(a), (b) and (d).

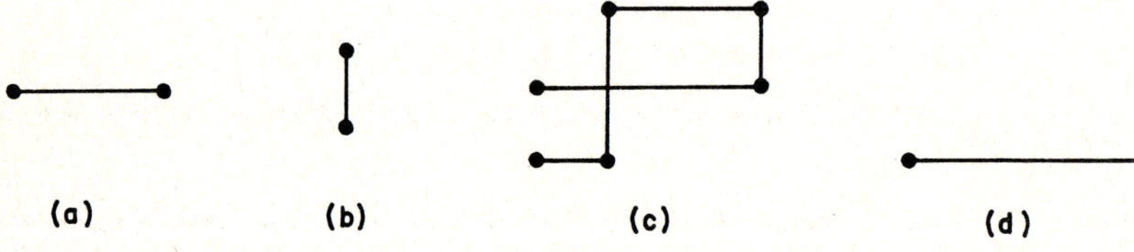

Figure 2.1

An *isothetic curve* is a connected sequence of isothetic line segments, see Figure 2.1(c).

Figure 2.2

An isothetic curve is *closed* if starting at any point p, on the curve, a path can be traced along the curve, visiting every point once and only once, which leads to p.

An isothetic curve is *simple* if:

(i) self-intersections only occur at endpoints, and

(ii) no more than two segments intersect at each endpoint.

A non-closed simple isothetic curve is a *staircase* if either for all points p and q on the curve, with p to the left of q, p is not above q, or for all points p and q on the curve, with p to the left of q, p is not below q. See Figure 2.2.

An *isothetic polygon* is a closed isothetic curve, which forms its boundary. We also consider a polygon to be the closed finite region of the plane enclosed by the closed curve. Note that this definition excludes the possibility of a polygon having *holes*. A *simple isothetic polygon* is a closed simple isothetic curve. See Figure 2.3(a) for a simple polygon and 3(b) for a non-simple polygon.

Figure 2.3

We often wish to distinguish polygons not only with respect to their simpleness, but also with respect to their isothetic convexity.

A simple isothetic polygon is *isothetically convex* if every isothetic line has either an empty intersection with the polygon or an isothetic line segment as its intersection. Figure 2.3(a) is a non-convex polygon, while Figure 2.4 displays a convex polygon. A special case of a convex polygon is a *rectangle*.

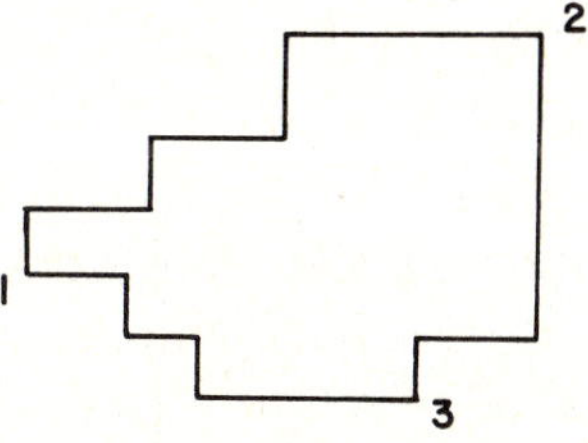

Figure 2.4

Observation *Every convex isothetic polygon can be decomposed into at most four staircases. In Figure 2.4 there are three:* 1 *to* 2, 2 *to* 3, *and* 3 *to* 1.

If a polygon is non-convex we often wish to refer to its convex hull, which is defined as follows:

The *isothetic convex hull* of a simple isothetic polygon is the smallest, in area, convex simple isothetic polygon containing the given polygon. Figure 2.5 displays the convex hull of Figure 2.3(a).

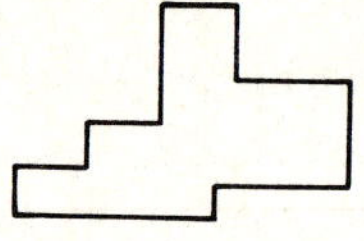

Figure 2.5

Already our definitions lead to three natural problems:

Problem 1 *Determine if a given isothetic polygon is simple.*

Problem 2 *Determine if a given simple isothetic polygon is convex.*

Problem 3 *Determine the convex hull of a given simple isothetic polygon.*

In the next section we will describe an algorithm to determine whether or not an isothetic polygon is simple, while the following section will discuss convexity.

3. INTERSECTION PROBLEMS

Throughout this and the following sections we will omit the adjectives *isothetic* and *simple.* All objects are isothetic and all polygons are simple unless it is explicitly stated otherwise. We first state:

Problem 4 *Given a collection of polygons determine if there are two* intersecting *polygons in the collection, that is two polygons having a common point.*

Problem 5 *Given a collection of polygons determine all pairwise intersecting polygons.*

Problem 6 *Given a collection of polygons and an integer* $k \geq 2$, *determine if there is a* k-intersection, *that is* k *polygons having a common point. In particular is there a p-intersection?*

Problem 7 *Given a collection of polygons determine the maximal* k *for which there is a* k-*intersection, that is the* thickness *of the given polygons.*

Using the *sweep line* paradigm, Shamos and Hoey [60] gave an $O(p \log p)$ time and $O(p)$ space algorithm for Problem 4 for the case of p line segments. This was extended by [4] and [8] to provide an $O(p \log n + k)$ time, where there are k pairwise intersections, and $O(p)$ space algorithm for Problem 5 for line segments.

The algorithm of [4] with some minor modifications gives us a fast algorithm for simplicity testing, that is Problem 1. There are two checks to be carried out. First, we need to check whether or not there are two edges which intersect at some point which is not an endpoint. Such non-endpoint intersections are caused (a) by a vertical edge crossing a horizontal edge, (b) by two vertical edges overlapping, or (c) by two horizontal edges overlapping. The algorithm of [60] will detect intersections of type (a), while those of types (b) and (c) can be found by sorting the edges. Second, we need to check that each endpoint intersection is formed by a unique pair of edges. Clearly there should only be n such pairs for an n-vertex simple polygon. Hence we count the endpoint intersections. If their number exceeds n then the given polygon is not simple. The [4] and [8] algorithm can be modified to perform this count, without affecting its time and space bounds, hence we obtain:

Theorem 3.1 *Given a polygon with* n *vertices it can be tested for simpleness in* $O(n \log n)$ *time and* $O(n)$ *space.*

The results for Problem 4 were extended to rectangles in [6] who gave an $O(p \log p + k)$ time algorithm. This was improved in [9] and [40] to also require only $O(p)$ space. Since it was already proved in [6] that these were lower bounds, this results in:

Theorem 3.2 [6], [9], [40]
Problem 4 for rectangles can be solved in $O(p \log p)$ *time and* $O(p)$ *space. Problem 5 for rectangles can be solved in* $O(p \log p + k)$ *time and* $O(p)$ *space, where there are* k *intersecting pairs. Moreover these are both time- and space-optimal in a comparison-based model of computation.*

Observe that $k = O(n^2)$ see Figure 3.1.

More recently time- and space-optimal divide-and-conquer algorithms for this problem have appeared, see [22] and [10]. While [14] have a new approach based on resolution.

These results can be extended to polygons fairly easily. We sketch the sweep-line approach for them.

The basic idea is to sweep, conceptually, a vertical line from the left of the polygons to the right of the polygons and, by this means, reduce a static two-dimensional problem to a dynamic one-dimensional problem. At each position of the sweep line there are polygons which it intersects, *active* polygons, polygons completely to its left, *dead* polygons, and polygons completely to its right, *inactive* polygons. The intersection of each active polygon with the sweep line results in a number of disjoint intervals, called *active intervals*. See position (b) of the sweep line in Figure 3.2. There are two crucial

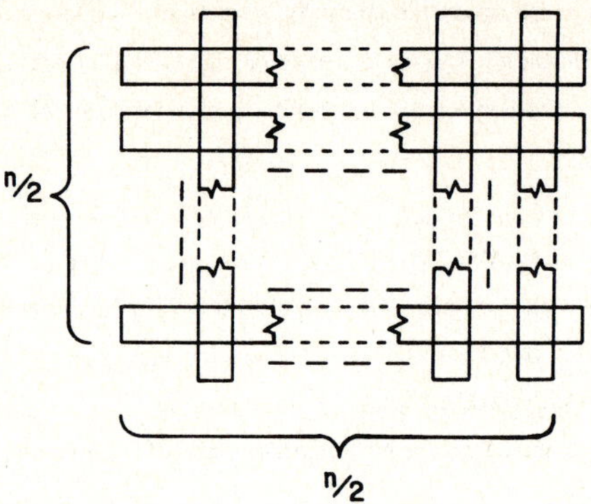

Figure 3.1

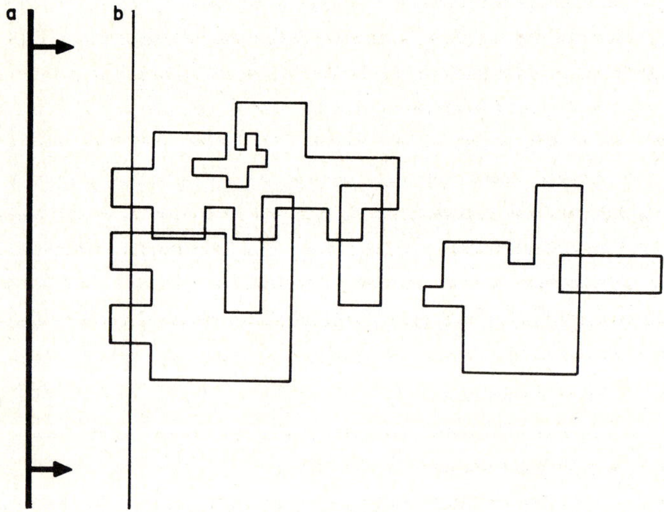

Figure 3.2

observations about the sweep line. First there are at most $2n$ different sets of active intervals, since the set of active intervals only changes when the sweep line meets a vertical edge of a polygon. Second two polygons intersect if and only if there is a position of the sweep line for which their corresponding active intervals intersect.

The first observation reduces a continuous sweep of the sweep line to $2n$ discrete jumps, the second reduces the two-dimensional problem to a one-dimensional one. Letting the x-projections of the vertices in ascending sorted order be known as *sweep points* we can describe the intersection algorithm as:

> **For** each sweep point x **do**
>
> **if** x corresponds to a left vertical edge **then**
>
> query active intervals with the edge for intersections, and insert the edge to form or extend an active interval **else**
>
> $\{x$ corresponds to a right vertical edge$\}$
>
> delete the edge from its corresponding active interval yielding zero, one or two new active intervals.

A *left* vertical edge is one on the left side of the polygon, while a *right* vertical edge is on the right side.

The dynamic one-dimensional problem can be solved with either a *segment tree*, requiring $O(n \log n)$ space, or a *1-fold interval tree* requiring $O(n)$ space. We choose the former, since it has many other applications, a description of the latter can be found in [9] and [40]. The segment tree is a binary leaf search tree, that is keys are to be found at external nodes while internal nodes contain *signposts, separating keys*, or *routers* as they are variously called. Moreover each external node not only represents a point in the key space, but it also represents the interval in the key space defined by the successor key in the tree. This implies that each internal node can also be considered to represent a key-space interval, namely the interval determined by its external nodes. So each interval specified by a pair of keys is associated with a unique node of the tree. In our setting the keys are y-values corresponding to the endpoints of vertical edges. In Figure 3.3 there are 7 vertical edges giving 7 y-values. The corresponding empty segment tree is displayed in Figure 3.4.

External node y_i represents the closed-open interval $[y_i, y_{i+1})$ for $0 \le i < 6$ and y_6 represents the closed interval $[y_6, y_6]$. Therefore the interval of node u, $int(u)$, is the interval $[y_0, y_4)$. To each node u is associated a set of intervals $cover(u)$; each interval I in it satisfies $int(u) \subseteq I$ and $int(\pi u) \not\subseteq I$, where πu is the parent of u. Initially $cover(u) = \varnothing$, for all nodes u in the tree. In contrast to binary search trees the structure does not change during insertions and deletions, only the cover sets of the nodes change. Hence we can construct the segment tree to be a minimal height tree initially. For example inserting the edge $[y_0, y_5]$ of B into the segment tree of Figure 3.4 causes $cover(u) := \{B\}$, $cover(v) := \{B\}$ and $cover(w) := \{B\}$. The latter assignment does not satisfy the rule given above since $int(w) = [y_5, y_6)$. However because $int(w) \cap [y_0, y_5] \neq \varnothing$ and w is an external node we assign B to $cover(w)$ anyway. If a segment tree has n external nodes, then it has height bounded by $\lceil \log_2 n \rceil$ and it is easily seen that the insertion of an interval only affects the cover sets of at most $4\lceil \log_2 n \rceil$ nodes. Deletion is not quite as simple, see [6] for details, but it can still be carried out in $O(\log n)$ time.

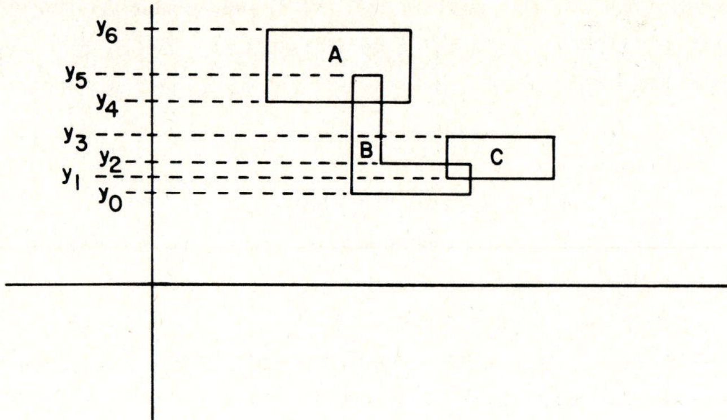

Figure 3.3

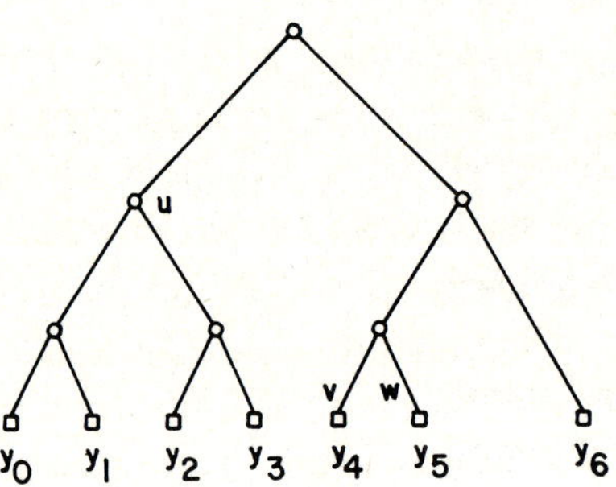

Figure 3.4

Disregarding the technical details of maintaining the active intervals of each polygon in the segment tree, consider a query. A query edge can intersect an active interval in one of four ways, see Figure 3.5.

Intersections of types (a) and (b) can be found by searching the segment tree for the external node corresponding to the y-value of the bottom endpoint of the query edge. An intersection of type (c) can be found by a similar query with the top endpoint y-value. In both cases all cover sets on the search path intersect the query edge. Now intersections of type (d) are not found by either of these queries. For this purpose we specify an additional field *downcover*(u) for each internal node u in the segment tree. This new field contains all intervals inserted in u's subtree but not at u, that is *downcover*$(u) = \bigcup$ *cover*(v), for all proper descendants v of u. Note that *downcover*(u) =

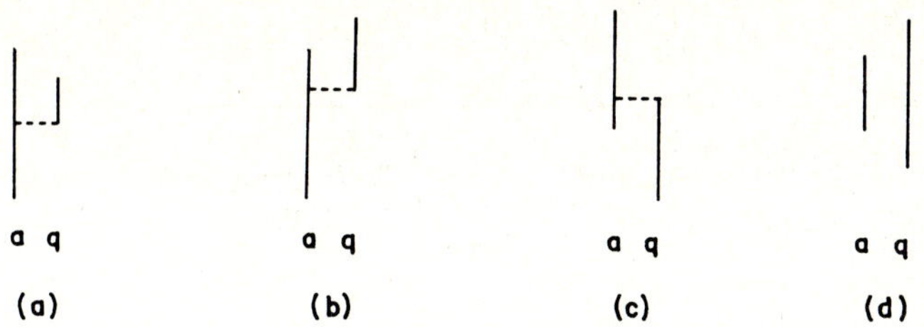

Figure 3.5

downcover(λu) $\cup$ *downcover*(ρu) $\cup$ *cover*(λu) $\cup$ *cover*(ρu) for all internal nodes u, where *downcover*(u) for u an external node is defined to be $\emptyset$. Maintaining *downcover* sets is no more difficult than maintaining cover sets, since an interval is inserted in at most $O(\log n)$ *downcover* sets. Hence insertion and deletion complexity is not affected. But an intersection of type (d) is now easily found. Perform a dummy insertion of the query edge. Now for each *cover*(u) to which it would be added all intervals in the corresponding *downcover*(u) intersect the query edge. Conversely every intersection of type (d) must be found in this way. Hence all pairwise intersections will be found and moreover each update requires $O(\log n)$ time and each query $O(\log n + k')$ time, where k' is the number of pairwise intersections found. Using the 1-fold interval tree rather than the segment tree gives an optimal space bound, thus we have:

Theorem 3.3 *Given p polygons with n vertices all k pairwise intersections can be found in $O(n \log n + k)$ time and $O(n)$ space, thus solving Problem 5.*

We now consider Problems 6 and 7. Again we first consider the case for p rectangles. Now for the p-intersection case we have:

Theorem 3.4 *Given p rectangles, there is a p-intersection, that is they all have a point in common, if and only if every pair of rectangles have a point in common.*

Proof: Clearly if there is a point common to all rectangles then every pair have a common point. Conversely, assume every pair have a common point. We argue by contradiction. If the p rectangles do not have a p-intersection, then there is a maximal k such that there is a k-intersection, where $2 \leq k < p$. Let $R_1,...,R_k$ be the k rectangles that have a k-intersection and R be one of the remaining rectangles. Consider a common point L of $R_1,...,R_k$ which is leftmost. Now L must lie on the leftmost edge of one of the $R_1,...,R_k$, let it be without loss of generality R_1, otherwise there would be another common point to its left. If R is to the left of L then R does not intersect R_1 - a contradiction. But a similar contradiction is obtained for each of the other three possibilities. Thus $R_1,\ldots,R_p$ have a p-intersection. □

This rectangle version of Helly's Theorem leads to the following stratagem for determining the existence of a p-intersection:

Determine if there is a pairwise non-intersection.

But this can be done by, essentially, the same algorithm which finds all pairwise intersections - the query at a sweeping point becomes: Is there an active segment which does not intersect the incoming segment?

An alternative approach is simply to count the number of pairwise intersections and if it is equal to $\binom{p}{2}$ then there is a p-intersection. Otherwise there is at least one pairwise non-intersection. Either way we obtain:

Theorem 3.5 *Given p rectangles the existence of a p-intersection can be determined in* $O(p \log p)$ *time and* $O(p)$ *space.*

Unfortunately Theorem 3.4 does not hold for polygons, see Figure 3.6. Indeed Figure 3.6 displays p *convex* polygons with $\binom{p}{2}$ intersections but no p-intersection, for $p \geq 3$.

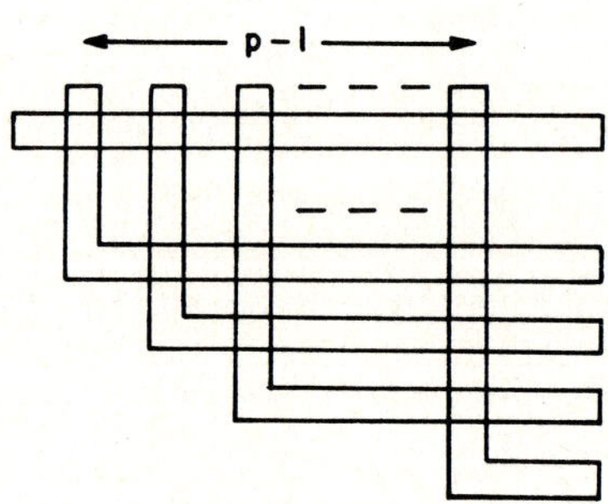

Figure 3.6

However for convex polygons Helly's Theorem holds directly, that is:

Theorem 3.6 *Given p convex polygons there is a p-intersection if and only if every three polygons have a common point.*

Although such a characterization does not hold in general we are still able to prove:

Theorem 3.7 *Given p polygons and an integer k then the existence of a k-intersection can be determined in $O(n \log n)$ time and $O(n)$ space, where the p polygons have n vertices.*

Proof Sketch: We use the sweep-line and segment tree approach described above to solve Problem 5. In other words we sweep the polygons from left to right so that at each position of the sweep line we have for each active polygon the disjoint intervals it forms with respect to the sweep line, that is the active intervals. Now there is a k-intersection of polygons if and only if at some position of the sweep line there is a k-intersection of active intervals. We represent the active intervals in a *counting segment tree*, that is $cover(u)$ for each node u is replaced by its size, $\#cover(u)$. Moreover for each node u in the tree define a new field $thickness(u)$ to be the maximal k-intersection in the subtree rooted at u. Observe that $thickness(u) = max(thickness(\lambda u), thickness(\rho u)) + \#cover(u)$, hence the thickness fields are easily updated after insertion and deletion. But this means that there is a k-intersection of active intervals exactly when $thickness(root) \geq k$. Moreover only insertions increase this value, so after each insertion $thickness(root)$ is checked. □

Note that the above proof sketch also yields a solution to Problem 7, therefore we have:

Theorem 3.8 *Given p polygons with n vertices their thickness can be determined in $O(n \log n)$ time and $O(n)$ space.*

4. CONVEXITY PROBLEMS

We have already defined the generally accepted notion of convexity for isothetic polygons. In this section we consider convexity testing, that is Problem 2, how the convex hull of a polygon is computed, that is Problem 3, how we might define the convex hull of a collection of polygons, and the convex skull problem.

Solving Problem 2 reduces to testing whether or not a given polygon can be decomposed into at least 2 and at most 4 staircases, which fit together appropriately. Ignoring the degenerate cases when they are only 2 or 3 staircases, observe that four staircases form a convex polygon if every staircase has the same orientation when they are rotated to place them in the north-west corner. This observation also holds, essentially, for the 2 and 3 case too. Thus testing a polygon for convexity reduces to testing if its north-west curve is a staircase of the form displayed in Figure 2.2(a). Since the polygon is specified by the sequence of its vertices in clockwise order, this reduces to five linear scans of them. The first determines the, at most, four curves, the second decides if the north-west curve is a staircase, the third tests the north-east curve, and so on, yielding:

Theorem 4.1 *Let P be a polygon with n vertices, then P can be tested for convexity in* $O(n)$ *time and space. Moreover both these bounds are optimal.*

Now solving Problem 3, that is computing the convex hull of a polygon, is similar to testing it for convexity. The only difference is that a smallest enclosing staircase for each curve is computed. This is straightforwardly computed, see [43], [44] and [48], yielding:

Theorem 4.2 *Let P be a polygon with n vertices, then the convex hull of P can be computed in* $O(n)$ *time and space, which is optimal.*

The convex hull of one polygon is easily defined, but when given a collection of polygons it is not even clear what definition should be used.

Thus we have:

Problem 8 *Given a collection of polygons define their convex hull.*

To begin to understand what is meant by the convex hull of a collection of polygons, we first make precise the notion of a *convex collection.* A collection C of polygons is convex if its intersection with every line is either empty or a line segment, in other words it is the same definition as is used for individual polygons. We now discuss three reasonable definitions of the convex hull of a collection of polygons.

Definition 4.1 Minimal Collection

Given a collection C of polygons its convex hull is the *minimal area convex collection* of disjoint polygons containing the given collection.

By the *area* of a collection we mean the area of its union, and by *containment* we mean a collection C_1 contains a collection C_2 if every polygon in C_2 is contained in the union of the polygons in C_1.

Definition 4.2 Minimal Polygon

Given a collection C of polygons its convex hull is a *minimal area convex polygon* P containing the given collection.

Definition 4.3 Half Plane

Given a collection C of polygons its convex hull is the intersection of all *closed isothetic half-planes* containing the collection.

There are four isothetic closed half-planes as displayed in Figure 4.1.

Figure 4.1

Definition 4.1 is the isothetic version of the classical definition of the convex hull and is used in [49]. Definition 4.2 is the one used in [43] and [44], while Definition 4.3 corresponds to finding the maximal vectors in a set of vectors [27]. We call these the *mc*-, the *mp*-, and the *hp-convex* hulls, respectively, and for a collection C we denote them by $MC(C)$, $MP(C)$ and $HP(C)$, respectively. It is not difficult to show that:

$$MC(C) \subseteq HP(C) \subseteq MP(C)$$

and, moreover, in many cases these containments are proper. See Figure 4.2, for example:

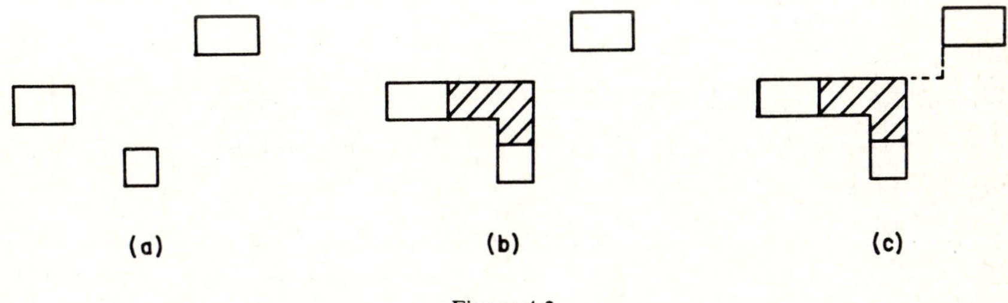

Figure 4.2

Figure 4.2(a) displays a collection of three rectangles and its *mc*-convex hull! Figure 4.2(b) and (c) show its *hp*-convex hull and an *mp*-convex hull, respectively.

For the usual non-isothetic convex hull of a point set there are a number of equivalent formulations, viz:

(1) The convex hull of a set is the intersection of all convex sets containing the given set.

(2) The convex hull of a set is the intersection of all half-planes containing the given set.

(3) A point is in the convex hull of a set if and only if it is in the convex hull of some three or fewer points from the given set.

(4) A point is in the convex hull of a set if and only if every line through the point divides the set into two non-empty sets.

Now (1) only holds for *mc*-convex hulls, (2) only holds for *hp*-convex hulls, when isothetic half-planes are meant, (3) only holds for *hp*- and *mc*-convex hulls, and (4) only holds for none of them, when isothetic lines are used.

For the usual non-isothetic convex hull we also require:

(5) The convex hull of a set to be *unique*.

(6) The convex hull of a set to be *stable* with respect to insertion and deletion of points.

Now (5) holds for both *hp*- and *mc*-convex hulls, but not for *mp*-convex hulls, see Figure 4.3. It is possible that there are infinitely many *mp*-convex hulls of a given collection, indeed Figure 4.3 only displays 3 possible *mp*-convex hulls for a collection of two rectangles. However from the point of view of stability *hp*- and *mp*-convex hulls are both stable, but *mc*-convex hulls are not, see Figure 4.4. In Figure 4.4(a) a collection and its *mc*-convex hull are displayed, while in Figure 4.4(b) a rectangle is added causing the *mc*-convex hull to grow tremendously.

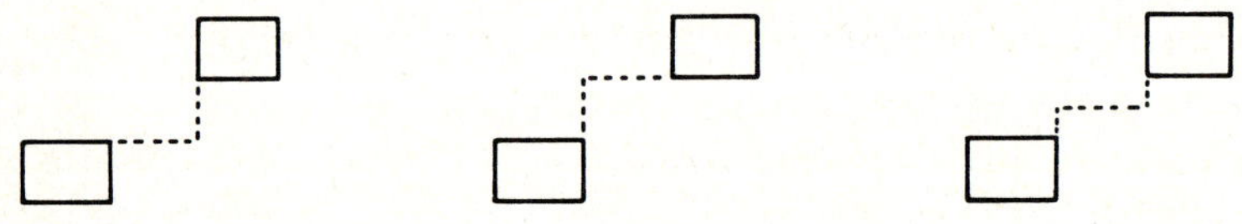

Figure 4.3

Each of these convex hulls can be computed in $O(n \log n)$ time and $O(n)$ space, where the collection contains p polygons with a total of n vertices. Moreover for the case of the *hp*-convex hull (and probably for the others) this can be improved to $O(n \log h)$ time, where h is the number of vertices in the *hp*-convex hull. To summarize:

Theorem 4.3 [48], [26]

Let C be a collection of p polygons with a total of n vertices. Then $MC(C)$ and $MP(C)$ can

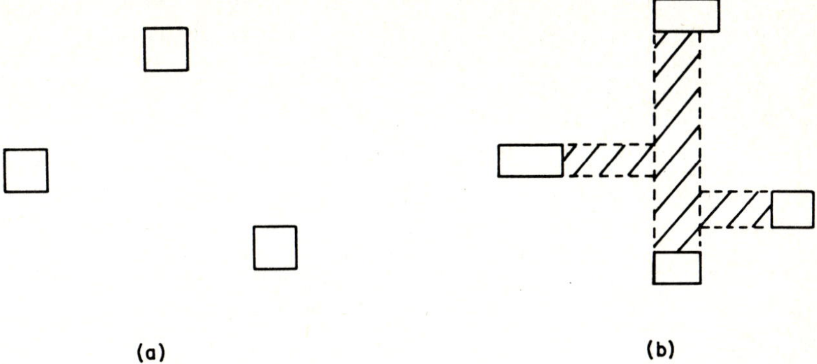

Figure 4.4

be computed in $O(n \log n)$ *time and* $O(n)$ *space, while* $HP(C)$ *can be computed in* $O(n \log h)$ *time and* $O(n)$ *space, where* h *is the number of vertices on the hp-convex hull.*

One interesting question for the *mp*-convex hull of a collection C is: Is $MP(C)$ unique? We have:

Theorem 4.4 [48]
Let C *be a collection of* p *polygons with* n *vertices. Then whether or not* $MP(C)$ *is unique can be decided in* $O(n \log n)$ *time and* $O(n)$ *space.*

Finally, the stability of the *hp*-convex hull means that *hp*-convex hulls can be maintained, efficiently, under both deletions and insertions.

Theorem 4.5 [53]
Let C *be a collection of* p *polygons with* n *vertices. Then* $HP(C)$ *can be maintained in* $O(nm \log^2 n)$ *time for each deletion or insertion of a polygon with* m *vertices.*

We now turn to the *convex skull*, which is in one sense a close relative of the convex hull. We have:

Problem 9 The Convex Skull Problem
Given a polygon P *find a largest area convex polygon* Q *with* $Q \subseteq P$. Q *is called a convex skull of* P.

This problem is also known as the *potato peeling* problem, namely peel the smallest area from P to leave a convex polygon [Go]. See Figure 4.5 for an example.

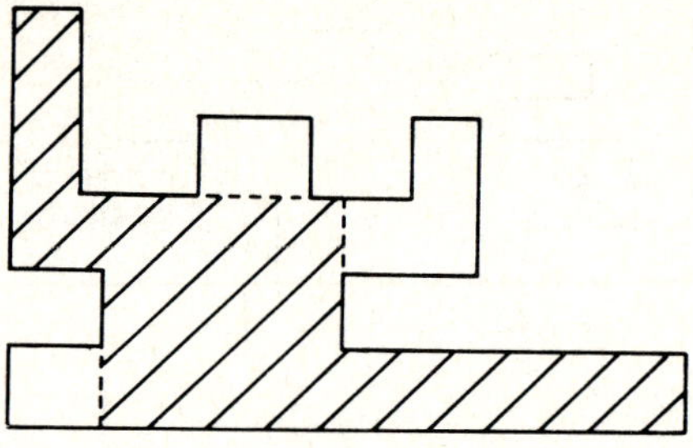

Figure 4.5

In [70] it is proved that:

Theorem 4.6 *A convex skull of a polygon with n vertices can be found in $O(n^3 \log n)$ time and $O(n)$ space.*

A simpler problem is:

Problem 10 *Given a polygon P find a largest area rectangle R with $R \subseteq P$.*

In this case we find:

Theorem 4.7 *A largest-area rectangle contained in a polygon with n vertices can be found in $O(n^2)$ time and $O(n \log n)$ space.*

Proof sketch: Each side of a largest-area rectangle with the given polygon must touch one edge, at least, of the polygon. For otherwise it could be enlarged. See Figure 4.6.

Hence, using the sweep line paradigm, sweep from left to right keeping track of possible largest-area rectangles. At each position of the sweep line there are at most n candidate rectangles; these can be represented in a segment tree. If the sweep point corresponds to a left vertical edge, then it starts a new candidate rectangle. If it corresponds to a right vertical edge then it terminates some candidates and restricts others, see Figure 4.7. Since there may be as many as n restrictions at each sweep point, see Figure 4.7, we obtain the result. □

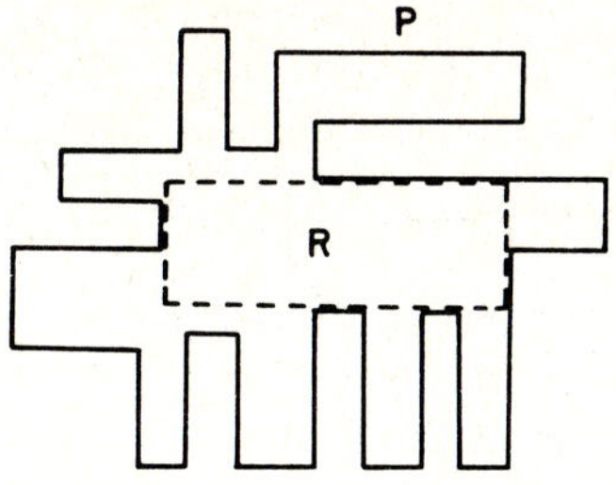

Figure 4.6

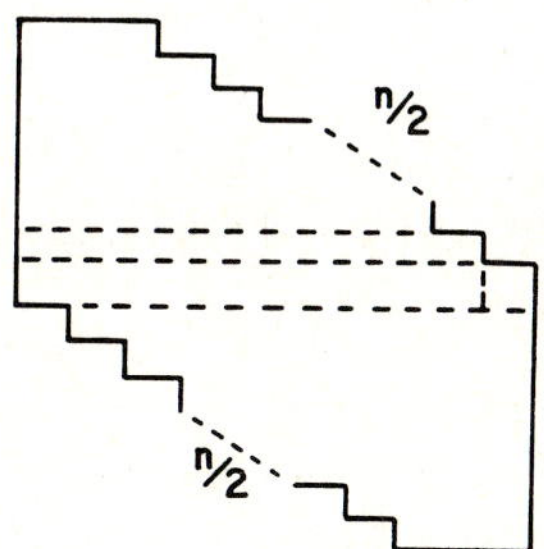

Figure 4.7

5. COMBINATIONAL PROBLEMS

By combinational we mean that the polygons are combined in some manner and, possibly, some properties of their combination are computed. For example the measure problem, that is finding the area of the plane covered by a collection, is one of the earliest of these problems, see [2] and [67]. We consider three combinational problems for a collection of polygons: the contour problem, the hidden line problem, and the boolean masking problem.

Problem 11 *Given p polygons with a total of n vertices, report the disjoint polygons obtained as their union; that is their* contour. See Figure 5.1.

Clearly the resulting polygon may have $O(n^2)$ edges, see Figure 3.1, for example. [36] gave the first algorithm for this problem, when the polygons are, in fact, rectangles. Their general approach is found in the subsequent work of [19], [20], and [69]. [36] observed that it is only necessary to compute the horizontal edges of the resulting polygons, since the vertical edges may then be deduced from

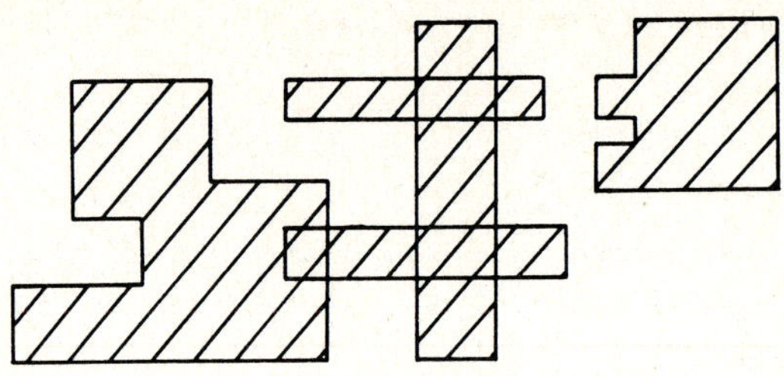

Figure 5.1

them. This gives a two-phase algorithm:

Phase 1:Compute the horizontal edges of the contour.

Phase 2:Compute the complete contour.

Phase 2 requires the horizontal edges to be sorted so for this reason it has a worst-case behavior of $O(n^2 \log n)$ or, more precisely, $O(r \log r)$ if there are r edges resulting from Phase 1. However Phase 1 can be solved in $O(n \log n + r)$ time and $O(n)$ space, as demonstrated in [19], [20], and [69]. In [69] a solution is given, based on the segment tree, which solves Problem 11 in full generality, not just for rectangles. The results proved in [69] are:

Theorem 5.1 *Given p polygons with a total of n vertices. Then the r horizontal edges of their contour can be computed in* $O(n \log n + r)$ *time and* $O(n)$ *space. Moreover these bounds are optimal.*

We now turn to our second problem:

Problem 12 The Boolean Masking Problem

Given p_r *disjoint red polygons with* n_r *vertices and* p_g *disjoint green polygons, with* n_g *vertices, report the polygons formed by their:*

(a) OR - *the union of the red and green polygons,*

(b) AND - *the intersection of the red polygons with the green polygons,*

(c) XOR - *their OR minus their AND,*

(d) NAND - *the intersection of the non-red portion of the plane with the green polygons.*

These operations are fundamental in mask design checking, see [28] and [42], where the red and green polygons correspond to polygons on two different layers.

Clearly Problem 12 is a special case of the contour problem, so Theorem 5.1 provides upper bounds for its solution where $p = p_r + p_g$ and $n = n_r + n_g$. Indeed this is the result in [55], which is also implied in [51], namely:

Theorem 5.2 *Given a collection of* $P = p_r + p_g$ *polygons with* $n = n_r + n_g$ *vertices, as above, their OR, AND, XOR, and NAND can be computed in* $O(n \log n + r)$ *time and* $O(n+r)$ *space, where* r *is the number of vertices of the resulting collection.*

Finally we turn to:

Problem 13 The Hidden Line Problem

Given p *opaque polyhedra in 3-space with* n *vertices and a given isothetic viewing direction, report the polygonal faces, and portions thereof, that can be seen from the viewing direction.*

In [50] a hidden line elimination algorithm is presented which deals with collections of arbitrary polyhedra viewed from an arbitrary direction. The restrictions to isothetic polyhedra and isothetic views leads immediately to:

Theorem 5.3 *Given a collection of opaque isothetic polyhedra with* n *vertices and given an isothetic direction, then the two-dimensional view from this direction can be computed in* $O(n \log^2 n + r)$ *time and* $O(n \log n)$ *space, where* r *is the number of vertices in the resulting view.*

In [58] the hidden line problem for a single polygon of n vertices is solved in linear time.

6. CLUSTERING PROBLEMS

We discuss three techniques for partitioning a collection of polygons into clusters. The first, and most obvious, is based on intersections and connectivity, that is connected components. The second is based on minimal area convex hulls, while the third is based on diagonal closure. The latter clustering technique occurs when viewing, from a geometrical point of view, deadlock and safety problems for locked transaction systems [54] and [36]. The second technique has been proposed as a method of partitioning components in a chip layout into macro-components, which are then more easily manipulated.

Problem 14 Connected Components Problem

Given a collection of polygons with n vertices determine their connnected components.

Given a collection of polygons, a subcollection is connected if its union is a connected set of points. A *connected component* of the collection is a subcollection which is connected and is also disjoint from the remaining polygons.

The connected components are the equivalence classes formed by taking the transitive closure of the pairwise intersection relation. Thus they can be found as a spin-off from Theorem 3.2. However this implies that a total of $O(n \log n + k \log k)$ time is required using the standard UNION-FIND method [1], where k is the number of pairwise intersections. Since $k = O(n^2)$ this method produces an inefficient algorithm. Fortunately a better approach is possible, as shown in [11] and [18].

Use the sweep line approach as we did in solving Problem 5, see Section 2. This gives rise to the notion of a *left-connected component*, with respect to the sweep line, that is a connected subcollection of the collection, where each polygon in the component is either dead or active. Clearly the connected components are the left components associated with the rightmost position of the sweep line. A crucial observation is that not only does the intersection of the sweep line with each left-connected component give at least one line segment, but also for two or more components these can be, at most, nested, they cannot overlap. In Figure 6.1(a) a valid configuration is displayed showing the nesting of two components A and B, while C is completely disjoint from both B and C. Figure 6.1(b) displays an impossible situation since the two appearances of B must be left connected as are the two appearances of A, hence A and B must intersect; a contradiction.

Since the one-dimensional projection of the left-connected components on the sweep line is so highly structured, it leads to an efficient algorithm to find connected components.

A high level algorithm can now be given:

For each sweep point x:

(1) x corresponds to a left edge.
Determine the left-connected components it joins, belongs to, or initiates. Modify left-connected components appropriately, forming unions where necessary.

(2) x corresponds to a right edge.
Determine the left connected component it belongs to and modify it appropriately.

Since we are forming equivalence classes, underlying this algorithm we have a classical UNION-FIND problem, since we need in both Steps (1) and (2) to perform FINDs, while in Step (1) we also

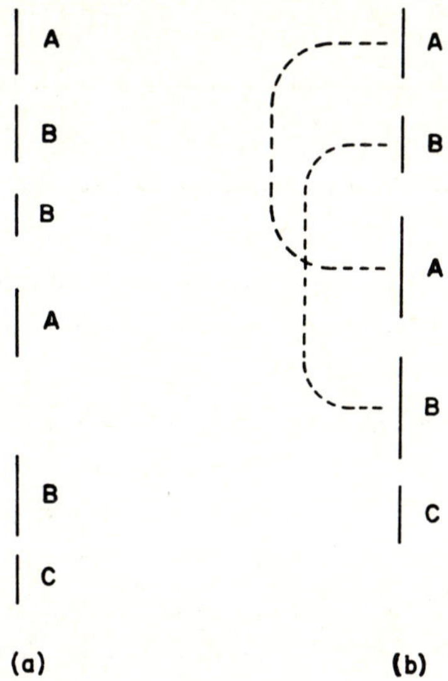

Figure 6.1

need to perform UNION. It is sufficient to use any one of the methods of solving UNION-FIND problems which requires $O(n \log n)$ time for $O(n)$ operations. Beyond this we need to represent each left-connected component as disjoint line segments, be able to insert and delete edges from them, and given an edge query determine which left-connected components it intersects. It turns out to be important to carry out the query in at most two phases. In the first phase it only reports the left-connected components it possibly intersects. For this purpose only the current topmost and bottommost points of each left-connected component are used. The result of the query is either it overlaps some of these endpoints or it is enclosed by some pair of endpoints. In the former case this immediately causes UNIONs to be carried out, while in the latter case the smallest enclosing pair should be investigated further to determine if one UNION should be carried out.

Data structures can be designed which yield the following result, generalizing the result for rectangles in [11] and [18]:

Theorem 6.1 *Given a collection of polygons with n vertices, its connected components can be found in $O(n \log n)$ time and $O(n)$ space.*

We now consider:

Problem 15 *Determine the minimal area convex hull of a collection C of polygons, that is*

MC(C).

As pointed out in Section 4 $MC(C)$ may consist of two or more disjoint convex polygons, each of which contains at least one of the original polygons. Now $MC(C) \supseteq \{MC(P) : P \text{ in } C\}$, and this collection can be found in time linear in the number of vertices. Therefore, without any loss of generality, we may assume that C is, in fact, a collection of convex polygons.

In [49] the following theorem is proved:

Theorem 6.2 *Given a collection C of polygons with a total of n vertices, $MC(C)$ can be computed in $O(n \log n)$ time and $O(n)$ space.*

The algorithm to compute $MC(C)$ is very similar to that for computing connected components, so we omit any further details.

Finally we turn to the notion of diagonal closure. Given a collection C of polygons we say it is *north-east closed* if for every pair of points $p_1 = (x_1,y_1)$ and $p_2 = (x_2,y_2)$ satisfying (1) and (2) below, the point (x_2,y_1) is in C, where:

(1) p_1 and p_2 are in the same connected component, and

(2) $x_1 < x_2$ and $y_1 > y_2$.

See Figure 6.2.

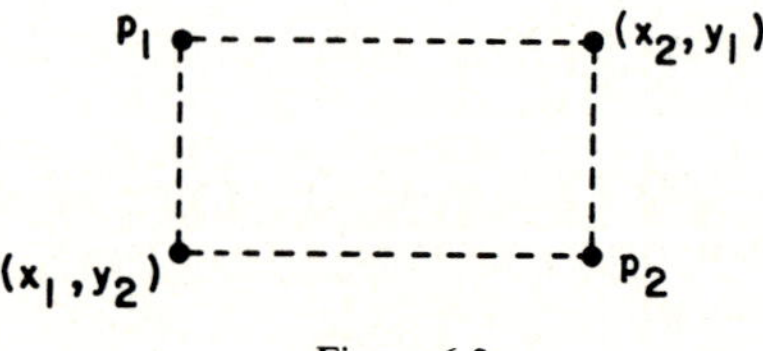

Figure 6.2

Similarly we say C is *south-west closed* if the point (x_1,y_2) is in C, and C is *diagonally-closed* if it is north-east and south-west closed.

Given a collection C of polygons we define its *north-east closure* as the minimal area collection C' of polygons, with $C' \supseteq C$ and C' is north-east closed. Again we obtain *south-west closure* and *diagonal closure*. For a collection C we denote these closures by $NE(C)$, $SW(C)$, and $D(C)$ respectively.

Problem 16 *Given a collection* C *of polygons determine* $NE(C)$, $SW(C)$, *and* $D(C)$.

In [63] the following theorem is proved:

Theorem 6.3 *Let* C *be a collection of polygons. Then* $D(C) = NE(SW(C)) = SW(NE(C))$.

Thus determining $D(C)$ reduces to determining the north-east and south-west closure of a collection. Let $rotate(C,\theta)$ denote the rotation of a collection C by θ degrees about the origin, hence $C = rotate(C,2\pi)$. Then we obtain:

Theorem 6.4 *Let* C *be a collection of polygons. Then* $NE(C) = rotate(SW(rotate(C,\pi)),\pi)$ *and* $SW(C) = rotate(NE(rotate(C,\pi)),\pi)$.

Thus we only need to determine north-east closure in order to be able to determine diagonal closure. In [63] it is shown that $NE(C)$ can be computed in $O(n \log n)$ time and space for a collection C of polygons having n vertices. The technique use is based once more on the sweep line paradigm. It is also similar to the connected components algorithm, needing some extra work, since regions need to be added to the original polygons. Thus:

Theorem 6.5 *Given a collection* C *of polygons with* n *vertices* $NE(C)$, $SW(C)$, *and* $D(C)$ *can be computed in* $O(n \log n)$ *time and space.*

The importance of diagonal closure stems from the work of [54] on locked transaction systems. Given two transactions, each of which can be considered to consist of a sequence of *lock* and *unlock* operations on at most n variables, we wish to determine whether they are deadlock-free and safe, when executed in a concurrent environment. Associating the two transaction sequences with the sequence of points 1, ... on the x- and y-axes, respectively, then they define rectangular forbidden regions. For example letting $T_1 = lock\ y,\ lock\ x,\ unlock\ x,\ unlock\ y,$ and $T_2 = lock\ x,\ lock\ y,\ unlock\ y,\ unlock\ x$ we obtain Figure 6.3.

The region labelled x corresponds exactly to the restriction that variable x may only be locked by one transaction at a time. A staircase from 0,0 to 4,4 which does not pass through any rectangle corresponds to a possible interleaving of T_1 and T_2, and vice versa, that is a *computation history*.

Consider the point $(2,2)$. A staircase which approaches close to this point (it cannot reach it since this would imply both T_1 and T_2 are locking variable y) cannot reach point (4, 4) at all, since it must go up or go right. The former is not possible since this requires T_2 to lock y also, and the latter is not possible since this requires T_1 to lock x also. In other words T_1 and T_2 are *deadlocked*. The rectangular area determined by (1, 1) and (2, 2) is a region of *deadlock*. Note

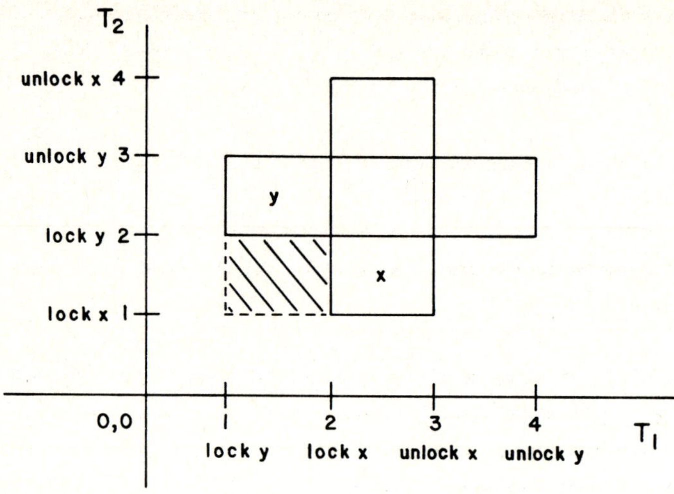

Figure 6.3

that the diagonal closure of the two rectangles fills in this region, the shaded region in Figure 6.3. To characterize deadlock we refer to the left edges of a collection as the left edges of its union, and to the lower edges as the lower edges of its union. Basically these edges, in the diagonal closure, must be in the original collection for deadlock freedom to occur.

Similarly, the notion of the *safety* of T_1 and T_2 can be characterized as the absence of any staircase which separates the forbidding rectangles into two non-empty sub-collections. Indeed [54] proves:

Theorem 6.6 *Let C be the collection of forbidding rectangles corresponding to two transactions T_1 and T_2 of n variables. Then T_1 and T_2 are* safe *if and only if $D(C)$ consists of a single polygon and T_1 and T_2 are* deadlock-free *if and only if the lower and left edges of $D(C)$ are (parts of) lower and left edges of C.*

7. VISIBILITY PROBLEMS

There are a number of visibility issues for polygons which either arise from some practical problem or arise from considerations of the complexity of polygons.

Examples of the latter situation are the various art gallery theorems.

Problem 17 *Given a polygon place a number of guards or watchmen on vertices, on edges, in the interior, or even in the exterior, so that each point on an edge, in the interior, in the exterior, etc., can be seen by at least one watchman.*

An elegant proof of one result of this type is:

Theorem 7.1 [46]

$\lfloor n/4 \rfloor$ *guards, stationed on vertices, are necessary and sufficient to see every point in the interior of a polygon of* n *edges.*

One can also assume that the guards have some mobility, for example along a line segment. This has been investigated by [47]. Other variations, which have not been investigated to the best of my knowledge, are only allowing guards isothetic mobility and isothetic sight. Assuming the walls are reflective is another possible assumption.

However we prefer to consider visibility problems arising from more practical situations. To this end we define:

Problem 18 *Given a collection of polygons we define a symmetric relation* vis, *for "is visible to," by: For two polygons* A *and* D, A vis D *if and only if there is a horizontal or vertical line segment joining* A *to* D *which intersects no other polygon, see Figure 7.1.*

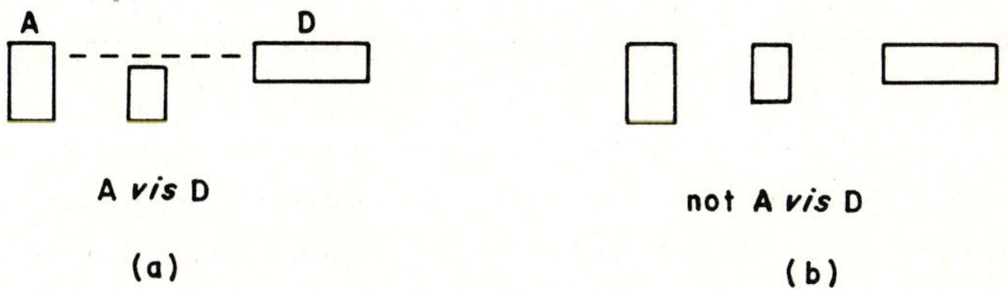

Figure 7.1

If we wish to connect components A and D with a wire, then when A *vis* D we have a straight connection, whereas if *not* A *vis* D then a connecting wire needs to be routed around the blocking obstacles.

It is convenient to split the *vis* relation into its horizontal and vertical components. Hence we say A *hvis* D if A *vis* D via a horizontal line segment and A *vvis* D if A *vis* D via a vertical line segment. In general we might have both A *hvis* D and A *vvis* D, since the given polygons are not necessarily convex. Note that *vis* may be the empty relation, see Figure 7.2.

We first prove a combinatorial result concerning the size of *vis*, namely the maximum number of

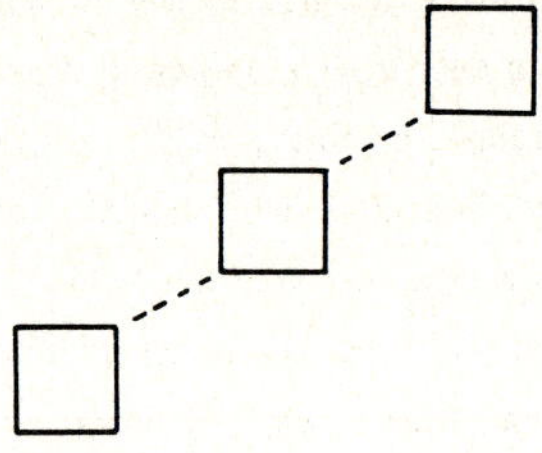

Figure 7.2

visibility pairs among n polygons.

Theorem 7.2 *Given a collection of* p *disjoint convex polygons, then* hvis *(and* vvis*) has size at most* $3p$ *and, therefore,* vis *has size at most* $6p$.

Proof: Consider *hvis* only. Each polygon has a bottommost and topmost edge. For every polygon, extend each of these edges to the right and left until they meet another polygon, otherwise they become rays or lines. In all three cases a partition of the plane is obtained, see Figure 7.3. Intuitively this process gives the horizontal *visibility regions* for each polygon. Now observe that each bottommost (topmost) edge of a polygon with its extension abuts three visibility regions as in Figure 7.4. This follows from the disjointness assumption and the construction of the regions.

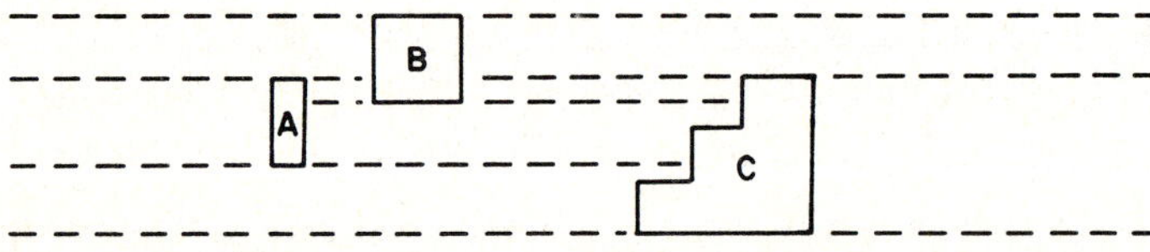

Figure 7.3

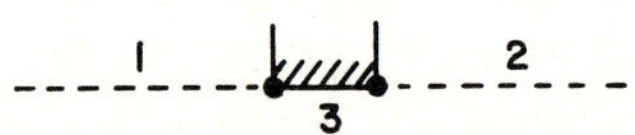

Figure 7.4

Each polygon together with its extension determines one horizontal edge for each of 6 regions. But each region, apart from the extremal ones, has two horizontal edges. Therefore there are, at most, $6p/2$ regions, that is $3p$ regions. □

This notion was used in [15] for printed circuit testing. In their application they have curves rather than polygons. Their problem is to determine a clustering of curves into disjoint groups, so that it is sufficient to test groups against groups for fabrication faults leading to short circuits. They show that there are at most 12 groups needed, in general, independent of the number of curves. In this setting a curve corresponds to a wiring and a visibility corresponds to the possibility of a short circuit. This is one application of the local relation *vis* to a global problem. There are bound to be others. For example a solution to the contour problem, see Section 5, depends on local visibility.

8. DISCUSSION

It is to be hoped that the survey presented here, although a biased one, will convince the reader that the isothetic viewpoint is a rich source of applications and problems. There are of course topics which have been omitted. Two of them, especially deserving of mention are: *motion* problems and *decomposition* problems. Elsewhere in this volume Sue Whitesides and Godfried Toussaint discuss motion problems in the context of robotics, see also [66]. Given a collection of polygons, some of which are fixed and are called *obstacles*, plan a sequence of movements to move the *movable* polygons to some given new positions, so that no two polygons collide and no two polygons touch. In [7] all polygons are movable and the aim is to separate them all, giving the game SEPARATION™.

Two examples of decomposition problems are to decompose the union of a collection of polygons into a minimal number of disjoint rectangles and into disjoint rectangles with minimal contour length. The first decomposition method occurs as a problem in databases [35], [39], the second as a problem in wire routing on VLSI chips [33].

Research into isothetic problems has: yielded new data structures, for example the segment tree, the 1-fold interval tree, and the priority search tree; resulted in new paradigms, for example the sweep line, dynamization, and paradigmatic transformations; and re-affirmed the utility of established paradigms, for example divide-and-conquer and dynamic programming.

Acknowledgements

Many of the results quoted in this survey stem from collaborative research in which I have been involved. The following are no longer just collaborators, but are also friends, without whom this survey could not have been written. I thank Jon Bentley, Bernard Chazelle, Herbert Edelsbrunner, Ralf Hartmut Güting, Joe O'Rourke, Thomas Ottmann, Mark Overmars, Eljas Soisalon-Soininen, Jan van Leeuwen, Peter Widmayer, and Chee Yap. Also Karel Culik II is responsible for the present form of Theorem 7.2, improving my own bound by p.

REFERENCES

[1] Aho, A.V., Hopcroft, J.E., and Ullman, J.D., *The Design and Analysis of Computer Algorithms* (Addison-Wesley, Reading, Mass., 1974).

[2] Bentley, J.L., Algorithms for Klee's Rectangle Problems. Unpublished (1977).

[3] Bentley, J.L., Kung, H.T., Schkolnick, M., and Thompson, C.D., On the Average Number of Maxima in a Set of Vectors and Applications. *Journal of the ACM 25* (1978) 536-543.

[4] Bentley, J.L., and Ottmann, Th., Algorithms for Reporting and Counting Geometric Intersections. *IEEE Transactions on Computers C-28* (1979) 643-647.

[5] Bentley, J.L., and Ottmann, Th., The Complexity of Manipulating Hierarchically Defined Sets of Rectangles. *Proceedings of the 10th International Symposium on Mathematical Foundations of Computer Science; Lecture Notes in Computer Science 118* (Springer-Verlag, 1981) 1-15.

[6] Bentley, J.L., and Wood, D., An Optimal Worst Case Algorithm for Reporting Intersections on Rectangles. *IEEE Transactions on Computers C-29* (1980) 563-580.

[7] Chazelle, B., Ottmann, Th., Soisalon-Soininen, E., and Wood, D., The Complexity and Decidability of SEPARATION™. *Proceedings of ICALP '84; Lecture Notes in Computer Science* (Springer-Verlag 1984) to appear.

[8] Brown, K.O. Comments on 'Algorithms for Reporting and Counting Geometric Intersections'. *IEEE Transactions on Computers C-30* (1981) 147-148.

[9] Edelsbrunner, H., A Time- and Space-Optimal Solution for the Planar All Intersecting Rectangles Problem. Technical University of Graz, IIG Technical Report F50 (1980).

[10] Edelsbrunner, H., and Overmars, M.H., Batched Dynamic Solutions to Decomposable Searching Problems. TU Graz IIG Technical Report No. F118 (1983).

[11] Edelsbrunner, H., van Leeuwen, J., Ottmann, T.A., and Wood, D., Connected Components of Orthogonal Geometric Objects. *RAIRO* (1984) to appear.

[12] Ferrari, L., Sankar, P.V., and Sklansky, J., Minimal Rectangular Partitions of Digitized Blobs. *Proceedings of the 5th International Conference on Pattern Recognition*, Miami Beach (1980) 1040-1043.

[13] Fowler, R.J., Paterson, M.S., and Tanimoto, S.L., Optimal Packing and Covering in the Plane are NP-Complete. *Information Processing Letters 12* (1981) 133-137.

[14] Gabow, H.N., Bentley, J.L., and Tarjan, R.E., Scaling and Related Techniques for Geometry Problems. *Proceedings of the Sixteenth Ann7ual ACM Symposium on Theory of Computing* (1984) 135-143.

[15] Garey, M.R., Johnson, D.S., and So, H.C., An Application of Graph Coloring to Printed Circuit Testing. *IEEE Transactions on Circuits and Systems CAS-23* (1976) 591-599.

[16] Goodman, J.E., On the Largest Convex Polygon Contained in a Non-Convex n-gon or How to Peel a Potato. *Geometriae Dedicata 11* (1981) 99-106.

[17] Groen, F.C.A., Verbeek, P.W., DeJong, N., and Klumper, J.W., The Smallest Box Around a Package. *Pattern Recognition 14* (1981) 173-178.

[18] Guibas, L., and Saxe, J., Problem 80-15. *Journal of Algorithms 4* (1983) 177-181.

[19] Güting, R.H., An Optimal Contour Algorithm for Iso-Oriented Rectangles. *Journal of Algorithms* (1984) to appear.

[20] Güting, R.H., Stabbing c-Oriented Polygons. *Information Processing Letters 16* (1983) 35-40.

[21] Güting, R.H., Doctoral dissertation, Dortmund, 1983.

[22] Güting, R.H., and Wood, D., Finding Rectangle Intersections by Divide-and-Conquer. *IEEE Transactions on Computers* (1984) to appear.

[23] Haralick, R.M., and Shapiro, L.G., Decomposition of Polygonal Shapes by Clustering. *Proceedings of the IEEE Computer Society Conference on Pattern Recognition and Image Processing* (1977) 183-190.

[24] Hwang, F.K., An $O(n \log n)$ Algorithm for Rectilinear Minimal Spanning Tree. *Journal of the ACM 26* (1979) 177-182.

[25] Imai, H., and Asabo, T., Finding the Connected Components and a Maximum Clique of an Intersection Graph of Rectangles in the Plane. University of Tokyo, Tokyo, Japan, Department of Mathematical Engineering and Instrumentation Physics; manuscript (1981).

[26] Kirkpatrick, D.G., and Seidel, R., The Ultimate Planar Convex Hull Algorithm? *Proceedings of the 20th Annual Allerton Conference on Communication, Control, and Computing* (1982) 35-42.

[27] Kung, H.T., Luccio, F., and Preparata, F.P., On Finding the Maxima of a Set of Vectors. *Journal of the ACM 22* (1975) 469-476.

[28] Lauther, U., An $O(N \log N)$ Algorithm for Boolean Mask Operations. *Proceedings of the 18th Design Automation Conference,* Nashville (1981) 555-562.

[29] Lee, D.T., and Preparata, F.P., An Improved Algorithm for the Rectangle Enclosure Problem. *Journal of Algorithm 3* (1982), 218-224.

[30] Lee, D.T., and Wong, C.K., Finding Intersections of Rectangles by Range Search. *Journal of Algorithms 2* (1981) 337-347.

[31] Lehert, P., Clustering by Connected Components in $O(n)$ Expected Time. *RAIRO Informatique/Computer Science 15* (1981) 207-218.

[32] Lingas, A., The Power of Non-Rectilinear Holes. *Proceedings of the 9th Colloquium on Automata, Languages and Programming; Lecture Notes in Computer Science 140* (Springer-Verlag 1982) 369-383.

[33] Lingas, A., Pinter, R.Y., Rivest, R.L., and Shamir, A., Minimum Edge Length Partitioning of Rectilinear Polygons. *Proceedings of the 20th Annual Allerton Conference on Communication, Control, and Computing* (1982) 53-63.

[34] Lipski, W. Jr., Finding a Manhattan Path and Related Problems. University of Illinois, Urbana, Illinois, Coordinated Science Laboratory; Report T-85 (1979).

[35] Lipski, W. Jr., Lodi, E., Luccio, F., Mugnai, C., and Pagli, L., On Two-Dimensional Data Organization II. *Fundamenta Informaticae II* (1979) 245-260.

[36] Lipski, W. Jr., and Papadimitriou, C.H., A Fast Algorithm for Testing for Safety and Detecting Deadlock in Locked Transaction Systems. *Journal of Algorithms 2* (1981) 211-226.

[37] Lipski, W. Jr., and Preparata, F.P., Finding the Contour of a Union of Iso-Oriented Rectangles. *Journal of Algorithms 1* (1980) 235-246.

[38] Lipski, W. Jr., and Preparata, F.P., Segments, Rectangles, Contours. *Journal of Algorithms 2* (1981) 63-76.

[39] Lodi, E., Luccio, F., Mugnai, C., and Pagli, L., On Two-Dimensional Data Organization I. *Fundamenta Informaticae II (1979) 211-226.*

[40] McCreight, E.M., Efficient Algorithms for Enumerating Intersecting Intervals and Rectangles. XEROX Palo Alto Research Center, Palo Alto, California; Report CSL-80-9 (1980).

[41] McCreight, E.M., Priority Search Trees. XEROX Palo Alto Research Center, Palo Alto, California; Report CSL-81-5 (1981).

[42] Mead, C., and Conway, L., *Introduction to VLSI Systems* (Addison-Wesley, Reading, Mass., 1980).

[43] Montuno, D.Y., and Fournier, A., Finding the $x-y$ Convex Hull of a Set of $x-y$ Polygons. University of Toronto CSRG Technical Report 148 (1982).

[44] Nicholl, T.M., Lee, D.T., Liao, Y.Z., and Wong, C.K., Constructing the X-Y Convex Hull of a

Set of X-Y Polygons. *BIT 23* (1983) 456-471.

[45] Nievergelt, J., and Preparata, F.P., Plane-Sweep Algorithms for Intersecting Geometric Figures. *Communications of the ACM 25* (1982) 739-747.

[46] O'Rourke, J., Galleries Need Fewer Mobile Guards: A Variation on Chvátal's Theorem. *Geometriae Dedicata 14* (1983) 273-283.

[47] O'Rourke, J., An Alternative Proof of the Rectilinear Art Gallery Theorem. Johns Hopkins University Technical Report 82-15 (1982).

[48] Ottmann, Th., Soisalon-Soininen, E., and Wood, D., On the Definition and Computation of Rectilinear Convex Hulls. *Information Sciences* (1984) to appear.

[49] Ottmann, Th., Soisalon-Soininen, E., and Wood, D., Rectilinear Convex Hull Partioning of Sets of Rectilinear Polygons. University of Waterloo, Computer Science Technical Report CS-83-19 (1983).

[50] Ottmann, Th., Widmayer, P., and Wood, D., A Worst-Case Efficient Algorithm for Hidden Line Elimination. University of Waterloo, Computer Science Technical Report CS-82-33 (1982).

[51] Ottmann, Th., Widmayer, P., and Wood, D., A Fast Algorithm for Boolean Mask Operations. University of Waterloo, Computer Science Technical Report CS-82-37 (1982).

[52] Overmars, M.H., *The Design of Dynamic Data Structures* (Springer-Verlag, Heidelberg/New York, 1983).

[53] Overmars, M.H., and van Leeuwen, J., Maintenance of Configurations in the Plane. *Journal of Computer and System Sciences 23* (1981) 166-204.

[54] Papadimitriou, C.H., Concurrency Control by Locking. *SIAM Journal on Computing 12* (1983) 215-226.

[55] Pracchi, M., Boolean Expressions of Rectilinear Polygons with VLSI Applications. University of Illinois-Urbana, Coordinated Science Laboratory Technical Report R-978 (1982).

[56] Rappaport, D., Eliminating Hidden-Lines from Monotone Slabs. *Proceedings of the 20th Annual Allerton Conference on Communication, Control, and Computing* (1982) 43-52.

[57] Sack, J.R., An $O(n \log n)$ Algorithm for Decomposing Simple Rectilinear Polygons into Convex Quadrilaterals. *Proceedings of the 20th Annual Allerton Conference on Communication, Control, and Computing* (1982) 64-74.

[58] Sack, J.R., A Simple Hidden-Line Algorithm for Rectilinear Polygons. *Proceedings of the 21st Allerton Conference on Communication, Control, and Computing* (1983).

[59] Shamos, M.I., Computational Geometry. Doctoral Dissertation, Yale University (1978).

[60] Shamos, M.I., and Hoey, D., Geometric Intersection Problems. *Proceedings of the 17th Annual IEEE Symposium on Foundations of Computer Science* (1976) 208-215.

[61] Sklansky, J., Measuring Concavity on Rectangular Mosaic. *IEEE Transactions on Computers C-21* (1972) 1355-1364.

[62] Sklansky, J., Chazin, R.L., and Hansen, B.J., Minimum Perimeter Polygons of Digitized Silhouettes. *IEEE Transactions on Computers C-21* (1972) 260-268.

[63] Soisalon-Soininen, E., and Wood, D., An Optimal Algorithm to Compute the Closure of a Set of Iso-Rectangles, *Journal of Algorithms* (1984) to appear.

[64] Srihari, S.N., Representation of Three-Dimensional Digital Images. *Computer Surveys 13* (1981) 399-424.

[65] Tompa, M., An Optimal Solution to a Wire-Routing Problem. *Journal of Computer and System Sciences 23* (1981) 127-150.

[66] Tousaint, G., and Sack, J.R., Some New Results on Moving Polygons in the Plane. *Proceedings of the Robotic Intelligence and Productivity Conference* (1983).

[67] van Leeuwen, J., and Wood, D., The Measure Problem for Rectangular Ranges in d-Space. *Journal of Algorithms 2* (1981) 282-300.

[68] Wesley, M.A., and Markowski, G., Fleshing Out Projections. Thomas J. Watson Research Center, Yorktown Heights, New York, Computer Science Department; Report RC 8884 (1981).

[69] Wood, D., The Contour Problem for Rectilinear Polygons. *Information Processing Letters* (1984) to appear.

[70] Wood, D., and Yap, C., Computing the Convex Skull of an Isothetic Polygon. In preparation (1984).

[71] Yannakakis, M., Papadimitriou, C.H., and Kung, H.T., Locking Policies: Safety and Freedom from Deadlock. *Proceedings of the 20th Annual IEEE Symposium on Foundations of Computer Science* (1979) 286-297.

CRANFIELD INSTITUTE OF TECHNOLOGY
LIBRARY